ADAPTING AGRICULTURE TO CLIMATE CHANGE

PREPARING AUSTRALIAN AGRICULTURE, FORESTRY AND FISHERIES FOR THE FUTURE

EDITORS: CHRIS STOKES AND MARK HOWDEN

CSIRO
PUBLISHING

National Library of Australia Cataloguing-in-Publication entry

Adapting agriculture to climate change : preparing
Australian agriculture, forestry and
fisheries for the future/Chris Stokes; Mark Howden.

9780643095953 (pbk.)

Includes index.
Bibliography.

Meteorology, Agricultural – Australia.
Bioclimatology – Economic aspects – Australia.
Bioclimatology – Social aspects – Australia.
Climatic changes – Australia.

Stokes, Chris
Howden, Mark.

630.2515094

Published by
CSIRO PUBLISHING
150 Oxford Street (PO Box 1139)
Collingwood VIC 3066
Australia

Telephone: +61 3 9662 7666
Local call: 1300 788 000 (Australia only)
Fax: +61 3 9662 7555
Email: publishing.sales@csiro.au
Web site: www.publish.csiro.au

Front cover photos by Gregory Heath/CSIRO (top), Richard Merry/CSIRO (bottom).

Set in 10/12 Adobe Palatino and Optima
Edited by Janet Walker
Cover and text design by James Kelly
Typeset by Desktop Concepts Pty Ltd, Melbourne
Index by Indexicana
Printed in Australia by Ligare

CSIRO PUBLISHING publishes and distributes scientific, technical and health science books, magazines and journals from Australia to a worldwide audience and conducts these activities autonomously from the research activities of the Commonwealth Scientific and Industrial Research Organisation (CSIRO).
The views expressed in this publication are those of the author(s) and do not necessarily represent those of, and should not be attributed to, the publisher or CSIRO.

CONTENTS

PREFACE   **V**

LIST OF CONTRIBUTORS   **VI**

1:   INTRODUCTION   *SM Howden and CJ Stokes*   **1**

2:   CLIMATE PROJECTIONS   *KJ Hennessy, PH Whetton and B Preston*   **13**

3:   GRAINS   *SM Howden, RG Gifford and H Meinke*   **21**

4:   COTTON   *MP Bange, GA Constable, D McRae and G Roth*   **49**

5:   RICE   *DS Gaydon, HG Beecher, R Reinke, S Crimp and SM Howden*   **67**

6:   SUGARCANE   *SE Park, S Crimp, NG Inman-Bamber and YL Everingham*   **85**

7:   WINEGRAPES   *L Webb, GM Dunn and EWR Barlow*   **101**

8:   HORTICULTURE   *L Webb and PH Whetton*   **119**

9:   FORESTRY   *TH Booth, MUF Kirschbaum and M Battaglia*   **137**

10: BROADACRE GRAZING   *CJ Stokes, S Crimp, R Gifford, AJ Ash and SM Howden*   **153**

11: INTENSIVE LIVESTOCK INDUSTRIES   *CJ Miller, SM Howden and RN Jones*   **171**

12: WATER RESOURCES   *RN Jones*   **187**

13: MARINE FISHERIES AND AQUACULTURE   *AJ Hobday and ES Poloczanska*   **205**

14: AGRICULTURAL GREENHOUSE GASES AND MITIGATION OPTIONS
JC Carlyle, E Charmley, JA Baldock, PJ Polglase and B Keating   **229**

15: ENHANCING ADAPTIVE CAPACITY   *NA Marshall, CJ Stokes, SM Howden and
RN Nelson*   **245**

16: SUMMARY   *CJ Stokes and SM Howden*   **257**

17: LOOKING FORWARD   *AJ Ash, CJ Stokes and SM Howden*   **269**

18: FREQUENTLY ASKED QUESTIONS   *CJ Stokes and SM Howden*   **275**

INDEX   **280**

PREFACE

The long-term sustainable production of food and fibre by our primary industries is imperative to all; not just living Australians, but also the billions of other people around the world who will need to be fed and clothed in the coming decades. Climate change presents perhaps the greatest economic and environmental challenge we have ever faced. Our primary industries will need to adapt to changes in temperature, extreme events, rainfall and its distribution, to an extent that we haven't experienced before. Further, our primary industries will need to adapt to the significant policy responses that will need to be made to substantially reduce our emissions in the coming years. It is increasingly apparent that changes to climate will have a greater impact on Australian primary producers than on most others: we already live in one of the harshest and most variable climates in the world and this is only likely to become more so under most current projections.

Two things are clear. Firstly, our adaptive response must be supported by good science. In this regard, we need to ensure our research, development and extension effort is well planned and well funded. The National Climate Change Research Strategy for Primary Industries, involving CSIRO, all Research and Development Corporations, all the States and the Commonwealth, is working hard to ensure efficient and effective targeted activity across all primary industry research agencies. Secondly, as the research strategy demonstrated, access to good quality information by all parties is essential to address the climate change challenge. This book is a result of both these drivers. It provides an excellent reference covering the latest research that will assist our many primary industries in meeting the challenges of climate change.

It also highlights just how much more needs to be done. Our innovative and adaptive producers must continue to push the envelope on the basis of good science combined with their intimate knowledge to feed and clothe the world in the decades ahead.

This book also demonstrates what can be done with collaboration and coordination. It is the result of many years of work by many scientists from many agencies. Importantly, it will assist broader collaboration among policy makers, industry, researchers and supporting agencies in finding and implementing practical solutions to the challenges of climate change.

Dr Michael B Robinson
Executive Director, Land & Water Australia
Chair, National Climate Change Research Strategy for Primary Industries

LIST OF CONTRIBUTORS

AJ Ash
CSIRO Climate Adaptation Flagship, 306 Carmody Rd, St Lucia, Qld 4067

JA Baldock
CSIRO Land and Water, PMB 2, Glen Osmond, SA 5064

MP Bange
CSIRO Plant Industry, Locked Bag 59, Narrabri, NSW, 2390; Cotton Catchment Communities Cooperative Research Centre, Locked Bag 1000, Narrabri, NSW 2390

EWR Barlow
School of Agriculture and Food Systems, University of Melbourne, Parkville, Vic 3121

M Battaglia
CSIRO Sustainable Ecosystems, CSIRO Sustainable Agriculture Flagship, Private Bag 12, Hobart, Tas 7001

HG Beecher
NSW Department of Primary Industries, Yanco Agricultural Institute, Yanco, NSW 2703

TH Booth
CSIRO Sustainable Ecosystems, CSIRO Climate Adaptation Flagship, GPO Box E4008, Kingston, Canberra, ACT 2604

JC Carlyle
CSIRO Sustainable Ecosystems, CSIRO Sustainable Agriculture Flagship, GPO Box 284, Canberra, ACT 2601

E Charmley
CSIRO Livestock Industries, CSIRO Sustainable Agriculture Flagship, Central Queensland Mail Centre, Qld 4702

GA Constable
CSIRO Plant Industry, Locked Bag 59, Narrabri, NSW 2390

S Crimp
CSIRO Sustainable Ecosystems, CSIRO Climate Adaptation Flagship, GPO Box 284, Canberra, ACT 2601

GM Dunn
School of Agriculture and Food Systems, University of Melbourne, Parkville, Vic 3121

YL Everingham
James Cook University, School of Engineering and Physical Sciences, Townsville, Qld 4811

DS Gaydon
CSIRO Sustainable Ecosystems, CSIRO Climate Adaptation Flagship, 306 Carmody Rd, St Lucia, Qld 4067

RG Gifford
CSIRO Plant Industry, CSIRO Climate Adaptation Flagship, GPO Box 1600, Canberra, ACT 2601

KJ Hennessy

CSIRO Marine and Atmospheric Research, CSIRO Climate Adaptation Flagship, PB 1, Aspendale, Vic 3195

AJ Hobday

CSIRO Marine and Atmospheric Research, CSIRO Climate Adaptation Flagship, GPO Box 258, Hobart, Tas 7000

SM Howden

CSIRO Sustainable Ecosystems, CSIRO Climate Adaptation Flagship, GPO Box 284, Canberra, ACT 2601

NG Inman-Bamber

CSIRO Sustainable Ecosystems, CSIRO Climate Adaptation Flagship, Private Mail Bag PO, Aitkenvale, Qld 4814

RN Jones

Centre for Strategic Economic Studies, Victoria University, Melbourne, Vic 3000

B Keating

CSIRO Sustainable Agriculture Flagship, 306 Carmody Road, St Lucia, Qld 4067

MUF Kirschbaum

Landcare Research, Private Bag 11052, Palmerston North 4442, New Zealand

NA Marshall

CSIRO Sustainable Ecosystems, CSIRO Climate Adaptation Flagship, PMB Aitkenvale, Qld 4814

D McRae

Queensland Climate Change Centre of Excellence, Environmental Protection Agency, PO Box 102, Toowoomba, Qld 4350

H Meinke

Centre for Crop Systems Analysis, Wageningen University, PO Box 430, NL 6700 AK Wageningen, The Netherlands

CJ Miller

CSIRO Sustainable Ecosystems, CSIRO Climate Adaptation Flagship, 306 Carmody Rd, St Lucia, Brisbane, Qld 4067

RN Nelson

CSIRO Sustainable Ecosystems, CSIRO Climate Adaptation Flagship, GPO Box 284, Canberra, ACT 2601

SE Park

CSIRO Sustainable Ecosystems, CSIRO Climate Adaptation Flagship, GPO Box 284, Canberra, ACT 2601

PJ Polglase

CSIRO Sustainable Ecosystems, CSIRO Climate Adaptation Flagship, GPO Box 284, Canberra, ACT 2601

ES Poloczanska

CSIRO Marine and Atmospheric Research, CSIRO Climate Adaptation Flagship, PO Box 120, Cleveland, Qld 4163

B Preston

CSIRO Marine and Atmospheric Research, CSIRO Climate Adaptation Flagship, PB 1, Aspendale, Vic 3195

R Reinke

NSW Department of Primary Industries, Yanco Agricultural Institute, Yanco, NSW 2703

G Roth

Cotton Catchment Communities Cooperative Research Centre Locked Bag 1000, Narrabri, NSW 2390

CJ Stokes

CSIRO Sustainable Ecosystems, CSIRO Climate Adaptation Flagship, PMB PO Aitkenvale, Qld 4814

L Webb

Centre for Australian Weather and Climate Research, a partnership between CSIRO and the Bureau of Meteorology, 107–121 Station St, Aspendale, Vic 3195; School of Agriculture and Food Systems, University of Melbourne, Parkville, Vic 3121

PH Whetton

CSIRO Marine and Atmospheric Research, PB 1, Aspendale, Vic, 3195; Centre for Australian Weather and Climate Research, a partnership between CSIRO and the Bureau of Meteorology, 107–121 Station St, Aspendale, Vic 3195

1
INTRODUCTION

SM Howden and CJ Stokes

KEY MESSAGES:

▪ The climate is changing and further change seems unavoidable, even if efforts are taken to reduce greenhouse gas emissions. For primary industries to continue to thrive in the future we need to anticipate these changes, be prepared for uncertainty, and develop adaptation strategies now.

▪ Some broad generalisations can be made about how plant growth, which underpins all the primary industries addressed in this book, will be affected by climate change. Warmer temperatures may benefit perennial plants in cool climates, but annuals and plants growing in hot climates may be negatively affected. Plant productivity would be expected to increase or decrease in accordance with any changes in rainfall, while the direct effects of CO_2 in stimulating plant growth and increasing water use efficiency could help by partly offsetting increases in evaporation or decreases in rainfall.

▪ While there are some general principles about how impacts of climate change will vary geographically, regional climate change projections are currently more useful for describing the wide range of uncertainty and for probability-based risk assessment than serving as precise estimates for predictive planning and decision making.

▪ Adaptation will need to take a flexible, risk-based approach that incorporates future uncertainty and provides strategies that will be able to cope with a range of possible changes in local climate. Initial efforts in preparing adaptation strategies should focus on equipping primary producers with alternative adaptation options suitable for the range of uncertain future climate changes and the capacity to evaluate and implement these as needed, rather than focussing too strongly yet on exactly where and when these impacts and adaptations will occur.

▪ In the short term, a common adaptation option will be to enhance and promote existing management strategies for dealing with climate variability. This will automatically track early stages of climate change until longer term trends become clearer.

A changing climate for agriculture

Australia's climate has many influences: seasonal synoptic circulations and frontal systems, the El Niño-Southern Oscillation (e.g. Pittock 1975), the Indian Ocean Dipole (Saji *et al.* 1999), the Southern Annular Mode (Marshall 2003), the Madden-Julian Oscillation (Donald *et al.* 2006), and the Inter-decadal Pacific Oscillation (Power *et al.* 1999) among others (see Table 1.1). Jointly, these have provided Australia with the world's most variable climate. Managing the impacts of climate variability on agricultural systems has thus been a major challenge since European settlement but has been improving gradually. Now, in addition to this highly variable and challenging climate, there is increasing evidence that the climate is changing and that humans are likely to be the cause of this change (Solomon *et al.* 2007). Climate change will likely cause a range of impacts on

Australian agriculture with a consequent need for adaptation responses to emergent risks and opportunities. This book is intended to be a first step towards effective climate change adaptation responses across Australian agriculture.

It is very likely that human-influenced emissions of greenhouse gases are affecting the global climate (Solomon *et al.* 2007). Global mean temperatures have risen approximately 0.76°C since the mid-1800s. The last decade is the warmest ever recorded instrumentally (0.42°C above the 1961–1990 baseline: Brohan *et al.* 2006) followed by the previous two decades (0.18 and 0.05°C respectively) while the last 100 years were the warmest of the millennium. This warming plus changes in continental-scale temperatures, rainfall patterns, wind fields, climate extremes and sea levels cannot be explained by natural causes alone: there is a strong human 'fingerprint' (Solomon *et al.* 2007). Additionally, rates of glacial and ice-field retreat and many other observations of physical and biological responses are consistent with expectations of 'greenhouse' climate change. It seems likely that these changes will continue for the foreseeable future due to ongoing, and even accelerating (Canadell *et al.* 2008) emissions of carbon dioxide and other greenhouse gases. Indeed, past greenhouse gas emissions alone are estimated to have committed the globe to a warming of about 0.2°C per decade for the next several decades (Solomon *et al.* 2007). The most up-to-date climate projections are for an increase in global average temperatures of 1.1–6.4°C by the end of the present century along with a large range of other climate changes (see Chapter 2). To place these temperature rises in perspective, a 1°C rise in average temperature will make Melbourne's climate something like that currently experienced by Wagga Wagga, a 4°C rise like that of Moree and a 6°C rise like that just north of Roma in Queensland (corresponding to shifts in latitude of 2.5°, 8° and 11° towards the equator respectively). Intuitively, it is hard to conceive that such changes will not have implications for Australia's agricultural industries. Unfortunately, at the moment the rate of greenhouse gas emissions, the build-up of atmospheric carbon dioxide, the global temperature increase and the rate of sea level rise are all at or above the worst-case scenario of the IPCC (Rahmstorf *et al.* 2007; Canadell *et al.* 2008) and thus the higher end of the range of change seems more likely than the lower.

Agricultural adaptation to climate change: a new need

Agricultural systems in Australia are well known to be sensitive to both long-term climatic conditions and year-to-year climate variability. This is evident in the systems used in a given geographic location or season type, average production and production variability, product quality, relative preferences for different agricultural activities, preferred soil types, the management systems and technologies used, input costs, product prices and natural resource management. Consequently, if the climate changes, there are likely to be systemic changes (or adaptations) in agricultural systems. Here we use the term 'adaptation' to include the actions of adjusting practices, processes and capital in response to the actuality or threat of climate change, as well as responses in the decision environment, such as changes in social and institutional structures, or altered technical options, that can affect the potential or capacity for these actions to be realised (Howden *et al.* 2007).

Adaptation is not new. Australian farmers have always adapted to past changes in prices, technologies and climate variations as well as institutional factors (e.g. McKeon *et al.* 2004). The rationale for having a focus on adaptation to climate change is that the changes are likely to be far-reaching, systemic and to some extent, able to be foreseen, and so there is a case for advanced preparation. For example, the recent IPCC Fourth Assessment Report concludes that Australian agriculture and the natural resource base on which it depends has significant vulnerability to the changes in temperature and rainfall projected over the next decades to 100 years (Hennessy *et al.* 2007). Climate change will add to the existing, substantial pressures on Australia's agricultural industries and will interact strongly with the food security challenge over the next decades: that is, to effectively double food production while reducing greenhouse gas emissions, reducing impact on biodiversity and the natural resource base while facing competition for land and water from urban encroachment and biofuel use. To be prepared for this challenge, we argue here that there is a need to start developing and implementing adaptation strategies now. This book is aimed at assisting such efforts.

Table 1.1: Major components of the climate system relevant to climatic variability and land management in Australia's agricultural systems. The cited reference indicates examples of the influence of components on Australian rainfall, controlling climate systems and vegetation response (after Meinke and Stone 2005; McKeon *et al.* 2009).

Component of climate system variation	Time period	Literature cited
Madden-Julian Oscillation (MJO)	Intra-seasonal (30–60 days)	Donald *et al.* (2006)
Quasi-biennial Oscillation	2½ years	White *et al.* (2003)
El Niño-Southern Oscillation (ENSO)	Inter-annual (2–7 years)	Pittock (1975)
Southern Annular Mode (SAM)	Inter-annual and trends	Marshall (2003)
Indian Ocean Dipole (IOD)	Inter-annual and decadal	Saji *et al.* (1999)
Pacific Decadal Oscillation (PDO) or Inter-decadal Pacific Oscillation (IPO)	Inter-decadal	Power *et al.* (1999)
Multi-decadal	30–100 years	Hendy *et al.* (2003)
Global Warming and Greenhouse	Since late-1800s	Nicholls (2006)
Stratospheric Ozone Depletion	Since 1970s	Syktus (2005)
Asian Aerosols	Since 1980s	Cai and Cowan (2007)
Land Cover Change	Since mid-1800s	McAlpine *et al.* (2007)
Very Long-term Oscillations (e.g. Milankovitch cycles or Ice Ages)	Thousands of years	De Deckker *et al.* (1988)

The importance of developing effective strategies for adapting to climate change has been recognised by the governments of the Commonwealth, States and Territories. Initiatives such as the Garnaut Climate Change Review (http://www.garnautreview.org.au), the National Climate Change Adaptation Research Facility (www.nccarf.edu.au) and the CSIRO Climate Adaptation Flagship (www.csiro.au/org/ClimateAdaptationFlagship.html) are seeking to more fully understand the implications of climate change and the actions that could be taken to address this challenge. It is now recognised that in order to assess the costs (and benefits) of climate change, we need to include the costs (and benefits) of mitigation, the costs (and benefits) of impacts and the costs (and benefits) of adaptation (Howden *et al.* 2007). Several of these interact with each other. For example, we would expect the size of the adaptation task to be lower if there is effective, but perhaps costly, mitigation and higher if mitigation is foregone. Similarly, the benefits of effective adaptation are likely to be greater if the climate change itself is large. Achieving this complex task of effectively informing public policy development will be challenging in its own right and this book is also a step towards that goal. In this we are building on several decades of research into the management of climate variability and approximately two decades of intermittent research into the impacts of climate change on Australian agricultural systems. However, there is relatively little

prior research on climate change adaptation, with only relatively few options analysed in a practical, participatory way with industry for their utility in reducing the risks or taking advantage of any opportunities arising from climate change. Even fewer adaptations have been evaluated in relation to the broader costs and benefits of their use. These few analyses show that practicable and financially viable adaptations will have very significant benefits in ameliorating risks of negative climate changes and enhancing opportunities where they occur. The benefit-to-cost ratio of undertaking research and development into these adaptations appears to be very large (indicative ratios greatly exceed 100:1, Howden and Jones 2004). In response to the growing demand from industry, government and the general public (including the recent National Climate Change Research Strategy for Primary Industries review), we prepared reports drawing together expertise from across Australia to identify prospective adaptation options (Howden *et al.* 2003; Stokes and Howden 2008). This book updates and expands these options drawing on more recent studies.

Scope of the book

In this book we cover all the major agricultural industries in Australia: grains (Chapter 3), cotton (Chapter 4), rice (Chapter 5), sugarcane (Chapter

6), winegrapes (Chapter 7), horticulture and vegetables (Chapter 8), forestry (Chapter 9), broadacre grazing (Chapter 10), intensive livestock (Chapter 11) and fisheries and aquaculture (Chapter 13). For each of these industries we consider the likely implications of projected climate change (as described in Chapter 2) and what actions industries might consider taking to adapt to these new challenges and opportunities. The total gross value of production of these industries is about $40 billion per year and it has been increasing at about 3.3% per year over the past decade (ABARE 2007). In the year 2006–7, exports from these industries were about $31.4 billion (about 15% of total exports). Hence, agriculture makes a substantial contribution to the national balance of trade. Throughout this chapter terms such as 'agriculture', 'enterprise', 'land use' and 'primary producer' are used to refer more broadly to the full range of industries that Australia's renewable natural resources support.

Past experience demonstrates that all these agricultural sectors have sensitivity to climate variations ranging from minor to substantial. Consequently, there are many management responses to climate variability and these will likely provide the basis of many initial adaptation strategies. This aspect is covered in each of the industry chapters along with other adaptation options these industries might use in responding to climate change and the uncertainties that need to be addressed to start developing adaptation strategies. We have also included cross-cutting issues of water resources and pests/diseases as previous work has suggested that these are highly sensitive to potential climate changes and they have significant implications for components of the agricultural sector. The material on pests and diseases is integrated into each sector as this will be how the impacts are largely expressed, while water resources are dealt with in a separate chapter (Chapter 12). Greenhouse gas emissions from agriculture and how to reduce these are dealt with in Chapter 14, and human dimensions of adaptation are covered in Chapter 15. The concluding chapters include a summary and an analysis of prospective ways ahead to make adaptation effective in a variable and changing climate.

There are several impacts of climate and atmospheric composition that are common across agricultural sectors as they impact on plant production, the primary driver for agriculture and

fisheries. To reduce repetition in the subsequent chapters we deal with these common responses below. We then address issues of managing uncertainty.

Direct impacts of atmospheric and climatic change

The almost certain increases in atmospheric CO_2 concentrations, the high likelihood of increases in temperatures and the possibility of changes in rainfall averages, seasonality and intensity (Chapter 2) are all likely to affect plant production directly. The broad nature of likely changes in these three factors (CO_2, temperature and rainfall) are described below followed by the likely impacts on plant production and marine systems.

Direct effects of rising atmospheric CO_2

Pre-industrial atmospheric concentrations of CO_2 were about 280 parts per million (ppm) and these have risen almost exponentially to their current level of about 389 ppm (a 37% increase). The IPCC developed sets of emissions scenarios about a decade ago, describing different economic, population and technology trajectories (Nakicenovic and Swart 2000). The low-emissions scenarios indicated a small rise in emissions followed by a sustained reduction, whereas the high-emissions scenarios involved almost exponential increases in emissions. Even the lowest emissions scenario stabilised CO_2 concentration at about 550 ppm (almost double pre-industrial levels) whereas the high-end scenarios had CO_2 levels exceeding 900 ppm at 2100. As noted earlier, emissions are now higher than the worst-case scenario (Canadell et al. 2008) and are growing rapidly (3.5% p.a.) with limited immediate prospect for global reductions in emission rates. Consequently, plants will increasingly be growing in high CO_2 environments and this will directly affect the resource use efficiency, productivity, and product quality of plant production in agriculture.

Elevated atmospheric CO_2 concentration increases the efficiency of use of light and water (Gifford 1979; Morison and Gifford 1984), nitrogen (Drake et al. 1997) and possibly efficiency or effectiveness of uptake of other minerals like soil phosphorus (Campbell and Sage 2002). In Australia where water, nitrogen and phosphorus are major limiting factors in production, this is an important first order, and generally positive feature of the response of agriculture to global atmospheric change. It is sometimes termed the 'CO$_2$ fertilisation effect'. The other, less certain, components of climate change such as rising temperature and changed rainfall will be superimposed on this primary response. As such it is appropriate to understand it and to consider how the benefits might be maximised.

The increase in light use efficiency in 'temperate' C$_3$ species, like wheat, barley, rice, cotton, oats, oil seeds, trees, and cool-season pasture species, derives substantially from the suppression of the process of photorespiration by elevated CO$_2$: photorespiration is essentially an inefficiency. The increase in light use efficiency can be substantial – 30% or more at the crop level with doubling of CO$_2$. The 'tropical' C$_4$ species (maize, sorghum, sugarcane and tropical grasses) lack photorespiration and the effect of CO$_2$ on increasing light use efficiency is correspondingly much lower in these species.

The increase in water use efficiency with elevated CO$_2$ results from partial closure of stomata (the small pores, mostly on the undersides of leaves, that allow CO$_2$ into the leaf but simultaneously let water vapour out). Higher concentrations of CO$_2$ in the atmosphere increase the rate that CO$_2$ diffuses into the leaf for photosynthesis relative to the rate of water loss through stomatal pores. This tends to increase water use efficiency expressed as: (1) an increase in photosynthesis while transpiration rates stay the same; (2) reduced transpiration while photosynthesis remains the same, or (3) an intermediate combination of increased photosynthesis and reduced transpiration. In the field, this may be observed as an increase in growth rate while soil water depletion remains unaltered (e.g. Samarakoon and Gifford 1995) or reduced soil water depletion with little growth effect or an intermediate combination (e.g. Stokes *et al.* 2008). This increase in plant dry matter production per unit of water used by plants occurs in both C$_3$ and C$_4$ species (Morison and Gifford 1984). In good growing conditions, under elevated CO$_2$ there is a tendency in some but not all plants for the decrease in water use per unit leaf area to almost match the increase in leaf area, resulting in almost identical time-course of water use to that found in lower CO$_2$ conditions (e.g. Samarakoon and Gifford 1995). However, in nutrient-poor situations, there is limited capacity to increase leaf area but the stomatal response still occurs, resulting in increased soil moisture contents in the subsoil and extension of the period when water is available for growth in the surface soil. This may significantly benefit trees and shrubs in savanna communities and also alter grass species dynamics (e.g. Stokes and Ash 2006). The increased efficiency with which plants use water can potentially be exploited in developing adaptation strategies and could be used to partly offset the effects of reduced rainfall or increased evaporative demand.

The increase in nitrogen (N) use efficiency (here referring to the capacity to grow more dry matter with the same amount of nitrogen) can occur in both C$_3$ and C$_4$ species. In C$_3$ species in particular this appears to occur as a result of the increased efficiency and hence lowered production of 'rubisco', a key photosynthetic enzyme that contains a large fraction of leaf nitrogen. Leaf and grain N-concentrations can also be reduced via passive 'dilution' whereby elevated CO$_2$ stimulates carbon fixation, and storage of carbohydrates but this is not matched by corresponding increases in uptake of N by plant roots. These changes in nitrogen (i.e. protein) and storage carbohydrates have implications for plant product quality such as herbage forage quality and grain quality. Adaptive management measures may be needed to compensate for these impacts where they are problematic, for example via application of nitrogenous fertilisers etc. (see Chapter 3), but in some circumstances these impacts can be beneficial (e.g. in livestock where growth is energy limited not N-limited, see Chapter 10).

In legume species that have symbiotic N-fixation in the roots, elevated CO$_2$ concentration has frequently been shown to increase the rate of N-fixation per plant or per unit ground area by increasing the size of the root system and mass of nodules (Chapter 10). Additionally, the growth response of legumes to elevated CO$_2$ concentration is generally greater than that of grasses. Thus in mixed farming systems the need for artificial fertiliser, to maintain grain protein levels for example, may be reduced by using legume-based leys (all else being equal).

Effects of increased temperature

The primary climatic effect of increasing concentrations of greenhouse gases is an increase in the average temperature of the terrestrial and ocean surface and the lower atmosphere. The rate of development of many agricultural plants is approximately linearly dependent on cumulative air temperature ('heat sum') above a base temperature at which development rate is essentially zero. In addition, plant growth rate often shows a flat bell-shaped response to temperature with each species having its own 'optimum' temperature characteristics. The optimum is, however, subject to acclimation such that plants within a species growing in high temperatures have a higher optimum than those growing in low temperatures. Generally speaking, most agricultural crops are grown in areas where average temperatures are below their acclimated optimum. Thus, as temperatures rise (and assuming all else being equal), we might expect both dry weight growth rates and rates of progression through developmental phases to increase with the effect on rate of development being the stronger of the two. However, plant responses to possible changes in frequency of occasional high temperature or frost stress make generalisation very problematic. For annual crops, warmer conditions tend to reduce yields owing to any faster growth rate not being sufficient to compensate for the earlier attainment of maturity reducing the opportunity to accumulate sunlight and hence biomass. Higher temperatures also increase respiration, decreasing yields (e.g. Manunta and Kirkham 1996).

For perennials such as trees and pasture species in regions where cool winters slow growth, it might be anticipated that warming would increase winter growth and extend the more rapid growth period. However, for the perennial subterranean clover, it was found that 3.5°C continuous warming of the atmosphere in a field experiment did not increase winter growth and for the whole year decreased herbage growth by almost 30%, offsetting a positive response to concurrent elevated CO_2 concentration (Lilley *et al.* 2001). The temperature responses of productivity are clearly complex, involving interdependent effects on photosynthesis, respiration, transpiration, nutrition and plant development.

In addition to the above effects, minimum temperature is inversely related to vapour pressure deficit (VPD: the 'dryness' of the air), which is in turn linearly related to evaporation rates. High vapour pressure deficits also result in lower water use efficiencies. Thus if VPD increases there are two compounding negative impacts (higher water demand and lower water use efficiency). VPD is likely to increase with higher temperatures provided that there is not a marked asymmetry between the increases in night-time and daytime temperatures. An additional effect on VPD and water use efficiency may arise from the influence of elevated CO_2 in reducing stomatal aperture, which would reduce evaporative cooling of the leaf from transpiration, increase the temperature differential between the leaf boundary layer and the air and thus increase effective VPD. Reduced transpiration (with reduced evaporative cooling) would also create an additional daytime warming of plant leaves above and beyond the leaf temperature increases associated with warming of the lower atmosphere.

Increases in night-time temperatures are known to affect numerous developmental and product quality attributes of protected horticultural crops such as on flowering, plant height, seed and flower set and fruit quality. Similarly, in rice, increasing night temperatures can increase spikelet sterility, reducing yields (Ziska and Manalo 1996). Thus adapting to higher night-time temperatures may need to take into account situation-specific responses created by species (or genotype) by environment interactions.

Low night-time temperatures are necessary for some annual and perennial plants via the process of vernalisation or provision of chilling units. Increases in temperature are already reducing the achievement of chilling units in industries such berry-growing and apple and stonefruits. Further warming will likely require much more significant adaptations such as changes in location, varieties or species and use of chemical or other control options. In contrast, for many industries such as viticulture, horticulture and some grains, frost is a significant risk. In some regions such as central Queensland, frost risk is already declining with warming conditions, but the reverse trend is occurring in southern Australia. There remains some uncertainty as to the likely changes in frost risk in southern Australia under climate change.

Interactions between elevated CO_2 concentration and temperature are complex. Although there seemed to be solid theoretical reasons why the magnitude of the CO_2 growth response of C_3 species would increase with temperature (Gifford

1992), synthesis of experimental evidence from the literature indicated no trend of increased CO_2 sensitivity with increasing temperature (Morison and Lawlor 1999). Hence, we cannot assume that the responsiveness of plant growth to CO_2 will become greater with global warming.

Increased intensity of the hydrological cycle

Implicit in the theory of global warming is a positive feedback of increasing greenhouse gas concentrations, increasing temperatures, increasing evaporation and thus atmospheric water vapour that will further increase temperatures and so on. This intensification of the hydrologic cycle means (on average across the globe) higher evaporation rates, higher absolute atmospheric humidity and higher rainfall (Solomon et al. 2007). However, the places where evaporation may increase are not necessarily the same as those where rainfall may increase. Across the globe, climate change scenarios indicate a tendency for increased rainfall in the equatorial and cool-temperate to polar regions but decreased rainfall in the subtropics and Mediterranean mid-latitude climate zones (Solomon et al. 2007). This tendency for lower rainfall and more droughts in southern Australia in particular is discussed in detail in the following chapter but, if realised, it will affect agriculture and water resource management across the continent. However, it is important to note that whereas we can have high confidence in CO_2 increases, there are major increases in the uncertainty as we consider temperature and particularly rainfall changes and as we move from global to local scales (Giorgi 2005). Consequently, local-scale rainfall changes in particular are fraught with uncertainty due to the large range of factors that affect them as well as the effectiveness of descriptions of the underlying climatic processes. Nevertheless, there appears to be more certainty in relation to changes in the intensity of rainfall. Even in regions where average rainfalls may decrease, rainfall intensities may increase through more rainfall falling as a result of thunderstorm-type activity rather than frontal rainfall (which tends to be of lower intensity). Similarly, high intensity rainfall may increase from cyclones as these are likely to increase in intensity and severity but not necessarily in frequency or latitudinal distribution (Solomon et al. 2007). Increased rainfall intensity will require management responses to reduce erosion risk across a broad range of agricultural industries (e.g. McKeon et al. 1988).

Marine system changes

The dominant climate change influences on marine environments will be warming of the oceans, sea level rises, changes in circulation patterns and changes in ocean chemistry. Increasing temperatures in oceans are likely to threaten coral reefs with more frequent bleaching episodes, cause fish species to migrate towards the poles, reduce the viability of southern kelp forests and threaten many important fisheries that have specific ecological, physiological or temperature requirements, such as abalone, rock lobster and Atlantic salmon (Hennessy et al. 2007; Hobday et al. 2008). Oceans serve as a strong buffer for atmospheric CO_2 levels and have been estimated to have absorbed about a third of all anthropogenic CO_2 emissions to date. However, this buffering already appears to be decreasing with the fraction of total CO_2 emissions absorbed by the ocean decreasing from about 31.5% in 1960 to only 25.5% now (Canadell et al. 2008). Further decreases seem likely as the oceans warm (CSIRO 2007). Rising levels of CO_2 in oceans are increasing their acidity and reducing the availability of calcium carbonate, which is required by many creatures with calcium carbonate-based shells: evidence from the Southern Ocean suggests this decline has already begun (Moy et al. 2009, Chapter 13). Increased stratification of oceans will potentially reduce overturning and nutrient cycling and this could alter productivity, particularly in upwelling regions with subsequent effects on fisheries (Chapter 13). Estuaries, which are important nurseries, are likely to be affected by rising sea levels and changes in flows of freshwater from rivers, often positively.

Adapting to uncertain changes

There are many sources of climate change uncertainties that form a cascade, each compounding one after another. As a result, there is a broad spread in the range of projected climate changes over the next decades to centuries (see Chapter 2). To illustrate this, there is substantial uncertainty about what actions humans will take globally to reduce future greenhouse gas emissions. For any given emissions scenario, there are uncertainties about how the greenhouse emissions will influence global climate and how feedbacks from the biosphere in response to these climate changes will further increase or decrease the atmospheric greenhouse gas levels. As we move from global climate to management scales (regional to local) the climate projections become less reliable. For any given climate change projection, there are then further uncertainties as to what the impact will be on current land (and ocean) use practices and the underpinning natural resources. There is further uncertainty as to the effectiveness of adaptations by affected industries, and whether their responses will or will not increase emissions, thus feeding back to climate change. How to deal with this cascade of uncertainty is a key challenge for adaptation.

There is a general view that the agricultural sectors in Australia are highly adaptive, developing management, technologies and other responses to a range of challenges and opportunities. Often, these have been in the face of fairly well-defined single factors such as changes in relative price of inputs or outputs, market access or change in consumer preference. In contrast, as indicated earlier and also in Chapter 2, climate change is, in many respects, highly uncertain in both the nature and degree of the change and it has multiple related dimensions. These are not only through potential changes in climate (and hence productivity) at local, regional and global scales, but also through the emerging carbon economy that may affect input prices and potential new products such as carbon storage (Keating et al. 2008). For example, it is important that adaptations do not increase emissions, thus increasing the size of the mitigation task (Howden et al. 2007). There is also the prospect of maladaptation that could arise from either under-adapting or over-adapting, or by adapting too quickly or too slowly.

The following chapters address these and other points such as costs, benefits and barriers to adaptation and likely adaptive capacity to provide an industry-by-industry perspective on adapting to climate change. These accounts are then synthe-

sised in the Summary (Chapter 16) into a region-by-region overview of adaptation options.

A consistent theme that emerges throughout this book is that the impacts of climate change are only partly certain. This theme is revisited towards the end of the book in considering the human dimension of adaptation, where a suggested approach for dealing with this uncertainty is presented (Chapter 15). A related theme is that adaptation should be viewed as an ongoing process that is part of good risk management, whereby drivers of risk are identified and their likely impacts on systems under alternative management are assessed. In this respect, adaptation to climate change is similar to adaptation to climate variability, changes in market forces (cost/price ratios, consumer demands, etc.), institutional or other factors. Isolating climate change from other drivers of risk may be helpful, especially during the initial stages of assessment when awareness of the relative importance of this risk factor is still low.

Operationally, however, translating adaptation options into adaptation actions requires consideration of a more comprehensive risk management framework. This would allow exploration of quantified scenarios dealing with all the key sources of risk (not just climate change), providing more effective decision making and learning for farmers, policy-makers and researchers: an increase in the 'climate knowledge' (Howden et al. 2007). Importantly this framework should not be limited to technical approaches to risk assessment, which tend to assume that people are 'locked in' to the existing enterprise mix or management systems. It is highly likely that incremental adaptations applied to existing systems will have limits to their effectiveness under more severe climate changes. Hence, more systemic, transformational changes in resource allocation need considering, such as targeted diversification of production systems and livelihoods. We hope that this book is a step on the pathway to an Australian primary industries sector that is flexible, forward-looking and more able to adapt to a variable and changing climate.

References

ABARE (2007). 'Australian commodity statistics.' Australian Bureau of Agriculture and Resource Economics, Canberra.

Brohan P, Kennedy JJ, Harris I, Tett SFB, Jones PD (2006). Uncertainty estimates in regional and global observed temperature changes: a new dataset from 1850. *Journal of Geophysical Research* **111**, D12106.1-D12106.21, doi:10.1029/2005JD006548.

Cai W, Cowan T (2007). Impacts of increasing anthropogenic aerosols on the atmospheric circulation trends of the Southern Hemisphere: an air–sea positive feedback. *Geophysical Research Letters* **34**, L23709.1-L23709.5.

Campbell CD, Sage RF (2002). Interactions between atmospheric CO_2 concentration and phosphorus nutrition on the formation of proteoid roots in white lupin (*Lupinus albus* L.). *Plant Cell and Environment* **25**, 1051–1059.

Canadell J, Le Quéréc C, Raupach MR, Field CB, Buitenhuis ET, Ciais P, Conway TJ, Gillett NP, Houghton RA, Marland G (2008). Contributions to accelerating atmospheric CO_2 growth from economic activity, carbon intensity, and efficiency of natural sinks. *Proceedings of the National Academy of Sciences* **104**, 18866–18870.

De Deckker P, Kershaw AP, Williams MAJ (1988). Past environmental analogues. In *Greenhouse, Planning for Climate Change*. (Ed. GI Pearman) pp. 473–488. CSIRO Australia, East Melbourne.

Donald A, Meinke H, Power B, Maia de HN, Wheeler MC, White N, Stone RC, Ribbe J (2006). Near-global impact of the Madden-Julian Oscillation on rainfall. *Geophysical Research Letters* **33**, L09704.

Drake BG, GonzalezMeler MA, Long SP (1997). More efficient plants: A consequence of rising atmospheric CO_2? *Annual Review of Plant Physiology and Plant Molecular Biology* **48**, 609–639.

Gifford RM (1979). Growth and yield of CO_2-enriched wheat under water-limited conditions. *Australian Journal of Plant Physiology* **6**, 367–378.

Gifford RM (1992). Implications of the globally increasing atmospheric CO_2 concentration and temperature for the Australian terrestrial carbon budget – integration using a simple model. *Australian Journal of Botany* **40**, 527–543.

Giorgi F (2005). Interdecadal variability of regional climate change: implications for the development of regional climate change scenarios. *Meteorology and Atmospheric Physics* **89**, 1–15.

Hendy EJ, Gagan MK, Lough JM (2003). Chronological control of coral records using luminescent lines and evidence for non-stationary ENSO teleconnections in northeast Australia. *The Holocene* **13**, 187–199.

Hennessy K, Fitzharris B, Bates BC, Harvey N, Howden SM, Hughes LSJ, Warrick R (2007). Australia and New Zealand. In *Climate Change 2007 – Impacts, Adaptation and Vulnerability – Contribution of Working Group II to the Fourth Assessment Report of the International Panel on Climate Change*. (Eds ML Parry, OF Canziani, JP Palutikof, PJ van der Linden, CE Hanson) pp. 507–540. Cambridge University Press, Cambridge.

Hobday AJ, Poloczanska ES, Matear RJ (2008). 'Implications of climate change for Australian fisheries and aquaculture – a preliminary assessment.' CSIRO, Hobart.

Howden SM, Ash AJ, Barlow EWR, Booth T, Charles S, Cechet R, Crimp S, Gifford RM, Hennessy K, Jones RN, Kirschbaum MUF, McKeon GM, Meinke H, Park S, Sutherst R, Webb L, Whetton PJ (2003). 'An overview of the adaptive capacity of the Australian agricultural sector to climate change – options, costs and benefits'. Australian Greenhouse Office, Canberra.

Howden SM, Jones RN (2004). Risk assessment of climate change impacts on Australia's wheat industry. In *New Directions for a Diverse Planet: Proceedings of the 4th International Crop Science Congress*. 26 September–1 October 2004, Brisbane, Queensland. p. 6. Regional Institute, Gosford.

Howden SM, Soussana JF, Tubiello FN, Chhetri N, Dunlop M, Meinke HM (2007). Adapting agriculture to climate change. *Proceedings of the National Academy of Sciences* **104**, 19691–19696.

Keating BA, Carberry PS (2008). Emerging opportunities for Australian agriculture? *Proceedings of the 14th Australian Society of Agronomy Conference.* http://www.regional.org.au/au/asa/2008/plenary/emerging_opportunities/5923_keatingb.htm

Lilley JM, Bolger TP, Gifford RM (2001). Productivity of *Trifolium subterraneum* and *Phalaris aquatica* under warmer, high CO_2 conditions. *New Phytologist* **150**, 371–383.

Manunta P, Kirkham MB (1996). Respiration and growth of sorghum and sunflower under predicted increased night temperatures. *Journal of Agronomy and Crop Science-Zeitschrift fur Acker und Pflanzenbau* **176**, 267–274.

Marshall GJ (2003). Trends in the Southern Annular Mode from observations and reanalyses. *Journal of Climate* **16**, 4134–4143.

McAlpine CA, Syktus J, Deo RC, Lawrence PJ, McGowan HA, Watterson IG, Phinn S (2007). Modelling the impact of historical land cover change on Australia's regional climate. *Geophysical Research Letters*, L22711.1-L22711.6 doi:10.1029/2007GL031524.

McKeon GM, Hall WB, Crimp SJ, Howden SM, Stone RC, Jones DA (1998). Climate Change in Queensland's grazing lands: I. Approaches and Climatic Trends. *Rangelands Journal* **20**, 151–176.

McKeon GM, Hall W, Henry B, Stone G, Watson I (2004). 'Pasture degradation and recovery in Australia's rangelands: learning from history.' Queensland Department of Natural Resources, Mines and Energy: Brisbane.

McKeon GM, Stone GS, Syktus JI, Carter JO, Flood N, Fraser GW, Ahrens DG, Bruget D, Crimp SJ, Cowley R, Johnston PW, Stokes C, Cobon D, Ryan JG, Howden SM (2009). Climate change impacts on rangeland livestock carrying capacity: a review. *Rangelands Journal* (submitted)

Meinke H, Stone RC (2005). Seasonal and inter-annual climate forecasting: the new tool for increasing preparedness to climate variability and change in agricultural planning and operations. *Climatic Change* **70**, 221–253.

Morison JIL, Gifford RM (1984). Plant growth and water use with limited water supply in high CO_2 concentrations .II. Plant dry weight, partitioning and water use efficiency. *Australian Journal of Plant Physiology* **11**, 375–384.

Morison JIL, Lawlor DW (1999). Interactions between increasing CO_2 concentration and temperature on plant growth. *Plant Cell and Environment* **22**, 659–682.

Moy AD, Howard WR, Bray SG, Trull TW (2009). Reduced calcification in modern Southern Ocean planktonic foraminifera. *Nature Geoscience* **2**, 276–280 doi: 10.1038/NGEO460.

Nakićenović N, Swart R (Eds) (2000). 'Special Report on Emissions Scenarios. A Special Report of Working Group III of the Intergovernmental Panel on Climate Change'. Cambridge University Press, Cambridge.

Nicholls N (2006). Detecting and attributing Australian climate change: a review. *Australian Meteorological Magazine* **55**, 199–211.

Pittock AB (1975). Climatic change and the patterns of variation in Australia rainfall. *Search* **6**, 498–504.

Power S, Casey T, Folland C, Colman A, Mehta V (1999). Inter-decadal modulation of the impact of ENSO on Australia. *Climate Dynamics* **15**, 319–324.

Rahmstorf S, Cazenave A, Church JA, Hansen JE, Keeling RF, Parker DE, Somerville RCJ (2007). Recent climate observations compared to projections. *Science* **316**, 709.

Saji NH, Goswami BN, Vinayachandran PN, Yamagata T (1999). A dipole mode in the tropical Indian Ocean. *Nature* **401**, 360–363.

Samarakoon AB, Gifford RM (1995). Soil water content under plants at elevated CO_2 and interactions with the direct CO_2 effects: a species comparison. *Journal of Biogeography* **22**, 1181–1190.

Solomon S, Qin D, Manning M, Chen Z, Marquis M, Averyt KB, Tignor M, Miller HL (2007). *Climate Change 2007: The Physical Science Basis. Contribution of Working Group I to the Fourth Assessment Report of the Intergovernmental Panel on Climate Change.* Cambridge University Press: Cambridge.

Stokes CJ, Ash AJ (2006). Impacts of climate change on marginal tropical animal production systems. In *Agroecosystems in a Changing Climate.* (Eds PCD Newton, RA Carran, GR Edwards, PA Niklaus) pp. 323–328. CRC Press, London.

Stokes CJ, Ash AJ, Holtum JAM, Woodrow I (2008). Savannas face the future: windows into a future CO_2-rich world. In *Proceedings of the 15th Biennial Australian Rangeland Society Conference: A Climate of Change in the Rangelands.* (Ed. DM Orr). Australian Rangeland Society, Charters Towers.

Stokes CJ, Howden SM (Eds) (2008). 'An overview of climate change adaptation in the Australian agricultural sector – impacts, options and priorities. Report prepared for the National Climate Change Strategy for Primary Industries.' CSIRO Sustainable Ecosystems, Canberra.

Syktus J (2005). Reasons for decline in Eastern Australia's rainfall. *Bulletin of the American Meteorological Society* **86**, 624.

Watson RT, Noble IR, Bolin B, Ravindranath NH, Verardo DJ, Dokken DJ (2000). 'Land use, land-use change, and forestry. A special report of the Intergovernmental Panel on Climate Change'. Cambridge University Press, Cambridge.

White WB, McKeon GM, Syktus JI (2003). Australian drought: the interference of multi-spectral global standing modes and travelling waves. *International Journal of Climatology* **23**, 631–662.

Ziska LH, Manalo PA (1996). Increasing night temperature can reduce seed set and potential yield of tropical rice. *Australian Journal of Plant Physiology* **23**, 791–794.

2
CLIMATE PROJECTIONS

KJ Hennessy, PH Whetton and B Preston

KEY MESSAGES:

- Warming of the climate system over the past century is beyond doubt, and it is partly due to anthropogenic increases in atmospheric concentrations of greenhouse gases.

- Since 1950, Australia has become warmer, with less rainfall in the south and east, and more rainfall in the north-west.

- Further global warming and climate change is expected due to projected increases in greenhouse gases.

- Australia is likely to become warmer and drier in the future.

- Small changes in average climate can have large effects on extreme weather events.

- Extremely hot years and days are likely to occur more often, with fewer frosts, more heavy rainfall in summer and autumn, less heavy rainfall in winter and spring in the south, more droughts, more fires, more intense tropical cyclones, less hail in the south and more hail along the east coast.

Introduction

Warming of the climate system over the past century is beyond doubt, and it is partly due to increases in greenhouse gases (IPCC 2007a). Further climate change is expected due to projected increases in greenhouse gases (IPCC 2007a). The consequences of such climate change will vary across regions and sectors (IPCC 2007b). While global warming can be slowed through large reductions in greenhouse gas emissions, some warming is unavoidable and adaptation will be needed to reduce damages and take advantage of opportunities (IPCC 2007b). A key part of planning for climate change is having access to reliable and relevant regional projections of climate change.

This chapter summarises the latest information about the climate change that has been observed in Australia since the early 20th century and its attribution to natural and anthropogenic causes. In addition, this chapter presents projections of global climate change over the 21st century as well as regional projections for Australia. This information is based on international climate change research, including conclusions and data from the Fourth Assessment Report of the Intergovernmental Panel on Climate Change (IPCC), and builds on a large body of climate research that has been undertaken for Australia in recent years. More detail is available in a technical report (CSIRO and Australian Bureau of Meteorology 2007).

Observed climate variability and change

Internal and external factors drive climate variability on a range of timescales. Internal factors are natural in origin and arise from complex interactions within the climate system, such as the

El Niño-Southern Oscillation (ENSO), the Indian Ocean Dipole, the Southern Annular Mode and the Inter-decadal Pacific Oscillation (CSIRO and Bureau of Meteorology 2007). A number of natural external factors also influence climate variability including the Earth's rotation, variations in the energy from the Sun, volcanic eruptions and changes in the Earth's orbital parameters. However, some external factors are human-induced, such as changes in land use, emissions of greenhouse gases and aerosols and stratospheric ozone depletion.

The main greenhouse gases are water vapour, carbon dioxide (CO_2), methane (CH_4) and nitrous oxide (N_2O). Since the Industrial Revolution (around 1750 AD) the atmospheric concentrations of carbon dioxide, methane and nitrous oxide have increased by 35%, 148% and 18%, respectively (IPCC 2007a). The increases in CO_2 are mainly due to fossil fuel use and land use change, while those of CH_4 and N_2O are mainly due to agriculture (IPCC 2007a).

The Earth's average surface temperature has increased by 0.7°C since the beginning of the 20th century (IPCC 2007a). Most of the warming since 1950 is very likely due to increases in atmospheric greenhouse gas concentrations associated with human activities, and it cannot be explained by natural variability alone (IPCC 2007a). The warming has been linked to more heatwaves, changes in precipitation patterns, reductions in sea ice extent and rising sea levels (IPCC 2007a).

Australian climate change

Australian average annual temperatures have increased by 0.9°C since 1910. Most of this warming has occurred since 1950 (see Figure 2.1, page 41), with the greatest increases in the east and a slight cooling in the far north-west (Figure 2.2, page 41) associated with increased rainfall. The warmest year on record for the whole of Australia is 2005, but 2007 was the warmest year for southern Australia (Australian Bureau of Meteorology 2008c). The number of hot days (maximum temperature greater than 35°C) and nights (minimum temperature greater than 20°C) has increased and the number of cold days (maximum temperature less than 15°C) and nights (minimum temperature less than 5°C) has declined (CSIRO and Australian Bureau of Meteorology 2007).

Since 1950, most of eastern and south-western Australia has become drier (see Figure 2.3, page 42). Across New South Wales and Queensland, rainfall trends partly reflect a very wet period around the 1950s, though recent years have been unusually dry. In contrast, north-western Australia has become wetter over this period, mostly during summer. Since 1950, the number of very heavy rainfall events of over 30 millimetres per day (mm/day) and the number of wet days (at least 1 mm/day) have decreased in the south and east but increased in the north (CSIRO and Australian Bureau of Meteorology 2007).

Australian rainfall shows considerable year-to-year variability, partly in association with the ENSO. El Niño events tend to be associated with hot and dry years in Australia, and La Niña events tend to be associated with mild and wet years (Power *et al.* 2006). There has been a marked increase in the frequency of El Niño events and a decrease in La Niña events since the mid-1970s (Power and Smith 2007). The frequency of tropical cyclones in the Australian region has decreased in recent decades, largely due to the increasing frequency of El Niños. However, the number of severe cyclones has not declined.

Causes of Australian climate change

Most of the Australian warming since the mid-20th century is very likely due to increases in greenhouse gases (IPCC 2007a). The 40% decline in south-east Australian snow depths in early spring over the past 40 years is due to warming rather than a decline in precipitation (Nicholls 2005). About 50% of the rainfall decrease in south-western Australia since the late 1960s is likely to be due to increases in greenhouse gases (Cai and Cowan 2006). The autumn rainfall decline in south-eastern Australia since the late 1950s may be partly due to increases in greenhouse gases (Cai and Cowan 2008). Increased rainfall in the north-west since 1950 may be due to increased aerosol particles produced by human activity (Rotstayn *et al.* 2008). However, the extent to which natural versus human causes have altered patterns of drought in Australia remains uncertain (Nicholls 2007; Hennessy *et al.* 2008).

Climate change projections

The IPCC developed forty scenarios of greenhouse gas and sulphate aerosol emissions for the

21st century based on various assumptions about demographic, economic and technological change (Nakićenović and Swart 2000). The scenarios are grouped into four 'storylines': B1, B2, A1 and A2, each of which has a number of variants. Six of these variants were utilised as 'illustrative' scenarios, designed to represent the broad range of uncertainty in emissions trajectories over the 21st century (see Figure 2.4, page 42). The climatic effects of projected changes in emissions can be simulated using climate models. These models are mathematical representations of the Earth's climate system based on well-established laws of physics, but involving simplifications of biophysical processes. Projections derived from climate models are not predictions because they are conditional on the emission scenarios and on the reliability of climate models.

Australian climate change simulations for the 20th and 21st centuries are currently available for 23 global climate models. Each model has been given a score based on its ability to simulate average observed (1961–1990) patterns of temperature, rainfall and mean sea level pressure in the Australian region (CSIRO and Bureau of Meteorology 2007). Models with higher scores are given greater emphasis in developing future climate projections.

The projections are relative to the baseline period 1980–1999 (i.e. 20 years centred on 1990). For convenience, this baseline is called 1990. Similarly, projections for 2030 and 2070 give an estimate of the average climate around these years. Individual years will, of course, show some variation from this average.

Uncertainties in projected climate change to 2030 are mostly due to differences between the results of the climate models rather than the different emission scenarios, so projections for 2030 are given for a mid-range emission scenario (the IPCC's A1B scenario). Beyond 2030, the magnitude of projected climate changes is strongly affected by the different emission scenarios, so results for 2070 are given for a low emission scenario (called B1) and for a high emission scenario (called A1FI). Since 1990, carbon dioxide emissions, global-mean temperatures and global-mean sea level have been tracking the upper edge of IPCC range of emission scenarios (Rahmstorf *et al.* 2007; Raupach *et al.* 2007).

To represent the uncertainty due to differences between climate model results, probability distributions are fitted to the range of results for temperature, rainfall and some other climate variables. This provides a central estimate based on the 50th percentile (the mid-point of spread of model results), and a range of uncertainty based on 10th and 90th percentiles (lowest 10% and highest 10% of the spread of model results). It also allows calculation of the probability of a change in climate being greater (or less) than a given threshold. Some uncertainties associated with projecting future global and regional climate change cannot be easily quantified. Therefore, changes outside the ranges given here cannot be excluded. In the following section, information is based on the report 'Climate Change in Australia – Technical Report 2007' (CSIRO and Bureau of Meteorology 2007) unless otherwise stated.

Australia's future climate

Annual and seasonal average changes in temperature

The central estimate of annual warming over Australia by 2030, relative to 1990, is about 1.0°C for the A1B emissions scenario (see Figure 2.5, page 43). Warming is likely to be a little less in coastal areas and a little more in the inland regions, with the least warming in winter and the most in spring. The 10th to 90th percentile range of uncertainty due to differences between models is about 0.6°C to 1.5°C for most of Australia, with the probability of the warming exceeding 1°C by 2030 being 10–20% for coastal areas, and over 50% for inland regions.

Later in the century, the warming is more dependent upon the emissions scenario. By 2070, the annual warming is around 1.8°C (range of 1.0°C to 2.5°C) for the low emission scenario and around 3.4°C (range of 2.2°C to 5.0°C) for the high emission scenario. By 2070, for the low emission scenario (B1), there is over 90% likelihood of warming exceeding 1°C in Australia, a 20–60% chance of exceeding 2°C over most inland areas, and about a 10% chance of exceeding 2°C in most coastal areas. For the high emissions (A1FI) case there is around a 30% chance of exceeding 3°C in southern and eastern coastal areas and a much greater chance inland, while the chance of exceeding 4°C is around 10% in most coastal areas and 20–50% inland.

Annual and seasonal average changes in rainfall

Unlike temperature, which is projected to increase, rainfall projections show increases in some regions and decreases in others. The direction of change (increase or decrease) is considered 'likely' where at least two-thirds of the models agree. Therefore, decreases are likely in southern areas for average annual and winter rainfall, in southern and eastern areas for average spring rainfall, and along the west coast for average autumn rainfall (see Figure 2.6, page 43). Outside of these regions, the models do not give a likely direction of rainfall change, although model ranges show a tendency to decrease in most cases. An increase in rainfall is not 'likely' in any region or season.

For 2030, central estimates of annual rainfall change indicate little change in the far north and decreases of 2% to 5% elsewhere. Decreases of around 5% occur in winter and spring, particularly in the south-west where they reach 10%. In summer and autumn, decreases are smaller and there are slight increases in north-eastern New South Wales (NSW) in summer.

The range of rainfall change in 2030, allowing for differences between models, is large. The projected annual average change is around -10% to +5% in northern areas and -10% to little change in southern areas. Winter and spring changes range from -10% to little change in southern areas in the continental south-east, -15% to little change in the south-west, and -15% to +5% in eastern areas. In summer and autumn, the range is typically -15% to +10%. Decadal-scale natural variability in rainfall is comparable in magnitude to these projected changes and may therefore mask, or significantly enhance, the greenhouse-induced changes. This is important when considering the possible causes of the rainfall decreases in southern and eastern Australia over the past decade.

In 2070 for the low emission scenario, the range of annual rainfall change is -20% to +10% in central, eastern and northern areas, with a best estimate of little change in the far north grading to around -7.5% elsewhere. The range of change in southern areas is from -20% to little change, with a central estimate of around -7.5%. Seasonal changes follow the pattern described for 2030, but are correspondingly larger. There is a 40 to 50% chance of an annual rainfall decrease of at least 10% in western and central areas, and a 10 to 20% chance of rainfall decreases of at least 20% in these areas. There is a 10 to 20% chance of rainfall increases of at least 10% in some northern areas.

In 2070 for the high emission scenario, the range of annual average rainfall change in central, eastern and northern areas grows to -30% to +20%, with a central estimate of little change in the far north grading to around -10% in the south-west. The range of change in southern areas is from -30% to +5%, with a central estimate of around -10%. Seasonal changes may be larger, with projected winter and spring decreases in the south-west of up to 40%. There is a 40 to 50% chance of rainfall decreases of at least 20% in the south-west. There is a 10 to 20% chance of rainfall increases of at least 20% in the north.

Annual and seasonal average changes in other climate variables

Other changes in climate include an increase in potential evaporation, higher wind speeds in many coastal locations, less snow and humidity, more sunshine in the south, warmer and more acidic oceans, and higher sea levels (Table 2.1).

Extreme events

Small changes in average climate can have a significant effect on extreme events. This section summarises projected changes in the frequency of extremely hot days, daily temperatures over 35°C and below 0°C, heavy daily rainfall intensity, extremely dry years, fire weather, oceanic storm surges, tropical cyclones, hail storms and tornadoes.

Extremely hot years were defined as occurring once every 20 years on average (Hennessy *et al.* 2008). Seven regions were considered: Queensland, NSW, Victoria and Tasmania (Vic&Tas), the Murray–Darling Basin, north-western Australia, south-western Australia, and the far south-west near Perth. Over the period 1900–2007, climate models show that extremely hot years affect about 4.5% of the area in each of the seven regions. By 2010–2040, for a mid-range emission scenario (A1B), the mean area may increase to 60–80%, with a low emission estimate of 40–60% and a high emission estimate of 80–95%. By 2010–2040, the frequency of extremely hot years may increase to once every one to two years.

Table 2.1: Projected changes to climate variables other than temperature and rainfall.

Variable	Projection
Solar radiation	Little change, although a tendency for increases in southern areas of Australia is evident, particularly in winter and spring.
Relative humidity	Small decreases over most of Australia.
Potential evaporation	Increases 2–6% by 2030 (central estimate), with largest increases in the north and east in autumn and winter.
Wind speed	Increases of 2–5% by 2030 (central estimate) in most coastal areas, except for a band around latitude 30°S in winter and 40°S in summer where there are decreases of 2–5% by 2030.
Snow	10–40% less annual average snow-covered area by 2020, with shorter snow-cover duration and smaller peak snow depths.
Sea level rise	A global-average rise of 18–59 cm by 2095, plus an allowance of another 10–20 cm for a potential disintegration of the ice sheets (IPCC 2007a). However, further ice sheet contributions may substantially increase the upper limit of sea level rise.
Ocean temperature	An increase of 0.6–0.9°C by 2030 (central estimate) in the southern Tasman Sea and off the north-west coast and 0.3–0.6°C elsewhere.
Ocean acidity	Increases in ocean acidity due to continued absorption of carbon dioxide, with the largest increases in the high and mid latitudes.
El Niño-Southern Oscillation	Significant uncertainty exists regarding changes in the frequency and intensity of El Niño and La Niña events.

The average warming projections were added to observed daily temperature data for 1971–2000, giving estimates of daily temperature data for 2030 and 2070. This assumes no change in daily temperature variability. Substantial increases in the frequency of days over 35°C are expected (Table 2.2). For example, the central estimate for Melbourne is a 25% increase by 2030, a 54% increase for the low emission scenario by 2070 and a 120% increase for the high emission scenario by 2070. Fewer frosts are also expected (Table 2.3). For example, the central estimate for Dubbo (NSW) is a 32% decrease by 2030, a 45% decrease for the low emission scenario by 2070

and a 79% decrease for the high emission scenario by 2070.

Heavy daily rainfall was defined as the highest 1% in a given year or season. The simulated intensity of heavy rainfall tends to increase in many areas, especially in summer and autumn. However, decreases are simulated in the south in winter and spring, associated with a strong decrease in average rainfall.

Droughts are likely to occur with greater frequency and severity in future, although details vary depending upon how drought is defined. For example, drought definitions based on rainfall

Table 2.2: Average number of days per year above 35°C at selected sites for the present climate (1971–2000) and central estimates for 2030 and 2070, with ranges of uncertainty in brackets. The mid, low and high emissions are A1B, B1 and A1FI, respectively. (State and Territory abbreviations are: Qld = Queensland; Tas = Tasmania; NSW = New South Wales; ACT = Australian Capital Territory; Vic = Victoria; NT = Northern Territory; SA = South Australia; WA = Western Australia.)

	Present average (1971–2000)	2030 average (mid emissions)	2070 average (low emissions)	2070 average (high emissions)
Brisbane (Qld)	1.0	2.0 (1.5–2.5)	3.0 (2.1–4.6)	7.6 (4–21)
Hobart (Tas)	1.4	1.7 (1.6–1.8)	1.8 (1.7–2.0)	2.4 (2.0–3.4)
Sydney (NSW)	3.5	4.4 (4.1–5.1)	5.3 (4.5–6.6)	8.2 (6–12)
Canberra ACT)	5.3	7.8 (7.6–10.5)	10.4 (8.2–13)	16.8 (11.7–25.1)
Melbourne (Vic)	9.1	11.4 (11–13)	14.0 (12–17)	20.0 (15–26)
Darwin (NT)	11.0	44.0 (28–69)	89.0 (49–153)	230.0 (140–308)
Adelaide (SA)	17.0	23.0 (21–26)	26.0 (24–31)	36.0 (29–47)
Perth (WA)	28.0	35.0 (33–39)	41.0 (36–46)	54.0 (44–67)

Table 2.3: Average number of days per year below 0°C at selected sites for the present climate (1971–2000) and central estimates for 2030 and 2070, with ranges of uncertainty in brackets. The mid, low and high emissions are A1B, B1 and A1FI, respectively.

	Present average (1971–2000)	2030 average (mid emissions)	2070 average (low emissions)	2070 average (high emissions)
Charleville (Qld)	12	6.1 (2.6–7.5)	3.0 (1.2–5.5)	0.5 (0–2)
Nuriootpa (SA)	12	7.0 (4.2–8.2)	4.8 (2.7–7.0)	1.6 (0.2–3.8)
Wandering (WA)	15	8.7 (6.7–9.5)	6.7 (3.8–8.5)	1.7 (0.5–3.9)
Dubbo (NSW)	22	15.0 (11–16)	12.0 (8–14)	4.6 (1.1–8.5)
Launceston (Tas)	35	26.0 (19–27)	21.0 (16–25)	12.0 (5–18)
Rutherglen (Vic)	44	32.0 (25–35)	26.0 (20–32)	14.0 (9–23)
Canberra (ACT)	61	49.0 (42–51)	43.0 (36–48)	29.0 (20–39)

deficiency do not account for changes in potential evaporation. A drought index based on soil moisture deficiency includes the effects of changes in rainfall and potential evaporation. Since long-term measurements of soil moisture are not available, historical variations can be simulated using a soil moisture model. Daily soil moisture data over the whole of Australia was simulated for a 50 year period (1957–2006) (Hennessy *et al.* 2008). In the seven regions considered for exceptionally hot years (above), about 6% of each region has extremely low soil moisture, defined as occurring once every 16 years on average, in the current climate. Soil moisture was simulated for 50 years centred on 2030 for a mid-range emissions scenario (A1B), allowing for projected changes in rainfall and potential evaporation. The area with extremely low soil moisture years increases to around 7% in most regions (uncertainty range of 4.5–10%), with larger increases in southern Australia particularly the far south-west. The return period for these events falls to 12–14 years (uncertainty range 9–22 years) in most regions, with shorter return periods in the southern Australia, particularly the far southwest.

An increase in fire weather risk is likely with warmer and drier conditions (Lucas *et al.* 2007). Simulations from two models show that the number of 'extreme' fire danger days in south-eastern Australia generally increases 5–25% by 2020 for the low emissions scenario and 15–65% for the high scenario. By 2050, the increases are generally 10–50% for the low scenario and 100–300% for the high scenario. It is likely that the fire season will extend, shifting periods suitable for prescribed burning toward winter.

Three recent studies have produced projections for tropical cyclone changes in the Australian region. The combined effects of changes in cyclone frequency, duration and the region of genesis determine the annual average number of cyclone days (defined as days on which a cyclone resides in a 200 km x 200 km box) (see Figure 2.7, page 44). More cyclone days are simulated off the Queensland coast, with little change near the Northern Territory and fewer days off the Western Australian coast. In one study, the simulated frequency of severe cyclones (Categories 3 to 5) increases 60% by 2030 and 140% by 2070 (Abbs *et al.* 2006). Other studies indicate that by 2050, the increase in frequency may be 22% (Leslie *et al.* 2007) or 56% (Walsh *et al.* 2004).

Changes in local storm activity are difficult to simulate. There is potential for significant increases in coastal inundation due to higher mean sea level and more intense storms, such as tropical cyclones and other low pressure systems. Simulations indicate a possible reduction in cool season tornadoes and hail storms in southern Australia, but hail risk may increase along the east coast from Brisbane to Melbourne.

Discussion

The climatic conditions that have challenged Australian agriculture over the past decade could be the most recent manifestations of a long-term shift in both the national and global climate. There are two primary lines of evidence that indicate this shift will persist and intensify in the years and decades ahead. The first is the attribution of observed global and Australian climate change to human activity and greenhouse gas emissions in particular. This link between emissions and the changing climate suggests additional climatic change will occur as greenhouse gas emissions continue to rise. The second line of evidence is based upon simulations of the response of global and regional climates to scenarios of increasing emissions and atmospheric concentrations of greenhouse gases. As detailed in this chapter, such simulations generally suggest that Australia's mean climate will trend toward warmer and drier conditions while global sea level rises.

While these physical changes in the environment are significant in their own right, they also have important implications. The performance of biological and ecological systems, including agricultural production systems, is directly influenced by climatic conditions and atmospheric CO_2 concentrations as well as water availability (i.e. soil moisture and irrigation). Meanwhile, changes in the performance of these systems will have a range of downstream social and economic implications, particularly for those with close ties to the agriculture sector. Nevertheless, there is a range of uncertainties that influence understanding about the dynamics of climate change including the accuracy and precision associated with climate projections and their consequent impacts. As scientific understanding of the climate system advances, it may be possible to constrain some of the objective uncertainties associated with climate projections so as to better inform decision-making. However, in those sectors such as agriculture, where delaying decision-making until significant reductions in uncertainty are available is not practical, adaptive actions will necessitate the management of that uncertainty to identify options that are robust over the likely range of future climate outcomes.

References

Abbs DJ, Aryal S, Campbell E, McGregor J, Nguyen K, Palmer M, Rafter T, Watterson I, Bates B (2006). 'Projections of Extreme Rainfall and Cyclones'. A report to the Australian Greenhouse Office, Canberra, Australia.

Australian Bureau of Meteorology (2008a). Australian Climate Change and Variability. http://www.bom.gov.au/cgi–bin/silo/products/cli_chg/

Australian Bureau of Meteorology (2008b). Tropical cyclone trends. http://www.bom.gov.au/weather/cyclone/tc–trends.shtml.

Australian Bureau of Meteorology (2008c). Annual Climate Summary 2007. http://www.bom.gov.au/climate/annual_sum/2007/

Cai W, Cowan T (2006). SAM and regional rainfall in IPCC AR4 models: can anthropogenic forcing account for southwest Western Australian winter rainfall reduction? *Geophysical Research Letters* **33**, L24708, p 1-5. doi:10.1029/2006GL028037.

Cai W, Cowan T (2008). Dynamics of late autumn rainfall reduction over southeastern Australia. *Geophysical Research Letters* **35**, L09708, pp. 1–5. doi:10.1029/2008GL033727.

CSIRO, Australian Bureau of Meteorology (2007). 'Climate change in Australia'. CSIRO and Bureau of Meteorology Technical Report, Melbourne. www.climatechangeinaustralia.gov.au

Hennessy KJ, Fawcett R, Kirono D, Mpelasoka F, Jones D, Bathols J, Whetton P, Stafford Smith M, Howden M, Mitchell C, Plummer N (2008). 'An assessment of the impact of climate change on the nature and frequency of exceptional climatic events.' CSIRO and the Australian Bureau of Meteorology, Melbourne. http://www.bom.gov.au/climate/droughtec/

IPCC (2007a). *Climate Change 2007: The Physical Science Basis. Contribution of Working Group I to the Fourth Assessment Report of the Intergovernmental Panel on Climate Change.* (Eds S Solomon, D Qin, M Manning, Z Chen, M Marquis, KB Averyt, M Tignor HL Miller) Cambridge University Press, Cambridge.

IPCC (2007b). *Climate Change 2007: Impacts, Adaptation and Vulnerability. Contribution of Working Group II to the Fourth Assessment Report of the Intergovernmental Panel on Climate Change.*

(Eds ML Parry, OF Canziani, JP Palutikof, PJ van der Linden, CE Hanson) Cambridge University Press, Cambridge.

Leslie, LM, Karoly DJ, Leplastrier M, Buckley BW (2007). Variability of tropical cyclones over the southwest Pacific Ocean using a high resolution climate model. *Continuum Mechanics and Thermodynamics* **19**, Issue 3–4, 133–175.

Lucas C, Hennessy KJ, Mills G, Bathols JM (2007). 'Bushfire Weather in Southeast Australia: Recent Trends and Projected Climate Change Impacts'. A report prepared by the Bushfire CRC, Australian Bureau of Meteorology and CSIRO Marine and Atmospheric Research for The Climate Institute of Australia. http://www.climateinstitute.org.au/index. php?option=com_content&task=view&id=88&Itemid=41

Nakićenović N, Swart R (Eds) (2000). 'Special Report on Emissions Scenarios. A Special Report of Working Group III of the Intergovernmental Panel on Climate Change'. Cambridge University Press, Cambridge.

Nicholls N (2005). Climate variability, climate change and the Australian snow season. *Australian Meteorological Magazine* **54**, 177–185.

Nicholls N (2007). 'Detecting, understanding and attributing climate change'. Report for the Australian Greenhouse Office, Canberra.

Power S, Haylock M, Colman R, Wang X (2006). The predictability of interdecadal changes in ENSO activity and ENSO teleconnections. *Journal of Climate* **19**, 4755–4771.

Power SB, Smith IN (2007). Weakening of the Walker Circulation and apparent dominance of El Niño both reach record levels, but has ENSO really changed? *Geophysical Research Letters* **34**, L18702, pp. 1–4. doi:10.1029/2007GL30854.

Rahmstorf S, Cazenave A, Church JA, Hansen JE, Keeling RF, Parker DE, Somerville RCJ (2007). Recent climate observations compared to projections. *Science* **316**, 709–710.

Raupach MR, Marland G, Ciais P, Le Quéré C, Canadell JG, Klepper G, Field CB (2007). Global and regional drivers of accelerating CO_2 emissions. *Proceedings of the National Academy of Sciences of the USA* **104**, 10288–10293.

Rotstayn LD, Keywood MD, Forgan BW, Gabric AJ, Galbally IE, Gras JL, Luhar AK, McTainsh GH, Mitchell RM, Young SA (2008). Possible impacts of anthropogenic and natural aerosols on Australian climate: a review. *International Journal of Climatology* **10**, 1002/joc.1729.

Walsh KJE, Nguyen KC, McGregor JL (2004). Finer-resolution regional climate model simulations of the impact of climate change on tropical cyclones near Australia. *Climate Dynamics* **22**, 47–56.

3
GRAINS

SM Howden, RG Gifford and H Meinke

KEY MESSAGES:

■ In addition to dealing with climate variability, adaptation to climate change is already part of cropping systems management in Australia. In some instances adaptation can turn a problem into an opportunity or at least reduce climate-related vulnerability. For the wheat industry alone, relatively simple adaptations to future climate change may be worth between $100 million to $500 million per year at the farm gate. Further benefits are likely if a wider range of adaptations is practised but these are yet to be evaluated.

■ There is a range of technical adaptations such as changed crop management practices, new varieties, altered rotations and improved water management that may, in some situations, have considerable benefit. Often these are consistent with existing best practice for climate risk. However, these practices may need to be modified, enhanced or integrated in different ways to cope with the likely challenges posed by climate change.

■ There is also a range of policy adaptations that form the decision environment within which farm and enterprise adaptations take place. These are industry and regional development policies, stewardship programs, infrastructure development, industry capacity development programs and other policies such as those relating to drought support, rural adjustment and trade among many others. Maintaining a flexible research and development base to inform policy adaptations as well as farm-level changes is essential to deliver potential adaptation benefits.

■ Maladaptation can occur either through over- or under-adaptation to climate changes or because of misalignment of policy and practice. Effective monitoring of change and adaptation at a range of temporal, spatial and sectoral scales could help reduce the risks of maladaptation by learning from experience and fine-tuning adaptation measures to the unique local conditions. Assessments to ensure that adaptations do not increase net greenhouse gas emissions or have other unintended consequences will become increasingly important.

■ The translation of climate change information into adaptation action requires participatory approaches across the agricultural value chain. Such studies will carry the analysis from climate to biophysical impacts on crops and cropping systems to enterprise level adaptation options to farm financial impacts to regional economic and social impacts (such as via livelihoods analysis) and then through to policy options. Integration and adaptive learning are critical and could occur through social and analytical links between the user communities and researchers.

Introduction

Cropping of various grains, oilseeds and pulses (referred to generically here as grains) is the major agricultural activity in Australia with a gross value of about $11 300 million per year. Cropping occurs over an area of some 24 million hectares (Mha) distributed from the summer-dominated rainfall region of the central highlands of Queensland in an arc around the winter-dominated rainfall areas of southern Australia and around to Geraldton in Western Australia (WA). In the western and southern regions, the predominantly cool season rainfall (i.e. autumn to spring) allows cropping of wheat, barley, canola, lupins, oats and other cool season crops to take place on a variety of soil types, from the sands of WA to the heavy clay soils of the Wimmera in Victoria. In contrast, in the northern regions, cropping is largely restricted to heavier soils with high moisture-holding capacity which can store the predominantly summer rainfall so that it is available for the cool season crops. Summer crops such as sorghum and maize can also be grown in these regions. In all regions, the industry is highly sensitive to climate with both wet and dry years causing substantial fluctuations in regional yield and grain quality. The difference in production between a dry year (low production) and a wet year (high production) is typically a factor of two or more (Figure 3.1). In some regions, irrigation is practised so as to reduce the fluctuations in production caused by dry years. However, even irrigation is exposed to climate risks as reduced water allocations for irrigation are occurring frequently in Australia's many over-allocated river systems.

High rainfall years also can cause problems with waterlogging, flooding, rain and hail damage, higher pest and disease loads and intermittent recharge of water to groundwater tables and associated leaching of nutrients and movement of salts.

Grain-cropping systems have comparatively high levels of management input compared with extensive grazing or forestry. This, in combination with

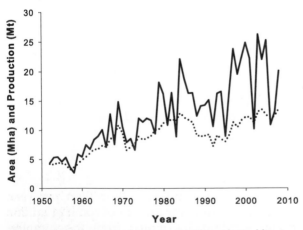

Figure 3.1: Area cropped to wheat (Mha – dotted line) and wheat production (million tonnes (Mt) – solid line) in Australia from 1952 to the present. ABARE 2009

the sub-annual time steps involved in the management of most crops allows considerable, but not unlimited, latitude for adaptation options to climate change. However, few of these options have previously been comprehensively analysed. The following sections outline the key adaptation options for cropping systems. The information in these is drawn from the literature, from a survey sent out to 26 of Australia's top grain-cropping farmers and from a range of participatory action research workshops (Crimp *et al.* 2006).

Impacts of projected climate change

Australian grain-cropping systems are likely to be impacted by combinations of higher levels of CO_2, increased temperatures and changes in rainfall, in order of diminishing confidence. These impacts have been reviewed recently by Tubiello *et al.* (2000). Elevated CO_2 has long been known to enhance plant growth under most environmental conditions due to increased photosynthetic rates and increased water use efficiency through decreased stomatal conductance. The degree of response depends on other variables such as temperature, soil moisture and soil nutrient availability, especially nitrogen. Grain yields may increase by about 21% at CO_2 concentration of 550 ppm and by 30% at 700 ppm when compared with a year 2000 baseline of 370 ppm. However, there is likely to be a concomitant decrease in grain protein content.

Temperature increases will accelerate plant developmental rates. For varieties adapted to existing temperature regimes, higher temperatures will result, all else being equal, in reduced yields due to the reduced time for accumulation of solar radiation. Increased temperature may increase evapotranspiration depending on changes in wind speed and absolute humidity, resulting potentially in more rapid depletion of soil moisture, particularly in spring when the grain is filling and ripening. This can reduce grain number, reduce harvest index, and under high soil nitrogen may result in 'haying off' with subsequent major reductions in grain yield. However, there remains some uncertainty as to how evapotranspiration will change in the future (e.g. Rayner 2007).

Rainfall is a key determinant of yields at both farm and national levels (e.g. Stephens and Lyons

1998), with much of the variability in national crop production arising from variability in rainfall. There remains considerable uncertainty as to likely changes in rainfall with climate change (Chapter 2) but on balance there appears to be a considerable risk of lower rainfall and hence lower yields across the cropping belts of southern Australia although small increases cannot be ruled out in certain areas. In the northern cropping zones, either increases or decreases in rainfall and hence yield appear to be about equally possible although there is some evidence that climate change may result in the mean state of the Pacific becoming more El Niño-like with consequent impacts on eastern Australian rainfall (IPCC 2007). In general, yields are likely to be highly sensitive to reductions in mean rainfall but less so to increases, as at present. Any reductions in rainfall in the Mediterranean cropping zones are likely to favour the use of light-textured soils rather than heavy soils with high clay content (van Ittersum *et al.* 2003) and may result in both shifts away from cropping in the dry margins of the cropping lands and expansion into the wet margins that previously were too wet for effective cropping. Lower rainfall may reduce the risks of dryland salinisation notwithstanding some opposite effects arising from slight increases in soil moisture from elevated CO_2. Increased rainfall intensity could increase already problematic rates of soil erosion leading to longer-term yield decline

Many pests, diseases and weeds are strongly influenced by climatic factors and so are anticipated to alter their abundance and distributions with the changing climate (Chakraborty and Datta 2003). For example, elevated CO_2 concentrations may increase the fecundity and evolution rate of anthracnose (*Colletotrichum gloeosporioides*) and other pathogens while temperature increases can increase pathogen development and survival rates, disease transmission and host susceptibility. Consequently, warming may increase the amount of stripe rust (*Puccinia striiformis*). In contrast, Take-all (*Gaeumannomyces graminis*), a fungal disease that can cause major crop losses when there are extended periods of high soil moisture, may be reduced in severity where winters become drier. Viral diseases such as Barley Yellow Dwarf, which rely on transfer by aphids, may increase with warmer winter temperatures with potential additional risks via the positive effects of CO_2 concentration on aphid populations. Elevated CO_2 may also increase insect damage via compensatory feeding due to lower leaf nitrogen concentrations. In other cases, increased CO_2 enhances plant defensive compounds reducing insect herbivore weight gains, increasing mortality and lowering fecundity. Further, some weed species (especially summer-growing C_4 weeds) may increase their competitive advantage under elevated CO_2 and higher temperatures.

Options for dealing with climate variability

Climate in Australia varies over a large range of timescales resulting in a large variety of management responses (e.g. Table 3.1).

Recent studies with selected farm managers in Queensland indicate that by using climate information (e.g. seasonal forecasts) in conjunction with systems analyses, producers can reduce the impact of various climate risks (e.g. Crimp *et al.* 2006). By identifying climate-related decisions that positively influence the overall farm operation in either economic or environmental terms, these producers have gained a better understanding of the system's vulnerability and thus improved climate risk management in their operations. Examples for actions taken when a forecast is for 'likely to be drier than normal' include: maximising no-till area (water conservation), applying some nitrogen fertiliser early to allow planting on stored soil moisture but splitting the application so as to apply some later if a good season eventuates; planting most wheat later than normal to reduce frost risk and increasing row spacing. In seasons that are likely to be wetter than normal, management options include: sowing wheat earlier; applying nitrogen to a wheat cover crop grown on a dry profile after cotton (normally not expected to produce a harvestable yield) and applying fungicides to wheat crops to minimise leaf diseases (Meinke and Hochman 2000).

At the crop level, wheat plantings are now three to four weeks earlier than in the 1950s (Stephens and Lyons 1998). This is largely the result of a drastically reduced frost incidence in central Queensland (see Box 3.1) aided by the availability of new wheat cultivars that are well-suited to this

Table 3.1: Agricultural decisions matched with timescales of climate variation (also see Chapter 1 in this volume; Meinke and Stone 2005, Table II).

Decision type	Frequency (years)
Logistics (e.g. scheduling of planting / harvest operations)	Intraseasonal (>0.2)
Tactical crop management (e.g. fertiliser, row spacing)	Intraseasonal (0.2–0.5)
Crop type (e.g. wheat or chickpeas)	Seasonal (0.5–1.0)
Crop sequence (e.g. long or short fallows)	Inter-annual (0.5–2.0)
Crop rotations (e.g. winter or summer crops)	Annual/bi-annual (1–2)
Crop industry (e.g. grain or cotton)	Decadal (≈10)
Agricultural industry (e.g. crops or pastures)	Inter-decadal (10–20)
Land use (e.g. agriculture or natural systems)	Multi-decadal (20+)
Land use and adaptation of current systems	Climate change

Box 3.1: Example of the grain industry adapting to changing frost risk

Across Australia, over the last five decades there has been a strong trend of increasing minimum temperatures (Torok and Nicholls 1996) that have resulted in reductions in frost frequency and duration (Stone *et al.* 1996), at least in the northern half of the continent. This has contributed to earlier sowing in affected areas, allowing increased yields due to increased likelihood of avoidance of drought during grainfill but without increase in frost risk (Howden *et al.* 2003). Nicholls (1997) argues that this effect plus other more minor climate changes have contributed to 30–50% of the observed increase in national yields since 1950 although there are alternative views on this (e.g. Gifford *et al.* 1998). In an assessment of alternatives to managing frost risk in Central Queensland, Howden *et al.* (2003) tested three different strategies: (1) planting date based solely on soil moisture status and rainfall, (2) planting date based on a 10% risk of frost during anthesis using the full 100 years of climate records and (3) an adaptive strategy where planting date aimed at a 10% risk of frost but based on the previous 10 years of climate only. The 100-year strategy had the lowest simulated gross margin ($29 per ha per year) because ignoring the climate trend meant extra frost risk and more damaged crops early in the 100 year record and missing high yield opportunities later. The rainfall-only strategy performed only slightly better ($34 per ha per year). In contrast, the adaptive strategy almost doubled gross margins ($52 per ha per year) as it continually re-balanced risk and opportunities according to the changing climate, developing a much more opportunistic cropping system (Pollock *et al.* 2001).

environment. Although these changes started to happen in the 1970s and 80s it is only recently that climate trends were identified as one of the drivers (Howden *et al.* 2003).

At the national level there have been strong trends to earlier sowing times, with sowing progressing a day earlier per year on average but with greater changes in Queensland and Western Australia (Stephens and Lyons 1998). This appears to be related to the adoption of new herbicide and planting technologies that increase speed of soil preparation and reduce rainfall requirements to sow (Kerr *et al.* 1992). Earlier sowing dates may also be in response to the strong observed increases in minimum temperatures over this period as discussed in Box 3.1. Increases in atmospheric CO_2 levels may have also contributed to increased yields by an estimated 8% over the past 100 years (Howden *et al.* 1999a).

In addition to using information on climate variability in on-farm decision-making, there are also developing applications in terms of policy and marketing. For example, a regional commodity forecasting system has been devised for wheat (Potgieter *et al.* 2006) and sorghum (Potgieter *et al.* 2005). It allows an assessment from the beginning of the cropping season of the likelihood of exceeding the long-term median shire yield associated with different season types. This system is now

run operationally for Queensland and northern NSW by monthly updates based on the actual rainfall that has occurred and any change in the SOI (Southern Oscillation Index) phase from month to month (e.g. http://www.dpi.qld.gov.au/cps/rde/dpi/hs.xsl/26_8099_ENA_HTML.htm). The system was originally designed to inform government in Queensland of any areas that might be more likely to experience poor crops in any year, but now regularly forecasts production conditions and harvest size for commercial clients, but also – in a more general form – for the general public. This information provides an alert for 'Exceptional Circumstances' issues associated with potential drought in the same manner described for pasture systems in Queensland by Carter *et al.* (2000). Anecdotal information received from marketing agencies based on their experience with the regional wheat outlook indicate that seasonal crop forecasting in their decision-making processes can be beneficial when it is used in addition to current information systems. Possible decisions to be taken when the outlook is for 'likely to be drier (or wetter) than normal' are, for instance, forward buying (or selling) of grain or shifting of resources from good yielding areas to poor yielding areas.

Comprehensive assessments of the options for dealing with climate variability in cropping systems can be found in Muchow and Bellamy (1991) and Hammer *et al.* (2000) among other sources.

Options for dealing with projected climate change

Many of the management level adaptation options available for cropping systems are largely extensions or intensifications of existing climate risk management or production enhancement activities in response to a potential change in the climate risk profile (Howden *et al.* 2007). For grain-cropping systems there are many potential ways to alter management to deal with projected climatic and atmospheric changes. If widely adopted, these adaptations singly or in combination have substantial potential to offset negative climate change impacts and take advantage of positive ones. For example, in a modelling study for Modena in Italy (Tubiello *et al.* 2000), simple and feasible adaptations altered negative impacts on sorghum (-48 to -58%) to neutral to marginally positive ones (0 to +12%). In that case the adaptations trialled were to alter varieties and planting times to avoid drought and heat stress during the hotter and drier summer months predicted under climate change. Similar approaches are relevant for Australia too. When summarised across many adaptation studies globally, there is a tendency for most of the benefits of adapting the existing cropping systems to be gained under moderate warming (<2°C) then to level off with increasing temperature changes (Figure 3.2; Howden and Crimp 2005; Howden *et al.* 2007). Additionally, the yield benefits tend to be greater under

scenarios of increased rainfall than those with decreased rainfall, reflecting the fact that there are many ways of more effectively using more abundant resources, whereas there are fewer and less effective options for ameliorating risks when conditions become more limiting. Overall, the potential global benefits of management adaptations are substantial and are similar in temperate and tropical systems (17.9% versus 18.6%). The following sections address these management level adaptations in detail.

Varietal change

Temperature increases reduce the duration of phenological stages of crops, restricting the time they have to accumulate radiation and nutrients. This will generally reduce grain yield thereby tending to counter the yield increase deriving from the CO_2 fertilisation (see Chapter 1). In Australia it was estimated that, in the absence of adaptive measures, a 1.5–2°C increase in mean temperature would cancel out the grain yield increase in wheat deriving from a CO_2 doubling assuming that no varietal adaptation was practised (Gifford 1989; Wang *et al.* 1992; Howden 2002). Where there is adequate moisture (wet regions or where climate change increases rainfall), there is likely to be an advantage in breeding and adopting slower-maturing cultivars (greater thermal time requirements) that could capitalise on the earlier date of flowering and potentially longer photosynthetically-active period before seasonal drought forces maturity. Where there is likely to be both increases in temperature and reductions in rainfall (e.g. in the strongly Mediterranean climate cropping regions) it may be advantageous to either keep varieties with similar or earlier-flowering characteristics than are currently used as this will allow grainfill to occur in the cooler, wetter parts of the year (Howden *et al.* 1999a; van Ittersum *et al.* 2003) particularly if planting can occur earlier due to reduced frost risk. Characteristics such as higher response to elevated CO_2 conditions, rapid germination, enhancing drought tolerance (e.g. 'Staygreen' sorghum), early vigour and increased retention of flowers in hot/windy conditions may also need to be considered (e.g. Richards 2002). Adoption of the best varietal strategy needs careful evaluation on a site-by-site basis, taking into account changes in temperature, rainfall and management. The tools to undertake such assessments are available,

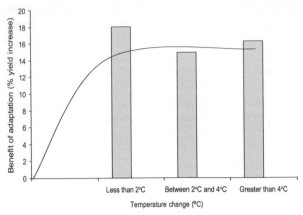

Figure 3.2: Mean benefit of adapting wheat cropping systems to the impact of temperature and rainfall changes calculated as the difference between % yield changes with and without adaptation. The temperature changes have associated changes in rainfall and CO_2 that vary between sites, scenarios and publications. The figures are drawn from a global synthesis of studies, with data from Howden *et al*. (2007). The line indicates the change in benefit including the scenario of no change.

but the plant breeding community needs to be better engaged on the issue.

Several of the adaptation strategies described below will require interactions with crop breeding groups. Particular issues are ensuring appropriate thermal time requirements, raising grain protein levels in higher CO_2 environments and maintaining pest and disease resistance. A further requirement may be for increased heat shock resistance. Heat shock occurs where there are high temperatures during grainfill (e.g. Blumenthal *et al*. 1991). These reduce dough-making quality of the grain. Heat shock incidence is likely to increase in northern Australia with climate change (1–50% increase; Howden *et al*. 1999a) perhaps requiring development of more heat-tolerant varieties so as to maintain capacity to produce high quality wheat. If there is increasing danger of more very hot days, then warnings of changed risk could be helpful in choosing crops that flower outside the key risk periods. The trend towards the incorporation of some vernalisation requirement into modern wheat cultivars will also tend to lock flowering into an appropriate relationship with a progressively earlier spring, reducing heat shock risk. Fine-tuning of this relationship would presumably occur over the several generations of varieties that will be developed as climate changes accumulate.

Hotter and drier conditions are likely to reduce the dry-down time prior to harvest and this also may affect final quality (e.g. grain cracking and small grains) requiring either breeding or management adaptation.

Given the importance of water as an increasingly limited resource, improving water use efficiency is of considerable interest to plant breeders. Attempts are now underway to genetically alter photosynthetic pathways in cereals by introducing aspects of the C_4 pathway into C_3 crops. While the biggest effort currently concentrates on rice (Mitchell and Sheehy 2006), other crops (e.g. wheat) are also being targeted. Using a simulation model, Meinke *et al*. (2008) found that grain yield advantage of a hypothetical 'C_4' rice varied strongly depending on climatic conditions and amounted to about half (average yield advantage: 23%) of the 50% yield increase hoped for by Mitchell and Sheehy (2006). For instance, considerably more nitrogen would need to be taken up to realise a 50% yield increase as C_4 photosynthesis responds more to leaf nitrogen than C_3 photosynthesis under high light conditions. Hence 'C_4' crops would need to develop a larger capacity to store carbohydrates produced before, during and shortly after flowering than existing C_3 crops. In other words, not only photosynthesis but also other mechanisms need to be genetically altered before such substantial yield increases can be realised (Meinke *et al*. 2008).

Species changes

Higher temperatures may enable the use of summer-growing grain and pulse species such as sorghum in temperate regions where these are not currently used in rotations. The negative impact of a reduction in rainfall is likely to be greater for rotation systems than for single crop systems as there is not a fallow in which to store soil moisture (Howden *et al*. 1999b). This will impact particularly on crops that show sensitivity to dry conditions such as canola and certain pulses. Furthermore, in rotations, options to vary planting windows are restricted, reducing flexibility to adapt management. Nevertheless, gradual adjustment of rotations will occur to minimise risk and maximise return. This will be aided by effective monitoring of soil moisture and nutrient levels, effective decision-support systems, improved seasonal climate forecasting and continuing improvements in crop management (including zero till, wide rows, variable planting density, canopy management, precision agriculture, etc.) that have relatively recently expanded planting options. If there is less frost risk, earlier spring plantings of warm season crops may be possible provided that moisture stress is managed.

Increased temperatures mean that cotton may also be able to be grown further south than currently (if adequate water is available) providing new rotation options in those areas with suitable soils (see Chapter 4). However, water availability is likely to be a key issue as the climate change forecasts are for greater reductions in mean flows in the catchments in the southern part of the Murray–Darling Basin than in the northern part (Chapter 12).

If there is reduction in rainfall and increased rainfall variability, it will make dryland cropping less attractive and farmers may well consider increasing the proportion of stock they have on-farm. Lowered production of annual pastures and crops may be partly offset by a greater planting of perennials such as lucerne that would be able to make use of summer rains and generally all available soil moisture. This will tend to extend the duration of rotations. The use of summer forage crops may increasingly be employed after summer rains; in regions with soils of low water-holding capacity (i.e. much of Western Australia) there may be a need for varietal development to increase the reliability of this option. However, livestock are also subject to the impacts of drought, requiring active management such as early removal of stock from pastures for grain-based finishing. These options may become more important under climate change.

Planting time variation

Higher temperatures are anticipated to reduce frost risks potentially allowing earlier planting (by a month or more) and consequently increased yields as grainfill is more likely to occur in the cooler months when the likelihood of water stress is lower (see Howden *et al.* 1999a, 2003; van Ittersum *et al.* 2003). This may require concurrent changes in thermal time requirements of the varieties used depending on any changes in planting dates. There is evidence that farmers are already planting earlier in response to lower frost risk (e.g. Stephens and Lyons 1998) and that this is enhancing yields.

The above adaptation of earlier planting assumes no change in the timing of 'autumn break' rains or of availability of stored soil moisture, both of which could be affected by climate change (Chapter 2). The effect of climate changes on the autumn break have not yet been investigated, however the greater likelihood of reductions than increases in future autumn rainfall in cropping areas in Western Australia, South Australia, Victoria and southern New South Wales (CSIRO 2007) suggests that the autumn break may be postponed compared with historical experience. Consequently, there may need to be ongoing reassessments of planting rules in these regions in conjunction with varietal changes. The rainfall scenarios for cropping regions in Queensland and northern New South Wales, where stored soil moisture is critical, have greater likelihoods of increased rainfall than those for southern Australia, suggesting that planting decisions need to continue to be sensitive to stored soil moisture levels and seasonal climate forecasts.

Crop management (spacing, tillage, fallows, rotations, irrigation)

There are many crop management practices that could be used in specific circumstances to lower risks from changed climate conditions (e.g. Easterling *et al.* 2007). These include:

- adopting zero-tillage practices (especially if there is increased rainfall intensity as greater infiltration will be needed with fewer but heavier events);

- develop more techniques that minimise disturbance (i.e. seed pushing, all-weather traffic lanes that allow planting while raining);

- using reactive strategies to track climate variation on daily or seasonal time steps (e.g. McKeon *et al.* 1993). For example crop planting decisions such as timing, cultivar and fertiliser rates can be based on soil moisture stores, nutrient concentrations, average seasonal rainfall (Hammer *et al.* 1996) and financial and commodity forecasts either on a single field or on a whole farm basis (e.g. Power *et al.* 2008);

- extending fallows to effectively capture and store more soil moisture (suitable mostly with heavy soils);

- dry sowing, later plantings or staggering planting times depending on the rainfall changes;

- widening row spacing, skip-row planting or lowering plant populations;

- developing efficient on-farm irrigation management with effective scheduling, application and transfer systems;

- reducing losses from irrigation systems during water transfers through improved channel lining, etc., and

- monitoring and responding rapidly to emerging pest, disease and weed issues, noting that support of effective research, development and extension would be needed.

Most of these have yet to be analysed for their benefits under climate change although such assessment is now underway with farmers (e.g. Crimp *et al.* 2006). It is likely that the benefits will be quite context-specific and involve trade-offs. For example, wide row and skip row plantings can increase yield stability and particularly increase yield in poor rainfall years. However, they also reduce ground cover, water storage at the end of the fallow and infiltration while increasing runoff and soil loss.

Nutrient management change (fertilisation and rotations)

There is a premium for high levels of protein in various crops (e.g. durum wheat, malting barley). Increases in atmospheric CO_2 levels are likely to result in declines in grain nitrogen and hence protein and flour quality (e.g. Rogers *et al.* 1998; Fangmeier *et al.* 1999). However, the amount of reduction varies with cultivar (Rogers *et al.* 1998) and N-supply to the crop. With ample N-status the decline in grain protein may be small (Kimball *et al.* 2001). However, in many cases, grain crops in Australia do not have ample nitrogen and so there is a risk of reductions in grain protein. Conversely, climate change seems likely to increase the risk of hot, dry finishes to the grainfill period that will tend to increase grain protein and may in part compensate for the above factors. To maintain grain nitrogen contents at historical levels, there may be a need to considerably increase the use of legume-based pastures (e.g. extending rotation length to have a longer pasture/legume phase), increase use of leguminous crops or further

increase nitrogen fertiliser application (extending an existing trend to higher applications: Hayman and Alston 1999). There will also be a need to continue monitoring soil nitrogen concentrations and to breed cultivars that limit decline in grain protein with increasing atmospheric CO_2 concentration. The adverse effects of higher CO_2 on grain protein will compound the protein reductions arising from long-term rundown of nutrient in Australian cropping soils (e.g. Dalal *et al.* 1991; Verrell and O'Brien 1996) and will likely require attention on better delivery technologies (e.g. direct delivery of ammonium solutions into the root zone with minimal soil disturbance), improved monitoring and enhanced education. Alternatively, some growers are already targeting a premium market in biscuit wheat, which requires low protein, soft wheat varieties. However, there are currently few other such market opportunities and low protein, hard wheat is generally being sold at the lower end of the market. Given that high CO_2 is global in its extent and impact, the prospects for maintaining a premium soft-wheat market seem to be limited as all producing nations are likely to be grappling with how to maximise returns from lower protein wheats.

The adaptations of fertiliser application and change in rotations will have impacts on soil acidification processes and water quality in some regions, but could also be a source of greenhouse gas emissions: production, packaging and distribution of nitrogenous fertiliser generate about 5.5 kg CO_2 per kg N (Leach 1976), while both fertilisation and legume rotations increase emissions of the potent greenhouse gas nitrous oxide (Prather *et al.* 1995).

Erosion management

Rainfall intensity is anticipated to increase with climate change even under scenarios where average rainfall may decrease continuing the current trends to higher intensity rainfall events in Australia (CSIRO 2007). This is likely to increase risks of soil erosion, particularly on soils with high erodability (e.g. solodised soils). Key adaptations may be to:

- increase residue retention and maintain crop cover during periods of high risk so as to reduce raindrop damage on the soil surface and allow for water to infiltrate;

- maintain erosion control infrastructure (e.g. contour banking, etc.), and

- adopt controlled traffic systems up and down slope.

These actions are already generally implemented in cropping systems but their importance is likely to increase over time. Improved warning of seasonal conditions with high erosion potential would enable improved risk management.

Management to reduce water-related soil erosion will also tend to reduce risks from wind erosion if this increases, although this is uncertain (Chapter 2).

Salinisation management

Increased rainfall intensity (and high CO_2 levels) may also increase drainage below the root zone, the driver for dryland salinisation by lifting the water table. This will be particularly prevalent on lighter-textured soils where indicative changes are for a 6–20% increase in drainage under a doubling of CO_2 alone (van Ittersum *et al.* 2003). This would represent a substantial potential change in landscape hydrology, which is likely to increase risks of salinisation in areas not yet affected and increase rates of this form of degradation in areas where it is already occurring. This increase may be more than offset in some regions if there is a reduction in rainfall in autumn and winter. For example, the strong drying trends across southern Australia would suggest a 30 to 80% reduction in drainage components depending on site (van Ittersum *et al.* 2003). Such large reductions in drainage would have strong implications for catchment management and policy development for addressing the dryland salinisation issue. In addition, the large reductions in farm profitability that may occur with climate change in Western Australia may reduce the capacity of individual farmers to implement practices (such as establishing perennial plants) to reduce dryland salinisation risk (John *et al.* 2005). The balance between these opposing tendencies is not well understood. In contrast, in north-east Australia, the tendency will be to increase drainage, increasing salinity risk (Howden *et al.* 1999c). Hence, policies relating to dryland salinisation need to take the potential impact of climate changes into account and may need to be adapted over time in conjunction with climate change and its effects on hydrological processes and farm profitability.

Moisture conservation

Potential evaporation changes under the influence of changes in wind speed, solar radiation and vapour pressure deficit (VPD: the difference between the moisture content of the air and its potential moisture content at that temperature). Potential evaporation is important for transpiration of water from plants, evaporation from soil and water storages and for the efficiency of water use by plants. Evaporation and vapour pressure deficit are affected by temperature (e.g. Tanner and Sinclair 1983). However, the way in which they may change in the future is highly dependent on the way in which daytime and night-time temperatures change as well as changes in wind speed and global atmospheric humidity, which is increasing. If we assume that the temperature change is symmetrical (i.e. the rate of change is similar for both night and day temperatures) then there will be, all else equal, a significant increase in evaporation (about 3% per °C) and VPD (about 6% per °C). Current global climate models do not indicate significant departures from symmetrical day/night change (CSIRO 2007). Many places in Australia (and elsewhere) have seen significant declines in potential evaporation with an overall continental decline over the last 30 years of about 2 mm per year (e.g. Jovanovic *et al.* 2008) due in large part to reductions in wind speed (Rayner 2007; McVicar *et al.* 2008)

If we assume that temperature changes will be symmetrical and wind speed ceases to decline (and hence evaporation and VPD will increase) then efficient moisture use can be enhanced by:

- increasing residue cover (maintaining stubble) particularly in association with minimal or no tillage;

- planting and phenology that maximises growth during the cooler, wetter months when VPD is low;

- developing varieties with higher water use efficiency;

- rapid establishment of crop canopy so that any water transfer at least results in crop growth;

- weed control, and

- maximising capture and storage of excess rainfall on-farm perhaps by incorporating raised bed technologies into controlled traffic operations and directing flows into storage zones. This may be especially important if rainfall intensity increases.

The effects of higher VPD on transpiration rates will be countered to some extent by the reduced stomatal conductance under elevated CO_2 concentration (see Chapter 1).

In some areas, such as the higher rainfall parts of the Western Australian wheatbelt, waterlogging and nitrogen leaching are currently problems. Consequently, the drier rainfall scenarios out to 2030 may result in beneficial impacts. However, further reductions in rainfall to those in the 2070 scenarios would result in soil moisture shortages requiring adoption of moisture conserving strategies even in these currently high rainfall areas.

In irrigated crops, higher evaporation rates and higher VPD will mean there is potential for greater water use per unit production, at the same time as there may be reductions in water allocation due to reduced river flows. Hence, key adaptations may be to ensure access to water and to increase water use efficiency. These are addressed below under 'Irrigation' and in Chapter 12.

Use of seasonal forecasting

The El Niño-Southern Oscillation system (ENSO) is a key source of variability in rainfall and wheat yield in Australia (e.g. Hammer *et al.* 1996). ENSO impacts are largest on Australian winter and spring rainfall and temperatures. El Niño events are generally associated with reduced rainfall across much of Australia and are known to adversely affect crop production, particularly in north-east Australia (e.g. Stone *et al.* 1993, 1996; Hammer *et al.* 1996). La Niña events tend to produce higher rainfall and hence higher yields but they may also result in greater incidence of waterlogging, crop spoilage and pest and disease problems (Meinke and Hochman 2000). There is a developing view that climate change may result in increased incidence of El Niño and possibly also La Niña events (e.g. IPCC 2007) but there remains considerable uncertainty.

Following early demonstrations of the value of using statistical seasonal forecasting in cropping management decisions (e.g. Clewett *et al.* 1991), there has been widespread but not ubiquitous adoption of this information (Meinke *et al.* 2006, 2007; Hayman *et al.* 2007). If the relationships between local weather and these broadscale factors (e.g. the Southern Oscillation Index or regional sea surface temperatures) remain largely stable, then the continued use of statistical seasonal climate forecasts provides a key way for agriculture to 'track' climate changes as first proposed by McKeon and Howden (1992) and McKeon *et al.* (1993). Process-based forecasts using coupled ocean-atmosphere models hold out the prospect of improved forecasts that will automatically incorporate the climate changes (Meinke *et al.* 2001; Power *et al.* 2007).

Decision-making using seasonal forecasts is improved if allied with on-ground observations such as soil moisture content at planting (e.g. Crimp *et al.* 2006, also see following section on monitoring/evaluation). Soil moisture is already a critical input into planting decisions especially in regions with heavier soils.

Irrigation

Irrigation, by removing water limitations to crop growth, is one way to reduce climate risks. However, changes to the way in which water rights are managed and traded may result in increasing levels of climate risk in the irrigated cropping sector. Furthermore, scenarios of climate change indicate substantial reductions in mean flows but higher flow variability in Australia (see Chapter 3) at the same time as possibly increased evaporative demand, indicating that climate change will also increase that risk. With water becoming more critical, there will be a need for further improvements in water distribution systems (to reduce leakage and evaporation), choice of crops to increase returns per litre and irrigation practices such as water application methods, irrigation scheduling and moisture monitoring.

Monitoring and evaluation

An important proactive step for producers to adapt to a changing climate is to maintain a thorough measurement and analysis program of their own local climate and production systems to compare with climate change scenarios (Gifford *et al.* 1996). Such proactive climate data acquisition and interpretation at the farm level could provide the capacity for reactive and opportunistic adaptive measures by farm managers. Parts of such a system already exist, for example with the Silo database that can provide an interpolated climate record for any point in the nation (Jeffrey *et al.* 2001), the AussieGrass project that provides spatial assessment of grazing systems across the nation (Carter *et al.* 2000), the seasonal crop outlook system referred to previously and the Australian RainMan decision-support package. There are many other activities that could contribute to this goal but largely remain uncoordinated and unlinked to climate change adaptation.

Management of pests, diseases and weeds

Pest impacts on crops are widespread and costly to industry, and include many trade access issues for grains and pulses. Many of the pests (such as *Heliothis* moths, armyworms, sucking bugs, diamond backed moths (on *Brassica*)) respond strongly to climate signals and their impacts are very dependent on climatic variability.

Adaptations to climate change are likely to happen via increased understanding of impacts and potential responses of recent climate variability manifestations (last 20 years) and may best be delivered via the two key emerging strategies: (1) integrated pest management and (2) area-wide management (i.e. coordinated responses of growers and policy-makers across an entire region).

Many of these tools rely on either intensive monitoring or on computer simulations of pest numbers to flag high-risk periods for each species of pest. The latter are poorly developed in Australia compared with our competitors overseas even where they use Australian software. In the absence of these, growers often choose to apply excessive amounts of chemical as 'insurance treatments' often because they do not have ready access to reliable information on the risks to the crop.

Summer rainfall could also be a problem with weed (volunteer cropping) species providing 'green bridges' for the diseases of our winter crops and this would necessitate their control in the summer months with spraying or grazing. However, climate change may also reduce the frequency of conditions suitable for spraying insecticides, herbicides and fungicides, requiring alterations in spray technologies and practices (Howden *et al.* 2007b) while elevated CO_2 may reduce the efficacy of products such as glyphosate for weed control (Ziska *et al.* 1999) and Bt for pest control (Wu *et al.* 2007).

Current management practices that respond to, or override, climatic variability include:

- genetic modification of crop plants to create insect- or disease-resistant and herbicide-tolerant varieties (via either conventional plant breeding or genetic modification);

- importation of exotic natural enemies of pests that were previously introduced without them. Also repeated, mass (inundative) releases of parasitic wasps to control insect pests;

- isolation and propagation of local natural enemies/diseases (e.g. *Metarhizium* on locusts, termites);

- cultural practices such as crop rotations, mixed crops, use of physical barriers to reduce disease transmission;

- chemical pesticides and increasing bio-pesticides (e.g. Bt) and bio-fumigation of soils using *Brassica* sp. as alternate crops;

- monitoring and use of predictive models to improve timing of interventions to coincide with high risk periods, and

- landscape-scale management involving groups of growers cooperating to reduce communal threats, e.g. when growing melons in rotation with soybeans or sugar, or chickpeas mixed with cotton.

These practices generally will need to be fine-tuned so as to cope with new challenges arising from climate change. One example is for better indicators of successful over-wintering of a wide range of insect pests and plant diseases. This information could then be used in phenological models and GIS producing regional scale outputs in real time. Associated with this will be a need for enhanced communication to make farmers aware of the nature of any imminent pest and disease risks and effective options for their control. A continuing commitment may be needed from Agricultural Protection Boards, State Governments and Shire Councils to extend their commitment to controlling listed weeds and pests and control volunteer crop species on road verges and Crown land to prevent disease build-up.

Research and development and education

Farmers cannot conduct controlled experiments to assess different management alternatives in an ever-changing environment. Instead, a key adaptation at the national level would be public sector support from a vigorous agricultural research and breeding effort, channelling experimental information into cultivar, breeds and technological and management alternatives. Complementing this would be an agricultural advisory network capable of interpreting property-specific climate records and production in terms of research findings and integrating this with farmer knowledge and aspirations followed by demonstration of the benefit. Mechanisms are needed also to ensure that farmer innovations for adapting to climate change are linked back to research and development groups for evaluation.

Farmers in 'core' cropping areas may also be able to learn much from those in currently marginal areas in terms of dealing with moisture limitations, nutrient and residue management and disease management.

Land use change (infrastructure, knowledge base)

Regional land use patterns are strongly affected by climate. Hence, changes in climate would indicate corresponding changes in land use (e.g. Reyenga *et al.* 2001; Ramankutty *et al.* 2002). In some cases, this could mean retreat of cropping zones from the dry margins and in others possibly expansion into either marginal zones in the north (i.e. Mitchell grass downs) or the wetter margins outside the current southern cropping zones (Howden *et al.* 1999a; Reyenga *et al.* 2001). The increasing water use efficiency from increasing atmospheric CO_2 is expected to have particularly strong effects on crop productivity at the dry margins (e.g. Gifford 1979; Ramankutty *et al.* 2002) but this effect can only compensate for lower rainfall to a limited extent and extremely dry conditions will still result in very low yields. If the more extreme climate change scenarios eventuate, it may become necessary to switch land use systems; for example from mixed farming to solely grazing or to water catchment or plantation forestry in some areas. Governments may have a role in monitoring land use and fostering change when necessary (e.g. via industry restructuring) taking into account potential competing uses (e.g. nature conservation, environmental stewardship) and dealing with potential conflict. For such large-scale adaptation, in addition to transitional support, there will likely be a need for a continuing education plan to retrain producers in new

enterprises and to maintain flexibility in adapting to new circumstances. These needs are only marginally different to those existing needs for increased management skills and fostering flexibility in agriculture. Infrastructure changes may also be needed to meet transport and processing changes if there are substantial land use changes (e.g. Hayman and Alston 1999).

Financial institutions and trade

Lending policies of financial institutions can greatly constrain or support options for producers to adjust their operations in the light of change. These institutions may have to change their policies to take account of predicted changes in the circumstances of their customers: information needs to be supplied to both the financial and farming industries. One possible adaptation is increased use of crop insurance. However, this does pose some actuarial challenges in assessing risks in a changing climate as distinct from the historical risk assessment approaches conventionally adopted in the industry. The conventional approaches by definition will provide a poor assessment of future risk.

Climate changes will also affect our trade competitors. The limited global analyses available so far on this suggest that there may be new cropping land viable in the northern hemisphere as a result of climate change but little in the tropics, that the negative effects of higher temperature (especially when it exceeds about 2°C) are likely to be pervasive across the equatorial to mid-latitude cropping belts and that the USA, South America, South Africa, India and China are all likely to be negatively impacted by rainfall reductions (Ramankutty *et al.* 2002; Easterling *et al.* 2007). The net effects of climate change interacting with improvements in productivity on global grain production are likely to approximately balance for some decades but then tend towards the negative (Easterling *et al.* 2007). A capacity to assess the ongoing production prospects of market competitors as well as the factors driving demand may assist Australian industry and government policy-makers in framing their own responses to climate change more effectively.

Risks of maladaptation

Maladaptation can occur through over- or under-adaptation to (potential) climate changes and also a misalignment between policy and practice, leading to unintended management responses or the provision of perverse incentives, issues that are currently discussed in relation to, for instance, the biofuel debate. Indeed, it could be argued that given the uncertainty associated with climate change, some degree of maladaptation is inevitable. Effective monitoring of adaptation and climate change at a range of scales could help reduce the risks of maladaptation by learning what adaptations work, which do not and why (Howden *et al.* 2007b).

Maladaptation can also arise through unintended negative consequences either inside or outside the agricultural value chain. For example, it is important to ensure that adaptations themselves do no lead to increased net greenhouse gas emissions such as might occur if there is increased application of fertiliser nitrogen to offset reductions in grain protein and to enable full response to rising atmospheric concentrations of CO_2. However, at the moment, there is no comprehensive analysis of these types of maladaptations. Consequently, an additional benefit from adaptation research on cropping may be understanding how short-term changes (e.g. peri-urban development in high rainfall areas potentially suitable for agriculture) may link to long-term adaptation options (e.g. land use change) so as to make sure that, at a minimum, management and/or policy decisions implemented over the next one to three decades do not undermine the ability to adapt to potentially larger impacts later in the century (Howden *et al.* 2007; Meinke and Nelson 2009).

Costs and benefits of adaptation

Many of the adaptations required for adapting to climate change are extensions of those currently used for managing climate variability (e.g. Howden *et al.* 2007b). The goals of such existing management strategies are usually to deliver one or more aspects of the 'triple bottom line' (i.e. economic, environmental and social outcomes). As such, the adaptations generally have immediate application as well as relevance to adapting to climate change. Much of what is known of the benefits of adaptation to productivity and sustainability of Australia cropping systems has been outlined in the previous section. However, in terms of making some assessment of the direct costs and benefits of adapting to climate change, there remain few studies to date in Australia. In particular, adaptations always have some costs

and these are often overlooked (Scheraga and Grambsch 1998).

The national financial benefit to the wheat industry of a subset of the possible adaptations to climate change has been assessed by Howden and Jones (2004) and Howden and Crimp (2005) using risk analysis approaches. The adaptations were varietal change and alteration of planting windows – key adaptations previously explored with farm level gross margin analyses by Howden *et al.* (1999a). Just these two adaptations could save the national wheat industry between $100 million and $500 million each year (in current dollar terms) by maintaining productivity in the face of change. These adaptations changed the mean result from being negative (on balance of probabilities) to positive. Clearly, investment in adaptation is extremely worthwhile for the wheat industry. However, in that study there remained a large negative 'tail' of results arising from very dry and hot climate change scenarios: hence adaptation cannot remove all the risk from climate change. Furthermore, since those analyses, the newer climate change scenarios have increased the probability and degree of serious negative climate change (CSIRO 2007). If such changes eventuate, additional adaptations such as those identified in the previous section of this chapter could further reduce the negative impacts of such changes, at least in most regions. However, comprehensive assessment of the benefits of adaptation options remains to be undertaken.

At the farm level, Howden *et al.* (1999c) assessed adaptations of fertiliser addition to maintain grain nitrogen contents at historical levels. They found that there will be a need to increase application rates by 40–220 kg/ha depending on the future climate and CO_2 scenario and location (Howden *et al.* 1999c). Optimum fertiliser application adaptation for a given scenario increased gross margin by about 20 to 25%. However, at higher levels of fertiliser applications, the increased cost of fertiliser was not offset by increased income (i.e. the strategy became maladaptive) with this level of fertiliser application being lower in drier regions and higher in wetter regions. Furthermore, such adaptations of increased fertiliser use will have their own impacts on greenhouse gas emissions (as discussed earlier), soil acidification processes and water quality in some regions and on farm economics. These were not costed.

Table 3.2: Priority climate change adaptation options for the grains sector (Howden *et al.* 2003). Priority 1 (high), 2 (medium) and 3 (low).

Adaptation option	Priority
Adaptation to climate change – crop and farm management	
Develop participatory research approaches to assist proactive adaptations on-farm and across the value chain	1
Develop further risk amelioration approaches (e.g. zero tillage and other minimum disturbance techniques, retaining residue, extending fallows, row spacing, planting density, staggering planting times, erosion control infrastructure, controlled traffic)	1
Alter planting rules to be more opportunistic depending on environmental conditions (e.g. soil moisture), climate (e.g. frost risk) and markets	1
Select varieties from the global gene pool with high levels of CO_2 responsiveness, appropriate thermal time and vernalisation requirements, heat shock resistance, drought tolerance (i.e. Staygreen), high protein levels, resistance to new pest and diseases and perhaps that set flowers in hot/windy conditions	1
Maximise utility of seasonal climate forecasts by research, develop and extension that combines them with on-ground measurements (i.e. soil moisture, nitrogen), market information and systems modelling	2
Research and revise soil fertility management (fertiliser application, type and timing, increase legume phase in rotations) on an ongoing basis including implications for off-farm impacts (e.g. greenhouse gas emissions)	3
Policy-related and other adaptations	
Continue training to enhance self-reliance via improved climate risk management, and to provide knowledge base for adapting	1
Provide public sector support for a vigorous agricultural research and breeding effort including undertaking further adaptation studies which include costs/benefits, adoption paths and maladaptations	2
Explore transformation options in the cropping zones that can provide positive production, environmental and social outcomes from these major changes	2
Establish an effective adaptation monitoring program to learn what works, what does not and why	2
Further improve water distribution systems (to reduce leakage and evaporation), irrigation practices such as water application methods, irrigation scheduling and moisture monitoring to increase efficiency of use	2
Further develop Area-wide Management operations, Integrated Pest Management and other innovative pest, disease and weed adaptations	3

Discussion

The sections above and Table 3.2 have detailed a large range of potential adaptation options across scales from farm-level to national policy and across the value chain. Some of these are largely new activities that may need to be implemented specifically in relation to climate and atmospheric changes (e.g. breeding to maximise the benefits of higher levels of CO_2). However, many of these options are currently being implemented to greater or lesser degrees as part of managing for climate variability, market vagaries, extant pest, disease and weed problems or the rundown of natural resources. Climate change is likely to raise the importance of these adaptations and require more widespread and rigorous implementation, often based on ongoing research, development and extension. Given the considerable uncertainties regarding rates and scale of climate change, there is a need for directed change in management and policy that is in turn monitored, analysed and learnt from, so as to iteratively and effectively adjust to the actual climate changes that will be experienced in coming decades (Howden *et al.* 2007).

Adaptation options for the grains sector should be undertaken via participatory engagement with decision-makers. By bringing their practical knowledge into the assessment, inclusion of decision-makers tends to identify a more comprehensive range of adaptations, as well as being able to assess the practicality of options and contribute to more realistic assessment of the costs and benefits involved in management or policy change. This process also focuses on the acceptability of adaptation options in terms of the factors important to the stakeholders and their perceptions of synergies and barriers. Therefore, involving stakeholders from project inception is critical if adaptation research is to be reflected in changed decisions and altered strategies and actions.

Within an existing grains cropping system, most of the benefits of adaptations accrue with moderate climate change and there are limits to their effectiveness under more severe climate changes. Hence, there is a need to consider more systemic changes in resource allocation, including livelihoods diversification. This may ultimately involve moving activities to new locations as has happened with the peanut industry, changing the enterprise to a new agricultural activity such as converting to grazing or in some cases cessation of that activity as looks possible with parts of the rice industry. These transformative changes will need integration across scales from farm to region to nation so as to look for synergies and to reduce maladaptive outcomes. Monitoring and evaluation of these adaptations in addition to those above is likely to be extremely instructive.

Finally, the above and other adaptations should be thought of as an ongoing process that is part of good risk management, whereby drivers of risk are identified and their likely impacts on systems under alternative management are assessed. In this respect, adaptation to climate change is similar to and can be integrated with adaptation to climate variability, changes in market forces (cost/price ratios, consumer demands, etc), institutional or other factors.

Acknowledgements

We would like to thank Diana Anderson, Barry Batters, Rod Birch, Peter Durkin, Andrew Lawson, Gerald Leach, Duncan McLelland, Rolf Meeking, John Mills, Darren Moir, Rod Morris, David Petering, David Pullar, Greg Rummery, Sandy Stump, Ian Sutherland and Bruce Watt for their input into an earlier version of this chapter, the CSIRO Climate Adaption Flagship and the (then) Australian Greenhouse Office for support.

References

ABARE (2007). Australian Bureau of Agriculture and Resource Economics AGSURF database. http://www.abareconomics.com/interactive/agsurf/

Blumenthal CS, Bekes F, Batey IL, Wrigley CW, Moss HJ, Mares DJ, Barlow EWR (1991). Interpretation of grain quality results from wheat variety trial with reference to high temperature stress. *Australian Journal of Agricultural Research* **43**, 325–334.

Carter JO, Hall WB, Brook KD, McKeon GM, Day KA, Paull CJ (2000). Aussie GRASS: Australian grassland and rangeland assessment by spatial simulation. In *Applications of Seasonal Climate Forecasting in Agricultural and Natural Ecosystems – The Australian Experience.* (Eds G Hammer, N Nicholls, C Mitchell) pp. 329–349. Kluwer Academic Press, The Netherlands.

Chakraborty S, Datta S (2003). How will plant pathogens adapt to host plant resistance at elevated CO_2 under a changing climate? *New Phytologist* **159**, 733–742.

Clewett JF, Howden SM, McKeon GM, Rose CW (1991). Optimising farm dam irrigation in response to climatic risk. In *Climatic Risk in Crop Production: Models and Management for the Semi-arid Tropics and Subtropics.* (Eds RC Muchow, JA Bellamy) pp. 307–328. CAB International, Wallingford, UK.

Crimp S, Gaydon D, Howden M, Hall, C, Poulton P, Hochman Z (2006). 'Managing natural resource issues in a variable and changing climate'. Final Report to Land and Water Australia, Managing Climate Variability Program, Canberra, Australia.

CSIRO, Australian Bureau of Metereology (2007). 'Climate change in Australia'. Technical Report. CSIRO and Bureau of Meteorology. http://www.climatechangeinaustralia.gov.au/

Dalal RC, Strong WM, Weston EJ, Gaffney J (1991). Sustaining multiple production systems. 2. Soil fertility decline and restoration of cropping lands in sub-tropical Queensland. *Tropical Grasslands* **25**, 173–180.

Easterling WE, Aggarwal PK, Batima P, Brander KM, Erda L, Howden SM, Kirilenko A, Morton J, Soussana J-F, Schmidhuber J, Tubiello FN (2007). Food, fibre and forest products. In *Climate*

Change 2007: Impacts, Adaptation and Vulnerability. Contribution of Working Group II to the Fourth Assessment Report of the Intergovernmental Panel on Climate Change. (Eds ML Parry, OF Canziani, JP Palutikof, PJ van der Linden, CE Hanson) pp. 273–313. Cambridge University Press, Cambridge.

Fangmeier A, De Temmerman L, Mortensen L, Kemp K, Burke J, Mitchell R, van Oijen M, Weigel HJ (1999). Effects on nutrients and on grain quality in spring wheat crops grown under elevated CO_2 concentrations and stress conditions in the European, multiple-ite experiment 'ESPACE-heat'. *European Journal of Agronomy* **10**, 215–229.

Gifford R, Angus J, Barrett D, Passioura J, Rawson H, Richards R, Stapper M (1998). Climate change and Australian wheat yield. *Nature* **391**, 448–449.

Gifford RM (1979). Growth and yield of CO_2-enriched wheat under water-limited conditions. *Australian Journal of Plant Physiology* **6**, 367–378.

Gifford RM (1989). Exploiting the fertilising effect of increasing atmospheric carbon dioxide. In *Climate and Food Security. Proceeds International Symposium on Climate Variability and Food Security in Developing Countries.* February 5–9 1987. pp. 477–487. International Rice Research Institute, Manila.

Gifford RM, Campbell B, Howden SM (1996). Options for adapting agriculture to climate change: Australian and New Zealand examples. In *Greenhouse: Coping with Climate Change.* (Eds WJ Bouma, GI Pearman, MR Manning) pp. 399–416. CSIRO Publishing, Melbourne.

Hammer G, Nicholls N, Mitchell C (2000). *Applications of Seasonal Climate Forecasting in Agricultural and Natural Ecosystems – The Australian Experience.* Kluwer Academic Press, The Netherlands.

Hammer GL, Holzworth DP, Stone R (1996). The value of skill in seasonal forecasting to wheat crop management in a region with high climatic variability. *Australian Journal of Agricultural Research* **47**, 717–737.

Harvell CD, Mitchell CE, Ward JR, Altizer S, Dobson AP, Ostfield RS, Samuel MD (2002). Climate Warming and Disease Risks for Terrestrial and Marine Biota. *Science* **29**, 2158–2162.

Hayman P, Crean J, Mullen J, Parton K (2007). How do probabilistic seasonal climate forecasts compare with other innovations that Australian farmers are encouraged to adopt? *Australian Journal of Agricultural Research* **58**, 975–984.

Hayman PT, Alston CL (1999). A survey of farmer practices and attitudes to nitrogen management in the northern New South Wales grain belt. *Australian Journal of Experimental Agriculture* **39**, 51–63.

Howden SM (2002). Potential global change impacts on Australia's wheat cropping systems. In *Effects of Climate Change and Variability on Agricultural Production Systems.* (Eds OC Doering, JC Randolph, J Southworth, RA Pfeifer) pp. 219–247. Springer: The Netherlands.

Howden SM, Ash AJ, Barlow EWR, Booth T, Charles S, Cechet R, Crimp S, Gifford RM, Hennessy K, Jones RN, Kirschbaum MUF, McKeon GM, Meinke H, Park S, Sutherst R, Webb L, Whetton PJ (2003). 'An overview of the adaptive capacity of the Australian agricultural sector to climate change – options, costs and benefits'. Australian Greenhouse Office, Canberra.

Howden SM, Crimp S (2005). Assessing dangerous climate change impacts on Australia's wheat industry. In *Proceedings of the International Congress on Modelling and Simulation.* (Eds A Zerger, RM Argent) pp. 170–176. Modelling and Simulation Society of Australia and New Zealand, Canberra.

Howden SM, Jones RN (2004). Risk assessment of climate change impacts on Australia's wheat industry. In *New Directions for a Diverse Planet: Proceedings of the 4th International Crop Science Congress.* 26 September – 1 October 2004, Brisbane, Australia. p. 6. Regional Institute, Gosford.

Howden SM, McKeon GM, Meinke H, Entel M, Flood N (2001). Impacts of climate change and climate variability on the competitiveness of wheat and beef cattle production in Emerald, north-east Australia. *Environment International* **2**, 155–160.

Howden SM, Meinke H, Power B, McKeon GM (2003). Risk management of wheat in a non-stationary climate: frost in Central Queensland. In *Integrative Modelling of Biophysical, Social and Economic Systems for Resource Management Solutions*. (Ed. DA Post) pp. 17–22. MSSANZ Inc., Canberra.

Howden SM, Reyenga PJ, Meinke H (1999a). 'Global Change Impacts on Australian Wheat Cropping. Report to the Australian Greenhouse Office'. CSIRO Wildlife and Ecology Working Paper 99/04, Canberra.

Howden SM, Reyenga PJ, Meinke H (1999b). Mixed wheat-sorghum cropping systems in eastern Australia under global change. In *Food & Forestry: Global Change and Global Challenges*. GCTE Focus 3 Conference, 20–23 September, Reading UK.

Howden SM, Reyenga PJ, Meinke H (1999c). 'Global change impacts on Australian wheat cropping: studies on hydrology, fertiliser management and mixed crop rotations. Report to the Australian Greenhouse Office'. CSIRO Wildlife and Ecology Working Paper 99/13, Canberra.

Howden SM, Soussana JF, Tubiello FN, Chhetri N, Dunlop M, Meinke HM (2007). Adapting agriculture to climate change. *Proceedings of the National Academy of Sciences* **104**, 19691–19696.

IPCC (2007). *Climate Change 2007: The Physical Science Basis. Contribution of Working Group I to the Fourth Assessment Report of the Intergovernmental Panel on Climate Change*. (Eds S Solomon, D Qin, M Manning, Z Chen, M Marquis, KB Averyt, M Tignor, HL Miller) Cambridge University Press, Cambridge.

Jeffrey SJ, Carter JO, Moodie KB, Beswick AR (2001). Using spatial interpolation to construct a comprehensive archive of Australian climate data. *Environmental Modelling and Software* **16**, 309–330.

John M, Pannell D, Kingwell R (2005). Climate change and the economics of farm management in the face of land degradation: dryland salinity in Western Australia. *Canadian Journal of Agricultural Economics* **53**, 443–459.

Jovanovic B, Jones DA, Collins D (2008). A high quality monthly pan evaporation dataset for Australia. *Climatic Change* **87**, 517–535.

Kerr NJ, Siddique KHM, Delane RJ (1992). Early sowing with wheat cultivars of suitable maturity increases grain yields of spring wheat in a short season environment. *Australian Journal of Experimental Agriculture* **32**, 717–723.

Kimball BA, Morris CF, Pinter PJ, Wall GW, Hunsaker DJ, Adamsen FJ, LaMorte RL, Leavitt SW, Thompson TL, Matthias AD, Brooks TJ (2001). Elevated CO_2, drought and soil nitrogen effects on wheat grain quality. *New Phytologist* **150**, 295–303.

Leach G (1976). *Energy and Food Production*. IPC Science and Technology Press: Guildford, Surrey.

McKeon GM, Howden SM (1992). Adapting the management of Queensland's grazing systems to climate change. In 'Climate change: implications for natural resource conservation.' (Ed. S Burgin) University of Western Sydney Occasional Papers in Biological Sciences No 1.

McKeon GM, Howden SM, Abel NOJ, King JM (1993). Climate change: adapting tropical and sub-tropical grasslands. *Proceedings of 17th International Grassland Congress*, Palmerston North, New Zealand, pp. 1181–90.

McVicar TR, Van Niel TG, Li LT (2008). Wind speed climatology and trends for Australia, 1975–2006: Capturing the stilling phenomenon and comparison with near-surface reanalysis output. *Geophysical Research Letters* **35**, L20403.

Meinke H, Hochman Z (2000). Using seasonal climate forecasts to manage dryland crops in northern Australia. In *Applications of Seasonal Climate Forecasting in Agriculture and Natural Ecosystems: The Australian Experience*. (Eds GL Hammer, N Nicholls, C Mitchell) pp. 149–165. Kluwer, The Netherlands.

Meinke H, Nelson R (2009). *Will global mitigation policy enhance or undermine local adaptation?* Conference on Integrated Assessment of Agriculture and Sustainable Development, 10–12 March, Egmond aan Zee, The Netherlands, in press.

Meinke H, Stone RC (2005). Seasonal and inter-annual climate forecasting: the new tool for increasing preparedness to climate variability and change in agricultural planning and operations. *Climatic Change* **70**, 221–253.

Meinke H, Pollock K, Hammer GL, Wang E, Stone RC, Potgieter A, Howden SM (2001). Understanding climate variability to improve agricultural decision making. *Proceedings of the Australian Agronomy Conference*, January 2001, Hobart.

Meinke H, Nelson R, Kokic P, Stone R, Selvaraju R, Baethgen W (2006). Actionable climate knowledge: from analysis to synthesis. *Climate Research* **33**, 101–110.

Meinke H, Sivakumar MVK, Motha RP, Nelson R (2007). Preface: Climate predictions for better agricultural risk management. *Australian Journal of Agricultural Research* **58**, 935–938.

Meinke H, Struik PC, Vos J (2008). Crop science for the future – adding value through modelling across scales and disciplines. *Proceedings of the 5th International Crop Science Congress*, 13–18 April, Jeju Island, Korea, published on CDROM and http://www.intlcss.org/congress–proceedings/

Mitchell PL, Sheehy JE (2006). Supercharging rice photosynthesis to increase yield. *New Phytologist* **171**, 688–693.

Muchow RC, Bellamy JA (1991). *Climatic Risk in Crop Production: Models and Management for the Semi-arid Tropics and Subtropics*. CAB International, Wallingford, UK.

Nicholls N (1997). Increased Australian wheat yield to recent climate trends. *Nature* **387**, 484–485.

Pollock KS, Meinke H, Stone RC (2001). The influence of climate variability on cropping systems in Central Queensland. *Proceedings of the 10th Australian Agronomy Conference*, Hobart.

Potgieter AB, Hammer GL, Doherty A, de Voil P (2005). A simple regional-scale model for forecasting sorghum yield across North-Eastern Australia. *Agricultural and Forest Meteorology* **132**, 143–153.

Potgieter AB, Hammer GL, Doherty A (2006). Oz-wheat: a regional-scale crop yield simulation model for Australian wheat. Queensland Department of Primary Industries & Fisheries, Brisbane.

Power B, DeVoil P, Payero J, Rodriguez D, Harris G (2008). Using the farm-scale system model APSFarm to improve profitability of irrigation cropping enterprises. In *Global Issues, Paddock Action. Proceedings of the 14th Australian Agronomy Conference.* 21–25 September, Adelaide, South Australia. Australian Society of Agronomy.

Power SB, Plummer N, Alford P (2007). Making climate model forecasts more useful. *Australian Journal of Agricultural Research* **58**, 945–951.

Prather MR, Derwent R, Ehhalt D, Fraser P, Sanhueza E, Zhou X (1995). Other trace gases and atmospheric chemistry. In *Climate Change 1994: Radiative Forcing of Climate Change and an evaluation of the IPCC IS92 Emission Scenarios*. (Eds JG Houghton, LG Meira Filho, J Bruce, H Lee, BA Callander, E Haites, N Harris, K Maskell) pp. 73–126. Cambridge University Press, Cambridge.

Ramankutty N, Foley JA, Norman J, McSweeney K (2002). The global distribution of cultivable lands: current patterns and sensitivity to possible climate change. *Global Ecology and Biogeography* **11**, 377–392.

Rayner DP (2007). Wind run changes: the dominant factor affecting pan evaporation trends in Australia. *Journal of Climate* **20**, 3379–3394.

Reyenga PJ, Howden SM, Meinke H, Hall WB (2001). Global change impacts on wheat production along an environmental gradient in South Australia. *Environment International* **27**, 195–200.

Richards RA (2002). Current and emerging environmental challenges in Australian agriculture – the role of plant breeding. *Australian Journal of Agricultural Research* **53**, 881–892.

Rogers GS, Gras PW, Batey IL, Milham PJ, Payne L, Conroy JP (1998). The influence of atmospheric CO_2 concentration on the protein, starch and mixing properties of wheat flour. *Australian Journal of Plant Physiology* **25**, 387–393.

Scheraga JD, Grambsch AE (1998). Risks, opportunities and adaptation to climate change. *Climate Research* **10**, 85–95.

Stephens DJ, Lyons TJ (1998). Variability and trends in sowing dates across the Australian wheatbelt. *Australian Journal of Agricultural Research* **49**, 1111–1118.

Stone RC, Nicholls, N, Hammer G (1996). Frost in NE Australia: trends and influence of phases of the Southern Oscillation. *Journal of Climate* **9**, 1896–909.

Stone RC, Hammer GL, Woodruff D (1993). Assessment of risk associated with climate prediction on management of wheat in north-eastern Australia. In *Proceedings 7th Australian Agronomy Conference, Adelaide*. pp. 174–177. The Regional Institute, Gosford.

Tanner CB, Sinclair TR (1983). Efficient water use in crop production: research or re-search? In *Limitations to Efficient Water Use in Crop Production*. (Eds HM Taylor, WR Jordan, TR Sinclair) pp. 1–27. ASA, Madison, Wisconsin.

Torok SJ, Nicholls N (1996). An historical annual temperature data set for Australia. *Australian Meteorological Magazine* **45**, 251–260.

Tubiello FN, Donatelli M, Rosenzweig C, Stockle CO (2000). Effects of climate change and elevated CO_2 on cropping systems: model predictions at two Italian locations. *European Journal of Agronomy* **13**, 179–189.

van Ittersum MK, Howden SM, Asseng S (2003). Sensitivity of productivity and deep drainage of wheat cropping systems in a Mediterranean environment to changes in CO_2, temperature and precipitation. *Agriculture, Ecosystems and Environment* **97**, 255–273.

Verrell AG, O'Brien L (1996). Wheat protein trends in northern and central NSW, 1958 to 1993. *Australian Journal of Agricultural Research* **47**, 335–354.

Wang YP, Handoko Jr., Rimmington GM (1992). Sensitivity of wheat growth to increased air temperature for different scenarios of ambient CO2 concentration and rainfall in Victoria, Australia – a simulation study. *Climate Research* **2**, 131–149.

Wright WJ, de Hoedt G, Plummer N, Jones DA, Chitty S (1996). Low frequency climate variability over Australia. In *Proceedings of the Second Australian Conference on Agricultural Meteorology*. 1–4 October 1996. University of Queensland Printer, Brisbane.

Wu G, Chen F, Ge F, Sun Y (2007). Effects of elevated carbon dioxide on the growth and foliar chemistry of transgenic Bt cotton. *Journal of Integrative Plant Biology* **49**, 1361–1369.

Ziska LH, Teasdale JR, Bunce JA (1999). Future atmospheric carbon dioxide may increase tolerance to glyphosate. *Weed Science* **47**, 608–615.

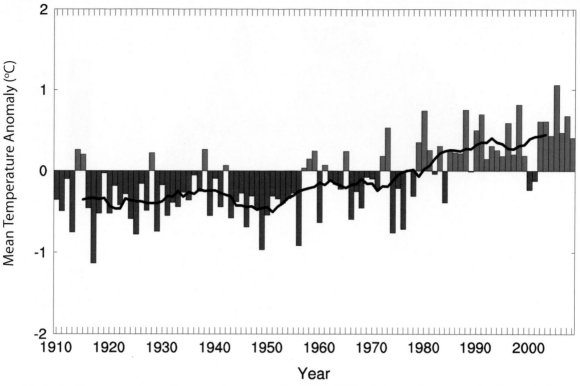

Figure 2.1: Australian average annual temperature anomalies (1910–2008 relative to the average for the period 1961–1990. Source: Australian Bureau of Meteorology 2008a.

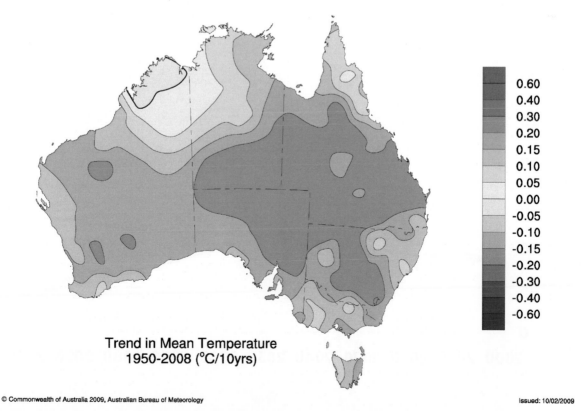

© Commonwealth of Australia 2009, Australian Bureau of Meteorology

Issued: 10/02/2009

Figure 2.2: Trend in annual mean temperature 1950–2008. Source: Australian Bureau of Meteorology 2008a.

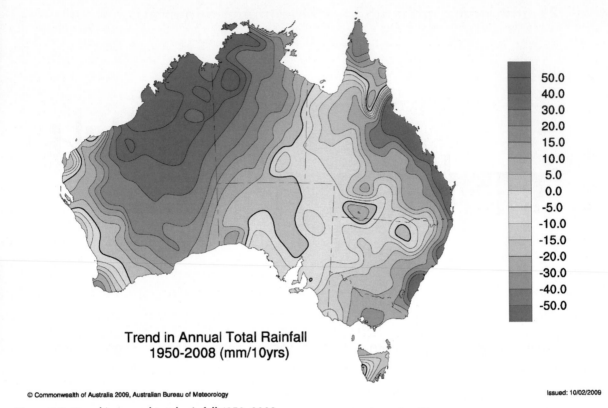

Trend in Annual Total Rainfall
1950-2008 (mm/10yrs)

© Commonwealth of Australia 2009, Australian Bureau of Meteorology

Issued: 10/02/2009

Figure 2.3: Trend in annual total rainfall 1950–2008. Source: Australian Bureau of Meteorology 2008a.

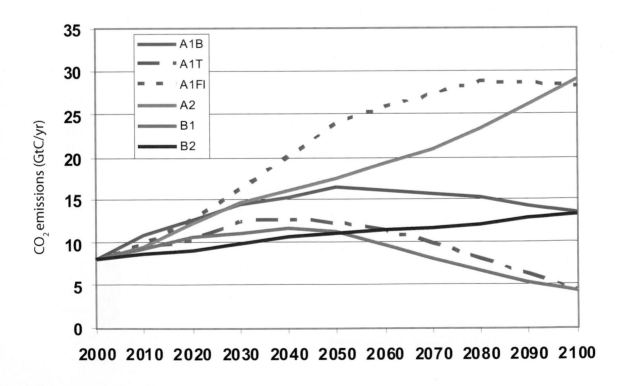

Figure 2.4: The IPCC's six 'illustrative' scenarios for 21st century CO_2 emissions.

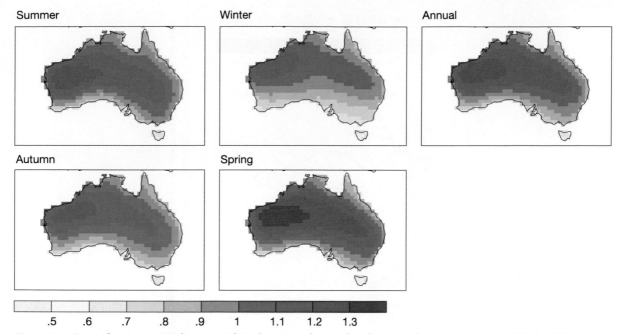

Figure 2.5: Central estimate (50th percentile) of projected annual and seasonal average warming (°C) by 2030, relative to 1990, for a mid-range emissions scenario (AIB).

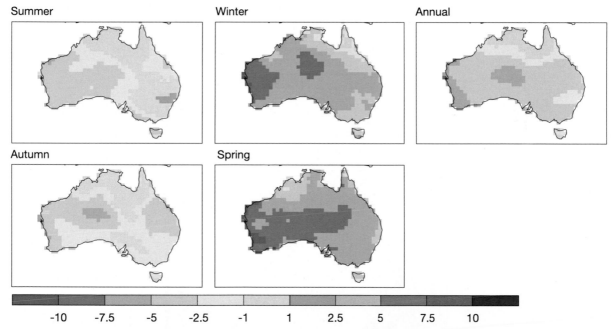

Figure 2.6: Central estimate (50th percentile) of projected annual and seasonal average rainfall change (%) by 2030, relative to 1990, for a mid-range emission scenario (AIB).

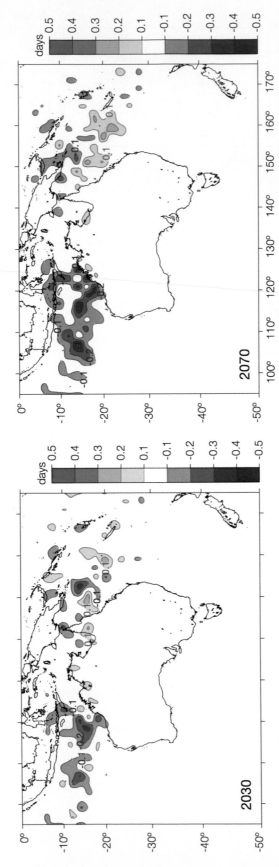

Figure 2.7: Changes in annual average numbers of tropical-cyclone days in the Australian region for 40 year periods centred on 2030 and 2070 (relative to averages for a 40 year period centred on 1980). Blue regions indicate a decrease in cyclone occurrence and red regions indicate an increase in occurrence. Source: Abbs et al. 2006.

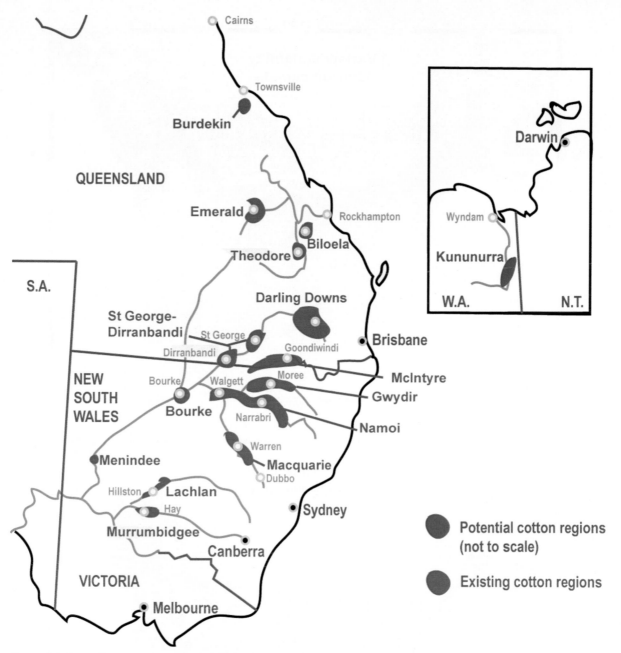

Figure 4.1: Australian cotton growing regions. Reproduced with permission from the Cotton Catchment Communities Cooperative Research Centre (2008).

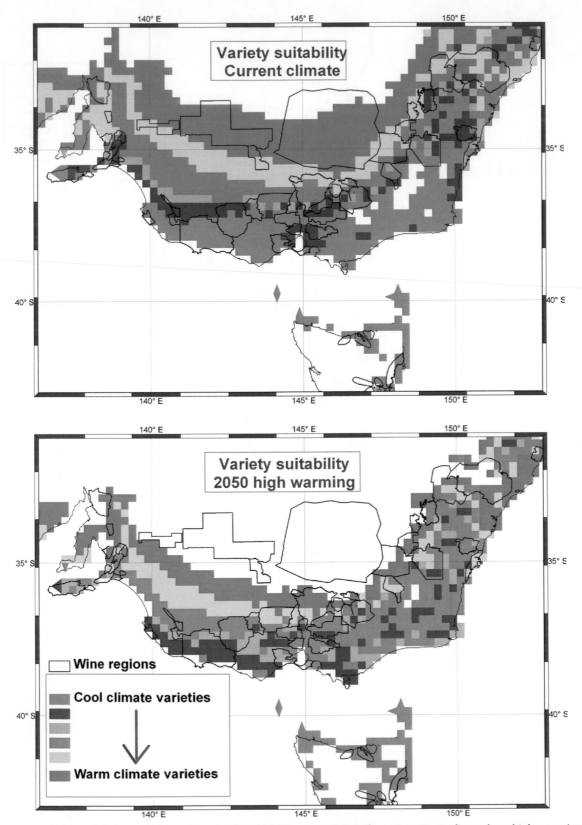

Figure 7.2: Projected shifting of grapevine variety suitability zones in south-eastern Australia under a high warming scenario by 2050.

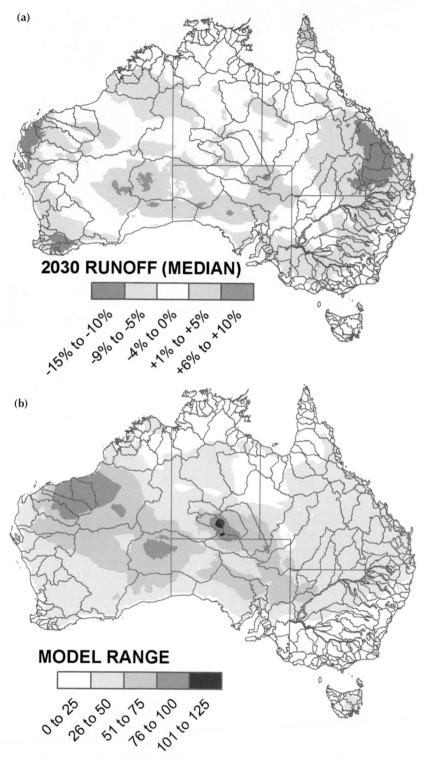

Figure 12.2: Estimated percentage changes in Australian runoff in 2030 produced using a simple hydrological model. (a) Median estimates based upon a series of 66 simulations using 11 different climate models, three different emissions scenarios, and low and high climate sensitivities. The 5th and 95th percentiles were calculated from the mean and standard deviations assuming normally distributed data. (b) Range between minimum and maximum result for all 66 simulations indicating levels of confidence. Smaller ranges indicated higher confidence in the projections. Catchment boundaries correspond to the 325 Surface Water Management Areas (SWMAs). After Preston and Jones 2008c.

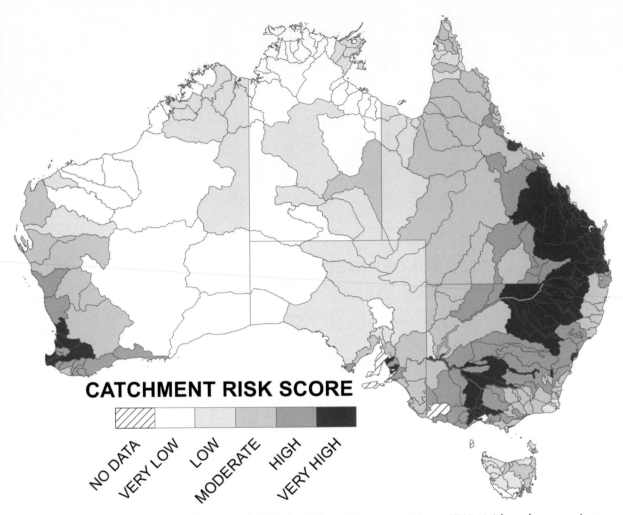

CATCHMENT RISK SCORE

NO DATA   VERY LOW   LOW   MODERATE   HIGH   VERY HIGH

Figure 12.3: Catchment risk scores for Australia's Surface Water Management Areas (SWMAs) based upon various indicators. Catchments at low risk of significant reductions in future runoff and with significant remaining development potential are at a lower vulnerability than those catchments projected to experience significant reductions in runoff and are already fully or over-developed. After Preston and Jones 2008c.

4

COTTON

MP Bange, GA Constable, D McRae and G Roth

KEY MESSAGES:

■ Climate change will have both positive and negative effects on cotton. Increased CO_2 may increase yield in well-watered crops, and higher temperatures will extend the length of growing season (especially in current short season areas). However, higher temperatures also have the potential to cause significant fruit loss, lower yields and reduced water use efficiencies.

■ Research into integrated effects of climate change (temperature, CO_2, and water stress) on cotton growth, yield and quality needs further analysis. This includes the development of cultivars tolerant to abiotic stress (especially for more frequent hot and water-deficit situations). Some consideration or allowance will be needed in these studies for both cotton cultivars and insect pests that have been naturally selected in rising CO_2 environments.

■ There will be a need to improve cotton farm resilience by maintaining and increasing cotton profitability through practices that increase both yield and fibre quality, while improving efficiency of resource use (especially energy, water and nitrogen). Although cotton is already well adapted to hot climates, continued breeding by conventional means as well as applying biotechnology tools and traits will develop cultivars with improved water use efficiency and heat tolerance. This can aid in developing the cotton farming system to be more resilient to climate change and climate variability. A better understanding of plant and crop physiology will be part of research identifying and assessing adaptation traits.

■ The declining availability of water resources resulting under climate change will increase competition for these resources between irrigated cotton production, other crops and environmental uses. These issues emphasise the need for continual improvement in whole farm and crop water use efficiencies and the need for clear information on water availability.

■ Region-specific effects will need to be assessed thoroughly as the cotton production regions span from southern NSW to north Queensland. This is necessary so that cotton growers can assess likely impacts at their business level. Continued research is justified into the development of sustainable cotton systems for northern Australia where water supply is more assured.

■ As cotton is a global commodity and all Australian cotton is exported it will be vital for Australia to understand global changes in cotton markets as part of its overall adaptation strategies.

Introduction

Cotton is a natural fibre produced by the cotton plant, a leafy, green perennial shrub, that is a member of the Hibiscus family. Although cotton grows naturally as a perennial plant to 3.5 metres in the tropics, commercially it is grown as an annual crop. Australia's cotton growing season usually starts in September to October with planting and finishes in March to April with picking.

Cotton is used every day in the form of clothing made from cotton fibre and products made from cotton-seed oil. Cotton is the most widely-produced natural fibre in the world and represents about 44% of the global textile market. By contrast, wool accounts for 3%, synthetics 51% and other fibres like silk, hemp and mohair make up a very small proportion (Cotton Australia 2007).

Cotton seed is a by-product of the more valuable cotton fibre, and can make up about 15% of the total financial returns to farmers. For every 227 kg bale of cotton lint, about 370 kg of cotton seed is produced. Cotton seed is a valued raw material for food oils for human consumption and high protein feed for livestock. Cotton-seed oil is one of the world's most popular vegetable oils.

The Australian cotton industry

Currently 70% of Australia's cotton is grown in New South Wales (NSW) with the remainder grown in Queensland (see Figure 4.1, page 45). In an effort to expand the cotton growing regions and address some of the challenges raised by our variable climate, cotton growing has been trialled in northern Australia (including the Ord River Irrigation Area (ORIA), Kununurra, and the Burdekin River Irrigation Area in north Queensland). Broadly the regions could be described as:

- Northern NSW (Namoi Valley, Gwydir Valley, Macquarie Valley, Bourke);

- Southern NSW (Hillston, Hay and Menindee districts);

- South Queensland (Macintyre Valley, Darling Downs, St George);

- Central Queensland (Theodore, Biloela and Emerald regions), and

- Northern Australia (ORIA and Burdekin Irrigation Areas).

Depending on water availability, about 400 000 ha of irrigated cotton is grown in Australia (Figure 4.2). The area of rain-fed cotton varies considerably from year to year depending on commodity prices, soil moisture levels and rain. The area of rain-fed crop can vary from 5000 to 120 000 ha.

Most Australian cotton farms are owned and operated by family farmers, are typically between 500 and 2000 ha, are highly mechanised, capital intensive, technologically sophisticated and require high levels of management expertise. About 80% of cotton farms are irrigated and as part of the enterprise mix generally grow other crops such as wheat and sorghum and/or graze sheep and cattle (Cotton Australia 2007). In 2005–06 cotton growing accounted for around $400 million towards gross regional product across all cotton regions (Stubbs *et al.* 2008).

On a global scale, Australia is a relatively small producer growing about 3% of the world's cotton. As with other agricultural commodities grown in Australia, the Australian cotton industry is a large exporter. Australia's reputation for producing high-quality cotton means it is internationally competitive. The gross value of cotton produced in Australia has increased rapidly since 1985, with the exception of the drought years 1986, 1992–94, and 2003–04. The gross value of production peaked at $1.9 billion in 2000–01 and was $1.2 billion in 2004–05 and $1.1 billion in 2005–06 making the industry important not only for the economy in cotton growing communities but also for the nation (Cotton Australia 2007). Drought conditions still plague the industry's access to water resources and continue to restrict cotton production (Figure 4.2).

Impacts of projected climate change

Cotton (*Gossypium hirsutum* L.) is a perennial plant with an indeterminate growth habit. Wild ancestors of cotton are found in arid regions often with high temperatures and are naturally adapted to surviving long periods of hot dry weather. Modern cultivars have inherited these attributes, making the cotton crop well adapted to the intermittent water supply that occurs with rain-fed (dryland) and irrigated production (Hearn 1990). Compared with other field crops, however, its growth and development are complex. Vegetative and reproductive growth occur simultaneously making interpretation of the crop's response to climate and management sometimes difficult. Climate change (Chapter 2) will influence cotton growth and development, and hence yield and fibre quality, through the combined net effects of increases in CO_2 concentration; reduced water availability and increased atmospheric evaporative demand as a result of lower rainfall and relative humidity; increases in daily temperature including marked

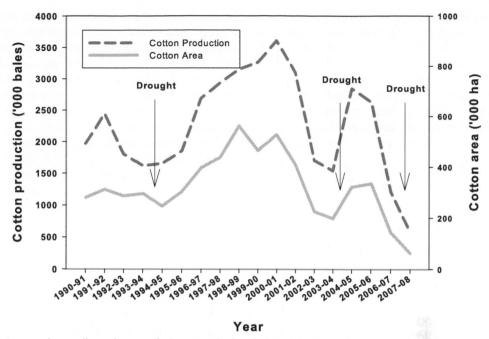

Figure 4.2: Area and overall production (bales – 227 kg lint/bale) of the Australian cotton industry. ABARE 2008

increase in the frequency of hot days and warm nights, but a less-marked decrease in the frequency of frosts; and changes in irrigation water availability resulting from a drier and hotter climate limiting runoff and replenishment of ground water. These effects on cotton growth and development are discussed in more detail below.

Increases in CO_2

Increased atmospheric CO_2 concentrations have been shown in cotton to increase leaf photosynthetic rates and crop radiation use efficiency (dry matter produced per unit of intercepted radiation by the crop) and reduce transpiration through reduced stomatal conductance, all leading to improvements in growth and yield. Research summarised by Reddy *et al.* (1996) in controlled glasshouse conditions showed that doubling CO_2 concentration in the atmosphere increased leaf photosynthesis by about 40% and lowered leaf transpiration by 30% which led to increased growth and yield (60% increase in lint yield) in well-watered environments. This work also showed that increasing CO_2 increased transpiration efficiency (CO_2/H_2O). From their studies they postulated that increased growth and yield would occur with higher CO_2 concentrations even in dry situations.

Field experiments growing cotton using free air CO_2 enrichment (FACE) showed that radiation use efficiency (dry matter per unit of intercepted radiation) was improved on average by 26% from 1.56 to 1.97 grams per megajoule of intercepted radiation resulting in increased biomass when CO_2 was increased to 550 ppm (Pinter *et al.* 1994; Mauney *et al.* 1994). They also found that regardless of irrigation treatment (wet or dry), radiation use efficiency was increased in CO_2-elevated treatments. This suggested that a rise in atmospheric CO_2 concentrations may partially compensate for plant stress caused by water shortages. As a consequence, lint yield on average was increased by 43% and was attributed to increased early leaf area (direct effect) and a longer flowering period (indirect effect).

Studies conducted by Samarakoon and Gifford (1996) in controlled environments also found that cotton growth, leaf transpiration and water use efficiency was improved when CO_2 was increased to 710 ppm. In contrast with the work of Reddy *et al.* 1996, transpiration was only reduced by 15%, which led to a smaller increase of transpiration efficiency in higher CO_2. This occurred concurrently with substantial increases in growth and leaf area resulting in water use per plant being 40–50% higher than other crop species. Samarakoon and Gifford (1996) also highlighted that in drying weather conditions, this resulted in soil drying faster under high CO_2 conditions, accelerating the onset of water stress compared with the low CO_2 treatment. This experiment was conducted in pots in a glasshouse. The potential for effects of water stress on growth to be exhibited earlier in pots compared with field studies (Jordan and Ritchie 1971) therefore needs to be taken into consideration.

The contrasting results of the effects of CO_2 on cotton transpiration and subsequent water use (Reddy *et al.* 1996; Samarakoon and Gifford 1996) needs resolution especially if cotton is subjected to the additional confounding effects of limited water availability and increased drying conditions caused by increased atmospheric evaporative demand. Substantially hotter temperatures could also potentially negate any benefits of increased CO_2 by reducing leaf photosynthesis (Reddy *et al.* 1998). This is discussed in more detail below.

Coviella *et al.* (2002) and Wu *et al.* (2007) have shown that cotton plants grown in elevated CO_2 have less expression of the *Bacillus thuringiensis* (Bt) toxin. Overall this effect did not change the performance of the cultivars to resist *Helicoverpa* spp. as there was evidence to suggest that *Helicoverpa* were adversely affected by feeding on cotton subjected to elevated CO_2. While their lifespan increased, their pupal weight, survival rate, fecundity, frass output, relative mean growth rates, and the efficiency of conversion of ingested and digested food was decreased (Chen *et al.* 2007; Wu *et al.* 2006).

In the case of weeds, there are concerns that some weed species will increase their competitive advantages under elevated CO_2 (IPCC 2007) and that it may increase tolerance to glyphosate in some weed species (Ziska *et al.* 1999).

Reduced water availability and higher evaporative demand

Water stress in cotton restricts both vegetative and fruit growth. Cotton's response to stress varies according to the stage of growth, the degree of stress, and the length of time imposed (Hearn 1979). Research in Australia has shown that to optimise yield, cotton crops generally require enough water to allow an average 700 mm of evapotranspiration (transpiration plus soil evaporation) (Tennakoon and Milroy 2003).

In situations where water is limited and there is high evaporative demand, crops may be unable to transpire enough to keep the canopies cool. Leaf temperatures then increase to a point where photosynthesis and growth are impaired (Hearn and Constable 1984). Higher evaporative demand in well-watered crops also has the potential to increase transpiration and lower water use efficiency (Rawson and Begg 1977). Research is currently underway to quantify the effects of evaporative demand on cotton growth and water use efficiency. Plant stress is increased when cotton growing on soil types with low water holding capacity encounters high evaporative demands (Neilsen 2008).

Increases in temperature

Temperature has two main influences on cotton growth and development. First, it determines rates of morphological development and crop growth (e.g. node development, rate of fruit production, photosynthesis and respiration) (Hearn and Constable 1984). Second, it also helps determine the start and end of a growing season (e.g. timing of frosts).

Consequently, rising temperatures under climate change may: (1) increase average daily temperatures warming both the start and end of cotton seasons, including the reduction of frost occurrence and plant chilling damage allowing for longer and better cotton growth (a positive effect); (2) cause quicker emergence and flowering (a positive effect); (3) reduce disease; (4) increase the number and severity of days with very high temperatures during the cotton season (negative effect); (5) increase average temperatures during boll filling, predisposing crops to high micronaire issues (both a positive and negative effect depending on the region), and (6) increase abundance and development and mortality of certain pests (may be a positive or negative effect).

As cotton is a perennial crop, warmer temperatures at the start and the end of cotton seasons will increase the length of time cotton has to grow and produce yield, provided adequate water and crop nutrition are available. For every extra week that the growth period (time between planting and maturity) is extended through warmer temperatures, the cotton crop has the potential to increase lint yield by 68 to 136 kg per ha (Bange and Milroy 2004a). Extending the cotton season has the potential to benefit the northern and southern NSW and southern Queensland cotton producing regions especially the eastern boundaries of these areas where cooler conditions prevail. It also provides growers with greater flexibility with planting dates at the start of the season.

Low temperatures after planting increase the time to emergence and reduce cotton seedling vigour often leading to poor establishment, poor early growth and increased risk of seedling diseases (Constable and Shaw 1988). A key indicator used in the Australian cotton industry of the degree of very cold temperatures that contribute to the conditions mentioned above is the 'cold shock' concept. A cold shock is defined as when minimum daily temperatures are ≤11°C and each event extends the duration to flowering by 5.2 day degrees (Hearn and Constable 1984). In some cotton producing regions in Australia the number of cold shocks can be as frequent as 40 (period 15 September to 30 November) (Bange and Milroy 2004b). Climate change has the potential to raise minimum temperatures and reduce the number of cold shocks.

The impacts of high temperature have been reviewed by Hearn and Constable (1984). Excessively high temperatures (greater than 35°C) during the day can decrease photosynthesis, while warm nights (above 25°C) mean that leaf temperature remains high, and respiration remains high, consuming stored assimilates. Maintenance respiration approximately doubles for every 10°C rise in temperature. Both situations reduce the amount of assimilates available for growth, and in turn reduce yield by increasing square and boll shedding and reducing seed number per boll. Loss of fruit may cause the crop to grow excessively vegetatively (rank) following the period of heat stress. In addition to reductions in assimilates available for growth, heat stress can also directly damage cotton plant tissue. During the day, very high relative humidity (which restricts evaporative cooling) in combination with clear skies can also increase tissue temperature to approach or exceed air temperature. A known consequence of direct tissue damage from severe heat stress is parrot-beaked bolls where high temperatures reduce the viability of the pollen at flowering. This results in small bolls with uneven seed numbers between the locks caused by poor pollination and seed set particularly in one lock. Where substantial numbers of bolls are affected, yield may be reduced. There are no known studies to show if the plant compensates for parrot-beak bolls by having other normal bolls grow bigger.

Large increases in temperature may also reduce the number of fruiting branches as well as the interval between flowering and boll opening, shortening the time to maturity and reducing yield. This may increase final micronaire by limiting the number of late set bolls that can have lower micronaire. Fibre length can also be affected by sustained periods of high temperatures as the time required for fibre elongation is reduced, preventing the genetic potential from being reached. Stockton and Walhood (1960) found that as boll temperatures increased above 32°C, fibre length was reduced. The consequences of hot conditions for yield and quality are exaggerated if water stress also occurs during these periods. These issues will be important for all cotton-producing regions.

A key factor affecting micronaire (a fibre quality trait) is the temperature during boll filling. Micronaire is an indirect measure of fibre fineness and maturity measured by passing air through a bundle of fibres using a HVI (high volume instrument). Growers in Australia have price discounts applied when cotton has too high or too low micronaire (optimum range 3.8 to 4.5). Cotton with low micronaire may be immature, while high micronaire cotton is considered coarse and provides fewer fibres in a cross-section of yarn. Fibre fineness is determined primarily at fibre initiation on a seed. Finer fibres mean that there are more fibres in the cross-section of a yarn when it is spun and so the yarn is stronger. Fibre maturity is the proportion of fibre cross-section occupied by cellulose and is influenced by variations in photosynthesis affecting assimilate supply to growing fibres. The degree of fibre maturity impacts dye absorbency and retention. As photosynthesis increases with temperature in the absence of water stress, more resources are available to mature the fibres thus increasing micronaire. High micronaire is more likely to occur in those seasons where there are frequent and sustained warm periods during boll filling (Constable and Bange 2007). High micronaire will be an issue for all but shorter growing season areas.

In the case of invertebrate pests, temperature directly affects insect development, survival, number of generations, timing and the duration of diapause, while rainfall affects the growth of plant hosts leading to differences in distribution and abundance of insect pests. Gregg and Wilson (2008) suggested the impacts of rising temperatures were increased chances and abundance of some insects in new and existing regions (they used Silver leaf whitefly (*Bemisia tabaci* B Biotype) as an example); increased development rates of some insects, and increased chances of overwintering of some pests. They cautioned, however, that these effects of warmer temperatures on insects could be offset by increased insect mortality. Furthermore, overwintering is influenced more by the amount of rainfall in autumn and winter (which affects the growth of plant hosts) than by temperature *per se*. The relative effect of warmer temperatures on the growth rates of crops and insects together also needs to be considered. In non-stressed conditions crop growth rates may increase in line with increased rates of insect development, potentially negating the effects of increased insect development and damage to the crop. This issue requires further study.

Options for dealing with climate variability

Although the rate of climate change is relatively slow, a small change in average temperature may have a relatively large impact. For example with days with 25°C maximum and 15°C minimum temperatures (equating to 8 day degrees) it takes about 63 days from planting to first square; add 1°C to mean temperature (the best estimate of annual warming to 2030 (CSIRO 2007)) and the number of days to first square is seven days quicker. Even so, the rate of climate change is still relatively small compared with the year-to-year variability in temperatures experienced in Australian cotton cropping regions. For example, mean monthly temperatures in some cotton regions were up to 2 to 3°C higher than their long-term monthly average at times in the 2006/2007 season and were of similar magnitudes less than their long-term monthly average in the 2007–08 season. Therefore practices that are adopted on the farm to deal with climate variability and improve sustainability offer opportunity to assist with climate change (Chapters 1 and 16). The wide geographic spread of the industry also means that some management practices differ as growers have adapted their practices for their various climates. Other than the direct effects of climate variability on physiology and growth of cotton, water availability, abundance of pests, weeds, and diseases are other impacts on cotton production that are also highly variable from season to season as a result of climate. Specific cotton crop management issues for dealing with climate variability for these issues are now discussed, as well as listing extension material and decision tools that the industry has currently available for use in managing this variability.

Water management

Most cotton in Australia is fully or partially irrigated which serves to reduce the impacts of rainfall variability. However, recent drought conditions coupled with reductions in water allocation across all major cotton production regions has reduced irrigation supply and placed significant emphasis on managing water resources more efficiently which will assist in adaptation to climate change. There are a number of practices that cotton growers can use to improve water use efficiency or to adapt to water-limited situations. These include:

- Implementing systems that monitor and assess whole farm water use efficiency to identify parts of the system which are inefficient. Growers consistently adopt practices to improve water storage and furrow irrigation efficiencies, and reduce transmission and application losses.

- Use of alternative irrigation systems such as lateral move, centre pivot, or drip irrigation systems.

- Better scheduling of irrigations utilising technologies that monitor weather (automatic weather stations) and crop soil water use (capacitance probes, neutron moisture meters) allowing for differences in soil types, demands of the crop (crop stage), and climatic conditions (e.g. temperature and humidity).

- Improving soil management by adopting controlled traffic and reduced tillage practices to minimise compaction, thereby improving soil structure and increases the rooting zone.

In addition, there are several management options that can be applied in limited water situations:

- Reduce the risk of crop failure by modifying the area of cotton grown to increase the ML per ha from irrigation water supplies before the season begins. Determining the area to plant is a decision that has to consider the yield and thus the water needed (accounting for climatic risk and system irrigation efficiencies) to break even. Simulation technologies such as those in HydroLOGIC (Richards *et al.* 2008) (which incorporates OZCOT (Hearn 1994)) that estimates yield with different water allocations and climatic impacts (including rainfall variability) can be used to assess cotton cropping areas. This concept is explained in more detail in an article by Hearn (1992).

- Better utilise stored soil water collected from crop fallows and employ practices to capture and retain soil moisture. Strategies such as reduced tillage and stubble retention are becoming standard practice for moisture conservation. New technologies such as films (e.g. polymers) deserve some research into how much soil moisture is saved and how much value is achieved. Use of rainfall to establish crops rather than pre-irrigation or watering-up are worth considering especially if there is flexibility in planting time.

- Avoid excess nitrogen fertiliser, which encourages extra vegetative growth, lowering crop water use efficiency. This also has the benefit of reducing nitrous oxide emissions (a greenhouse gas).

- Sow as rain-fed cotton, utilising supplemental irrigation strategies or modified row configurations (e.g. skip rows) to enhance crop access to soil moisture. These strategies are not necessarily the most water use efficient but offer significant risk mitigation in years where rainfall is limited (Montgomery and O'Halloran 2008). Skip-row configurations can also offer significant insurance against losses in both yield and quality in those regions and years where rainfall is highly variable, and can reduce input costs such as Bollgard II licence fees (Bange *et al.* 2005).

- Consider rain-fed production in those years where prices for cotton are profitable and there is a forecast for reasonable rainfall (Bange *et al.* 2005).

- Avoid water stress conditions, particularly during the flowering period.

- Shorten the time to maturity.

Pest, weed and disease management

One major production issue that growers face each season is the protection of the crop against a range of insect and mite pests. Key invertebrate pests of Australian cotton include *Helicoverpa armigera* and *Helicoverpa punctigera*, spider mites (*Tetranychus urticae*), aphids (*Aphis gossypii*) and mirids (*Creontiades dilutus*). The pest and beneficial insect complex is broadly similar for all cotton growing regions (including northern Australia). To control these pests they have historically relied strongly on intervention with chemical pesticides, which remain a significant component of the cost of production (Fitt and Wilson 2000). In addition, the use of chemical sprays gives rise to ecological problems from pesticide resistance in key pests, and environmental concerns about pesticide movement off-farm (Wilson *et al.* 2004; Fitt 2000).

The development of transgenic cotton with two Bt genes (Bollgard II) has reduced pesticide use for the control of major *Lepidopteran* pests (particularly *Helicoverpa* spp.) however, as the system is changing pests formerly suppressed by these sprays for *Helicoverpa* spp. are emerging as new challenges. The Australian industry is modifying integrated pest management (IPM) strategies to meet the challenges of emerging pests as well as maintaining viability of conventional and transgenic systems against pesticide resistance (Wilson *et al.* 2004). Implementation of IPM in the Australian industry has been supported by the use of industry IPM guidelines (Deutscher *et al.* 2004), decision tools (Hearn and Bange 2002), and specialist cotton consultants.

Seasonal climate variability (especially in relation to variations in temperature and rainfall) influences the distribution and abundance of insect pests. In the past the industry has experienced extremely wet and very dry years and the pest issues associated with these climates. Wet and warm years have seen abundant winter and summer weed hosts contributing to pest build up. Hot dry years when irrigation water is available are favourable for cotton growth and most likely have less insect pest build up also contributing to improved yields.

One of the fundamental elements of an IPM approach adopted by the Australian industry is the regular and accurate monitoring of the numbers of pest and beneficial insects and the use of economic thresholds to guide decisions on pest management. In doing this, growers are responding accordingly and adapting to regional and seasonal variation. Changes in climate will need to ensure industry monitoring strategies are in place to identify changes in the pest, weed and disease spectrums, especially those that prefer warmer climates. There is significant opportunity to further develop pest forecasting systems that can be used to predict the effects of climate change. As an example, a simulation model already used in the Australian cotton industry for *Helicoverpa* spp. is the HEAPS (HElicoverpa Armigera and Punctigera Simulation) model which has been used to assess movement of adult moths within a regional cropping system (Fitt *et al.* 1995). HEAPS includes modules for spatial representation in the region, moth movement, oviposition, pest and crop development and pest mortality.

Differences in weeds and crop diseases also occur across regions and seasons because of variation in rainfall and temperature. Australian cotton growers employ the use of transgenic cotton that allows over-the-top application of glyphosate for weed control to enable a rapid response to weed control along with integrated weed management. In the case of diseases, cold wet conditions are favourable for Black root rot (*Thielaviopsis basicola*), Fusarium Wilt (*Fusarium oxysporum*) and Verticillium Wilt (*Verticillium dahliae*). When conditions are warm and moist Alternaria Leaf Spot (*Alternaria macrospore* and *Alternaria alternate)* can be an issue. The industry employs an integrated disease management approach, and breeding efforts are heavily focussed on developing germplasm with in-built resistance.

Extension material and decision tools

Guidelines for crop management practices specific to Australian cotton systems are delivered through publications and decision-support tools. These publications make specific reference to managing climate variability and are made available through the Cotton Catchment Communities website (http://www.cottoncrc.org.au/). Some significant publications include WATERpak (CRDC 2008), NUTRIpak (Rochester *et al.* 2001b), WEEDpak (Charles *et al.* 2002), integrated

management guidelines for invertebrate pests (Deutscher *et al.* 2004) and diseases (Allen *et al.* 2002), and the rain-fed production guide (CRDC 1997). Several decision tools and modelling capabilities available to the Australian cotton industry integrate knowledge of cotton management and climate, including climate variability issues: (1) OZCOT cotton crop simulation model (Hearn 1994) is used for assessing the impact of climate variability and management on yield on both irrigated (Richards *et al.* 2008) and rain-fed cropping systems (Milroy *et al.* 2004). Current research for OZCOT is focussed on the model's ability to simulate high fruit retention transgenic cotton, impacts of changes in evaporative demand directly on plant stress, and enabling the model to predict fibre quality. (2) CotAssist (http://www.cottassist.cottoncrc.org.au) is a collection of interactive decision tools including a cotton day degree calculator, crop development tool, last effective flower predictor, NutriLOGIC (crop nutrition management), and *Helicoverpa diapause* induction and emergence tool. (3) CottBASE is a standalone software application based on the Whopper Cropper software concept (Nelson *et al.* 2002) that is used to analyse pre-run simulations of the OZCOT cotton simulation model to assess the effect of climate variability on cotton irrigation practice. (4) Cotton Greenhouse Gas Calculator is a web application that estimates greenhouse gas emissions from use of nitrogen fertiliser and fuel consumption (http://www.isr.qut.edu.au/tools/). Cotton growers also grow other crops so they have access to knowledge and technology available to other cropping industries.

Options for dealing with projected climate change

In addition to harnessing options for coping with climate variability, two other key considerations will be important in developing adaptation strategies. First, it will be important to develop a range of options that have synergistic effects in reducing the impact of climate change (e.g. cultivar, frequent irrigation, planting date strategies), rather than relying on just one approach. Second, these options will need to be flexible to cope with variability and uncertainty. It will also be vitally important that existing integrated management (insect, weed, and disease) systems discussed previously are maintained. Any management changes aimed at

'Neilo' is a 4700 ha mixed farming enterprise located west of Toobeah. It comprises approximately 1400 ha of land developed for surface irrigation on which cotton can be grown, 1100 ha of land developed for broadacre farming, with the rest of the property used for grazing running on average 350 breeding cows and calves. Dave Taylor from 'Neilo' believes that climate change will create a number of challenges and opportunities.

Dave explained that the focus at 'Neilo' on managing climate change and variability has taken the form of using modelling technology such as Rainman Streamflow, Howwet, Howoften, Whopper-Cropper and HydroLOGIC to investigate the effect that different rainfall and temperature patterns have on the production system and how they would alter planting opportunities, yield and harvest times. The output from these tools is also incorporated into a budget analysis to investigate any potential impact on gross margins and changes to the level of acceptable risk and equity growth.

Dave believes that regardless of climate change, the basics of any farming or production system still need to be carried out correctly since getting the basics right has the biggest influence on production. 'For example we will have to be very careful in our reliance on residual-type herbicides for weed management so as to not miss a cropping or cash-flow opportunity, but in doing so we must be mindful on any developing resistance to contact herbicides'.

As to adapting to changes in the future 'I believe marketing is going to be one of the best tools to offset some production or yield loss and it may change our risk paradigms if there is an opportunity that looks too good to miss. I feel climate change and looking for potential opportunities further enhances my wish to develop a permanent bed system in irrigation country based around minimum tillage and moisture retention'.

reducing the impacts of climate change should also be planned and chosen so as to avoid creating problems elsewhere in the farming system (for example pest resistance to pesticides). Here we

re-emphasise broad strategies and practices that can be adopted to help build resilient cotton systems in a changing climate.

Improving cotton yields

Australian cotton yields have been steadily increasing over the last 20 years at a rate of about 0.1 bales per ha each year despite expansion into new regions and including an increasing proportion of lower-yielding rain-grown production. Improvements in yield can be attributed to better cultivars, and considerable improvements in crop management, particularly factors such as soil management and crop rotation, irrigation scheduling and insect control (Constable *et al.* 2001). Baker and Hesketh (1969) of the University of Mississippi calculated that the maximum yield potential of cotton was 19.1 bales per ha. This value was estimated by using leaf photosynthesis and respiration data from their experiments. Constable and Bange (2006) reassessed this potential yield estimate and also found 19 bales per ha were possible, but only with a perfect climate and a long growing season.

There are many reports of very high yields at the field level, with up to 15 bales per ha recorded (80% of the potential yield). If the potential yield estimates are correct, then yields cannot continue increasing indefinitely unless there is a change in physiological yield potential of the cotton plant. There is research at a very basic level in plant physiology and biotechnology to raise photosynthesis and water use efficiency and lower respiration. But these processes are complicated and involve many genes, so advances in these areas will be many years away (Sinclair *et al.* 2004). As a consequence, we will most likely see that those already achieving high yields will see less significant increases in yield, and these growers should instigate practices to increase efficiency. Australia's average irrigated yield for the 2009 harvest was 8.2 bales per ha so there are many fields still producing yields less than this average. Therefore there are opportunities for some growers to realise yield improvements where climate and soil permits. These growers can harness practices of high yielding growers to raise their own yield and efficiency.

Cultivar choice is a strong component of realising both target yield and fibre quality levels on farm. A balance needs to be reached between yield,

Table 4.1: Comparison of the highest monthly average maximum and minimum temperatures for Narrabri and Emerald (Source: Bureau of Meteorology) and hot cotton production areas in Maricopa, Arizona (USA) and Multan (Pakistan) (Source: www.weatherbase.com) during their respective summer production season

Location	Summer maximum °C	Summer minimum °C
Multan, Pakistan	42.3 (Jun)	28.7 (Jul)
Maricopa, Arizona, USA	41.6 (Jul)	24.1 (Jul)
Emerald, Qld, Australia	34.3 (Dec)	21.0 (Jan)
Narrabri, NSW, Australia	33.9 (Jan)	18.5 (Jan)

fibre quality/price and other important considerations such as disease resistance, and insect and herbicide resistance. To meet this need, Australian cotton breeding programs are already developing cotton cultivars well suited to the environmental and climatic conditions experienced throughout all the cotton production regions (hot and cool), and therefore give growers options for selecting cultivars suitable to their conditions. Plant breeding programs have already developed cultivars with known high heat tolerance (Constable *et al.* 2001). As cotton is already grown in hotter climates than currently experienced in Australia (Table 4.1), accessing germplasm from those areas may offer genetic sources of improved heat tolerance and water use efficiency. In accessing this germplasm, plant breeding efforts, in combination with new biotechnology tools and screening techniques (Cottee *et al.* 2008), will improve cultivar tolerance to temperature and water stress.

In addition to these breeding efforts that account for regional climatic diversity, there are also specific breeding activities in place for rain-fed cotton production (Stiller *et al.* 2005). This means that there are already cultivars that are better adapted to less than fully irrigated conditions where plants are periodically stressed.

Planting time variation

Changing planting time offers a 'systems solution' that can provide benefits in maintaining yield, improving fibre quality, reducing the risk of adverse effects of high temperatures and low humidity (high evaporative demand), and reducing the incidence of seedling diseases early in the season. Research by Bange *et al.* (2008) showed that crops with higher fruit retention can maintain yield and improve fibre length and

micronaire for delayed planting dates in warmer seasons. Current studies are also investigating opportunities to utilise variations in planting date and cultivars with different growth habits to optimise yield, quality and water use efficiency. The intention is to optimise planting date in different regions to avoid periods of high temperatures and low humidity that lead to significant increases in crop water use. This approach needs consideration in those regions with longer growing seasons that offer opportunity to vary sowing time, especially if the climate change leads to less available water and lower humidity.

Overall higher temperatures in some regions may give more planting time options. In northern New South Wales and southern Queensland cotton regions, it may be possible to sow later in the season, while in shorter season regions, it may be possible to sow earlier. In a changing climate with the introduction of new cultivars there will need to be ongoing reassessment of planting time as a management option in the different cotton production regions. A key to understanding how these interactions affect other efficiencies and factors means that they will require careful investigation on cotton farms, and these efforts will need to involve a range of research and extension specialists working with growers.

Optimising efficiency of resource inputs

Improving production efficiencies can improve crop returns and can sometimes also allow for more resources (e.g. water from improved water use efficiency) to be accessed for production. Rising costs of fuel, fertiliser and chemicals, future costs associated with 'carbon pollution', environmental concerns, and possible reductions in available irrigation water (from reform in water regulation and climate change), will all impact on cotton production. Two key current focuses in cotton production on improving production efficiencies relate to the use of nitrogen fertiliser and water (Bange and Constable 2008).

Monitoring soil fertility and crop nutrient uptake is now more important within the Australian cotton industry, as growers realise the importance of avoiding nutrient deficiency and the economics and environmental concerns (including greenhouse gas emissions) of excess fertiliser use. Excess nitrogen fertiliser can also lower water use efficiency or yield by encouraging excessive vegetative growth and delaying maturity. NutriLOGIC or NUTRIpak decision-support systems (Rochester *et al.* 2001b) are used by growers to provide information for determining the appropriate rates for N fertiliser use and the need for other nutrients based on crop stage (utilising climate information) and performance. However, a survey of Australian cotton fields by Rochester *et al.* (2007) highlighted that a significant proportion (17 in 34 cotton crops) had low nitrogen use efficiency (kg lint per kg N uptake) as a consequence of excessive N fertiliser application. They calculated on average an excess of 40 kg N per ha was applied, increasing the chances of nitrogen being lost from the system and contributing to greenhouse gases.

An alternative approach for crop N nutrition is to supplement or even replace entirely the use of artificial nitrogen fertiliser with nitrogen fixed by legumes. Cotton crops can be grown with nitrogen provided entirely by legumes (Rochester *et al.* 2001a), with vetch (*Vicia villo*) especially being able to supply high yielding crops requiring up to 240 kg N per ha (Rochester and Peoples 2005). The additional benefits of utilising legumes in the cropping system include improvements in soil structure and soil health.

Since the effects of rising CO_2, warmer temperatures and changing rainfall on plant growth can all be modified by nutrient availability, it will be important to monitor and adapt fertiliser practices with climate change. Indeed monitoring approaches such as leaf and petiole tests will also need constant revision as climate changes since plant C:N ratio can be less in cotton grown in raised CO_2 environments (Wu *et al.* 2007).

Tennakoon and Milroy (2003) in their review of cotton water use efficiencies highlighted that there were significant opportunities to improve water use efficiency at all levels (whole farm to agronomic). Their analysis showed that cotton farms had an average farm irrigation efficiency of 57% (ranging from 37 to 68%) and that 40 to 45% of water losses from the farm system were through conveyance, storage and application losses or improper scheduling. Overall there continues to be a need for crop management and research efforts to remain diligent in challenging the paradigm of a cotton system that was developed assuming reasonable access to water, to one that accepts that water will always be limited. A range of current options already discussed with alternative irrigation and agronomic practices will be needed that can consistently produce profitable and high quality cotton crops with less irrigation water. Current research and extension efforts into improvements in water management at all scales (plant to catchment) are summarised by Roth (2007).

Matching crop maturity with season length

Crop maturity can be manipulated by choice of cultivar, insect management, nutrition, growth regulators, or late season irrigation management (Roberts and Constable 2003). Early crop maturity may avoid fibre quality downgrades, save water and reduce the need for late season insect protection. However, this needs to be balanced against the fact that in the Australian environment lint yield is reduced by 68 to 136 kg per ha per week of shorter maturity (Bange and Milroy 2004a). Longer growing seasons such as those experienced in northern New South Wales and southern Queensland have the potential for greater yields.

To cope with limited water availability, one option would be to reduce the time to maturity and then manage a crop to achieve a targeted economic yield threshold. Roberts and Constable (2003) and Bange *et al.* (2006) have shown that after cultivar choice, the main factor driving differences in crop maturity is fruit retention. The Bollgard II cultivar, with its ability to withstand early pest damage from *Helicoverpa* spp., thus maintaining higher fruit retention and avoiding plant terminal damage, can achieve similar yields to non–Bollgard II and use less water by maturing earlier (Richards *et al.* 2006).

Regional opportunities

There is significant opportunity for cropping regions in northern Australia to produce cotton (Yeates 2001). For the industry as a whole these regions have generated significant interest with the opportunity of more reliable water supply for production compared with existing cotton regions that have limited access to water supplies as a result of recurring droughts and reductions to allocations. Water availability will become further constrained with climate change (Chapter 12). Current regions being considered for cotton include the Ord and Burdekin Irrigation Areas.

Despite unsuccessful attempts in the 1970s at cotton production in the Ord region in northern Australia, the introduction of transgenic cotton (with in-built protection to *Helicoverpa* spp. and herbicide resistance), as well as adopting cropping practices tailored specifically to these regions, has meant cotton production may well be a viable alternative crop in these regions. For the Ord River Irrigation Area, research has demonstrated that cotton can be grown sustainably when grown with transgenic insect-protected cotton in the dry winter season. Management guidelines specific to this region have been developed (Yeates *et al.* 2007). In the Burdekin Irrigation Area research is currently underway to develop a sustainable system in the wet summer with key challenges relating to climate variability. The challenges are low radiation and high humidity during cotton growth and ensuring integration with sugar cane cropping systems (Grundy and Yeates 2007). Cotton may provide significant opportunities for further crop diversification in the region.

Improving the return for cotton

Generally, Australian cotton has a good reputation for having good strength, length, colour, and of being largely contaminant free and is purchased for ring spinning medium to fine count carded and combed yarns (Gordon *et al.* 2004). Cotton spinners require longer, stronger, finer, more uniform and cleaner cotton. To improve or maintain the returns for fibre, there are opportunities to develop niche markets that may attract premiums or to improve the quality and consistency of Australian cotton. It may be that a smaller area of a higher value product provides a greater return in situations of limited water. In discussion papers, Gordon (2008) and Morison and Tomkins (2008) proposed a number of other opportunities to add value to Australian cotton that may help to attract premiums or increase demand, such as developing an Australian long staple upland cotton market that could be used to blend with extra long staple cottons such as Pima; eco-labelling; reducing contamination by replacing the jute bagging of cotton bales; minimising neps through improvements in managing crop maturity and through improved ginning, and investing in research to identify and develop cotton fibre with unique and novel traits.

Cotton breeding programs have already developed a broad portfolio of fibre quality cultivars to provide a wider range of options for future market opportunities and environmental conditions. Choosing a cultivar with specific fibre properties can cover for some climate or management challenges. Examples include longer fibre cultivars reducing short fibre discounts in water stress environments, or high micronaire cultivars minimising low micronaire discounts as a result of cool or stress environments during boll maturation. However, to fully realise the benefits of improving quality, all sections of the industry need to work together to address the challenges and opportunities for improving quality. The task for cotton growers and the cotton industry is to optimise fibre quality in all steps from strategic farm plans, cultivar choice, crop management, harvesting and ginning (Constable and Bange 2007).

Crop substitution

Farming has always fundamentally been about effectively optimising and responsibly managing natural resources for generating livelihoods. Rising commodity prices of other crops (such as wheat, maize and sorghum) provides attractive options for generating returns similar to cotton production. Growers can utilise and substitute their farm resources (including water) for growing other crops. Importantly the increase in commodity prices means that it offers profitable crops to be grown in rotation with cotton that result in improvements in cotton growth and yield (Hulugalle *et al.* 2005, 2007). Cotton may simply become part of a more variable and viable crop rotation program.

Costs and benefits

No explicit investigation of the benefits of implementing adaptation practices for climate change has been conducted for the Australian cotton industry. Average expenses for cotton production have risen significantly as a result of increases in costs of chemical insecticides, fertiliser, fuel and oil, and water charges and purchases. Boyce (2007) has shown that the yield required to cover total growing costs of cotton is increasing by 0.2 bales per ha per year. This is double the rate at which the industry average yield is currently increasing (Constable *et al.* 2001). Therefore unless there are significant increases in the rate of yield increase, or reductions in the rate of increase of costs, the ability of cotton systems to make substantial returns will be limited not long after 2020 (Bange and Constable 2008). The appeal of grow-

Table 4.2: Specific adaptation options for the Australian cotton industry. Some feedback and prioritisation was assessed from the survey published by McRae *et al.* (2007). Priority 1 (high), 2 (medium) and 3 (low).

Adaptation options	Priority
Policy/industry	
Policy settings that encourage development of effective water-trading systems that allow for climate variability and support development of related information networks	1
Public sector support for vigorous agricultural research for adaptation studies and breeding efforts with access to global gene pools	1
Encouragement of diversification of farm enterprises (other crops and livestock)	1
Introduction of climate change adaptation (including minimising industry's greenhouse carbon footprint) into environmental management systems	1
Investigation of trends and extremes resulting from climate change both regionally and globally for production and exploration of implications for our markets and impact on competitors	2
Expansion of industry to other regions (including northern Australia)	2
Crop and farm management	
Improvement of nitrogen use efficiency of cotton crops through changes in fertiliser application, type and timing and increase use of legume phase in crop rotations	1
Improved management options in limited water situations (alternative irrigation systems; row configurations, irrigation scheduling strategies) to maximise water use efficiency	1
Selection of cultivars with appropriate heat-stress resistance, drought tolerance, higher agronomic water use efficiency, improved fibre quality, resistance to new pest and diseases (including introgression of new transgenic traits)	1
Ongoing evaluation of cultivar/management/climate relationships on both yield and quality with higher CO_2, increased temperature, and lower vapour pressure deficits)	1
Development of practices to take advantage of increased temperatures especially at the start and end of the growing season to raise yields	2
Linkage of on-farm adaptation with catchments impacts	2
Alteration of planting rules to improve yield and quality	3
Development of future cotton systems that are earlier maturing, use less water and allow more crops to be grown in rotation	3
Climate information and use	
Provision of information on the likely impacts at business level (downscaling climate change predictions to regional scales)	2
Tools and extension to enable farmers to access climate data and interpret the data in relation to their crop records and analyse alternative management options	2
Warnings of the likelihood of very hot days and heavy rain which may result in high erosion potential	3
Managing pests, disease and weeds	
Avoiding resistance of pests (both insects and weeds) through appropriate integrated pest and weed management systems to maintain transgenic technologies	1
Improvement in pest (weeds, invertebrate and diseases) predictive tools and indicators	2
Further development of integrated pest, weed and disease management including area-wide management operations	2

ing cotton would diminish considerably, especially if other commodity prices remain high.

Discussion

Climate change is a multifaceted and complex challenge for the cotton industry and will affect the sustainability of farms, ecosystems and the wider community. Overall though, in responding to some of the specific challenges, the industry is generally well positioned. The current industry is geographically dispersed which means that cultivars and systems have already been developed for a wide range of climatic conditions, and include options to cope with effects such as increased periods of high temperatures and water stress. In addition many of the potential adaptation responses available to the Australian cotton industry have immediate production efficiency benefits making them attractive options regardless of the rate and nature of future climate change. We reviewed potential challenges that climate change presents to the viability of the cotton industry. Table 4.2 summarises some of the adaptation options that the Australian cotton industry will need to consider in addressing several key challenges:

- There is a need to address the question of just how much climate change it would take to make it more appropriate to consider using land and water resources for purposes other than cotton or irrigated production.

- Region-specific impacts will need to be assessed thoroughly as the predominant cotton production regions range from southern NSW to north Queensland. This is necessary so that cotton growers can improve their capacity to assess likely impacts at their business level.

- Detailed integrative research is required to investigate a greater range of water stress and higher temperature scenarios (especially in Australian climatic conditions), to properly assess the influence of elevated CO_2 on cotton growth and insect pests, and to determine how these changes affect yield and quality.

- Research into the development of sustainable or low environmental impact cotton systems for northern Australia should be undertaken given the possibility that rainfall reductions may not be as great in the north as in southern Australia.

- Investigations are needed into the likely impact of climate change on cotton production worldwide. For example, a USA study (Doherty et al. 2003) suggested that while future climate change, with higher temperatures and decreased precipitation, will likely result in higher water-stress levels

and other water management issues for cotton growers, with appropriate management strategies including increased irrigation, cotton yields in the USA could increase in the future. It will be important for Australia to understand global changes in cotton markets as part of its adaptation strategies.

- Climate change is occurring against a background of naturally high climate variability in Australia and this variability is set to increase with climate change. It is important to distinguish between climate variability and climate change as there is the potential for maladaptation to occur if not identified.

- Introductions of transgenic technologies for pest management have delivered reductions in inputs and in future may offer opportunities to improve yields or stress tolerance. Research capability, particularly in crop physiology and agronomy will be needed to ensure their benefits are realised in the field.

Given that there will be no single solution for all of the challenges raised by climate change and variability, the best adaptation strategy for any industry will be to develop more resilient systems. Early implementation of adaptation strategies particularly in regard to enhancing resilience, have the potential to significantly reduce the negative impacts of climate change (Chapter 4). To meet these challenges there will be a greater need to incorporate other efficiencies into the analysis of systems, for example diesel or energy per bale, carbon emissions per bale in addition to bale per ML water and N applied per bale. Importantly, an integrated systems-based approach to research and extension is needed to identify management for minimising economic, social and environmental harm, while maximising new opportunities.

References

ABARE (2008). Australian Bureau of Agriculture and Resource Economics AGSURF database. http://www.abareconomics.com/interactive/agsurf/

Allen SJ, Nehl DB, Moore N (2002). 'Integrated Disease Management for Australian Cotton'. Australian Cotton Cooperative Research Centre, Narrabri, NSW.

Baker DN, Hesketh JD (1969). Respiration and the carbon balance in cotton (Gossypium hirsutum, L). In Proceedings Beltwide Cotton Conference. 7–8 January, New Orleans, Louisiana. pp. 60–64. National Cotton Council, Memphis, Tennessee.

Bange MP, Constable GA (2008). Cotton farming systems for a changing climate. In *Proceedings of the 14th Australian Cotton Conference.*12–14 August, Broadbeach, Queensland. Australian Cotton Growers Research Association.

Bange MP, Milroy SP (2004a). Growth and dry matter partitioning of diverse cotton genotypes. *Field Crops Research* **87**, 73–87.

Bange MP, Milroy SP (2004b). Impact of short-term exposure to cold night temperatures on early development of cotton (*Gossypium hirsutum* L.). *Australian Journal of Agricultural Research* **55**, 655–664.

Bange MP, Caton SJ, Milroy SP (2008). Managing yields of high fruit retention in transgenic cotton (*Gossypium hirsutum* L.) using sowing date. *Australian Journal of Agricultural Research* **59**, 733–741.

Bange MP, Carberry PS, Marshall J, Milroy SP (2005). Row configuration as a tool for managing rain–fed cotton systems: review and simulation analysis. *Australian Journal of Experimental Agriculture* **45**, 65–77.

Bange MP, Milroy SP, Roberts GN (2006). Factors influencing crop maturity. In *Proceedings Beltwide Cotton Conference.* 2–5 January, San Antonio, Texas. National Cotton Council, Memphis, Tennessee.

Boyce Chartered Accountants (2007). 'Australian cotton comparative analysis 2007 crop.' Cotton Research and Development Corporation and the Australian Cotton Catchment Communities Cooperative, Narrabri, NSW.

Charles G, Johnson S, Roberts G, Taylor I (2002). 'WEEDpak'. The Australian Cotton Cooperative Research Centre, Narrabri, NSW.

Chen F, Wu G, Parajulee MN, Ge F (2007). Long-term impacts of elevated carbon dioxide and transgenic Bt cotton on performance and feeding of three generations of cotton bollworm. *Entomologia Experimentalis et Applicata* **124**, 27–35.

Constable GA, Bange MP (2006). What is cotton's sustainable yield potential. *The Australian Cottongrower* **26**, 6–10.

Constable GA, Bange MP (2007). Producing and preserving fiber quality: from the seed to the bale. In *Proceedings 4th World Cotton Conference.* 10–14 September, Lubbock, USA.

Constable GA, Shaw AJ (1988). 'Temperature requirements for cotton'. Agfact P5.3.5. Division of Plant Industries, New South Wales Department of Agriculture.

Constable GA, Thomson NJ, Reid PE (2001). Approaches utilized in breeding and development of cotton cultivars in Australia. In *Genetic Improvement of Cotton: Emerging Technologies.* (Eds JN Jenkins, S Saha) pp. 1–15. Science Publishers, Enfield.

Cottee N, Bange MP, Tan D, Campbell LC (2008). Identifying cotton cultivars for hotter temperatures. In *Proceedings of the 14th Australian Cotton Conference.* 12–14 August, Broadbeach, Queensland. Australian Cotton Growers Research Association.

Cotton Australia (2007). http://www.cottonaustralia.com.au/

Coviella CE, Stipanovic RD, Trumble JT (2002). Plant allocation to defensive compounds: interactions between elevated CO_2 and nitrogen in transgenic cotton plants. *Journal of Experimental Botany* **53**, 323–331.

CRDC (1997). 'Australian dryland cotton production guide'. 2nd edn. Cotton Research and Development Corporation, Narrabri, NSW.

CRDC (2008). 'WATERpak'. 2nd edn. (Eds P Smith, E Madden). Cotton Research and Development Corporation, Narrabri, NSW.

CSIRO (2007). 'Climate change in Australia – observed changes and projections'. Australian Climate Science Program. CSIRO, Australia.

Deutscher SA, Wilson LJ, Mensah R (2004). 'Integrated pest management guidelines for cotton production systems in Australia'. The Australian Cotton Cooperative Research Centre, Narrabri, NSW.

Doherty RM, Mearns LO, Reddy KR, Downton MW, McDaniel L (2003). Spatial scale effects of climate scenarios on simulated cotton production in the southeastern U.S.A. *Climatic Change* **60**, 99–129.

Fitt GP (2000). An Australian approach to IPM in cotton: integrating new technologies to minimise insecticide dependence. *Crop Protection* **19**, 793–800.

Fitt GP, Wilson LJ (2000). Genetic engineering in IPM: Bt cotton. In *Emerging Technologies for Integrated Pest Management: Concepts, Research, and Implementation*. (Eds C Kennedy, TB Sutton) pp. 108–125. The American Phytopathological Society, St. Paul, Minnesota.

Fitt GP, Dillion ML, Hamilton JG (1995). Spatial distribution of *Helicoverpa* populations in Australia: simulation modelling and empirical studies of adult movement. *Computers and Electronics in Agriculture* **13**, 177–192.

Gordon S (2008). 'Development strategies for adding value to Australian cotton'. Discussion paper prepared for Cotton Research and Development Corporation. CSIRO, Australia.

Gordon SG, van der Sluijs MHJ, Prins MW (2004). Quality issues for Australian cotton from a mill perspective. The Australian Cotton Cooperative Research Centre, Narrabri, NSW.

Gregg PC, Wilson LJ (2008). The changing climate for entomology. In *Proceedings of the 14th Australian Cotton Conference*. 12–14 August, Broadbeach, Queensland. Australian Cotton Growers Research Association.

Grundy P, Yeates S (2007). Is a sustainable cotton industry possible in the Burdekin? *The Australian Cottongrower* **28**, 16–20.

Hearn AB (1979). Water relationships in cotton. *Outlook on Agriculture* **10**, 1159–1166.

Hearn AB (1990). Prospects for rain-fed cotton. In *Proceedings of the 5th Australian Cotton Conference*. 8–9 August, Broadbeach, Queensland. pp. 135–144. Australian Cotton Growers Research Association.

Hearn AB (1992). Risk and reduced water allocations. *The Australian Cottongrower* **13**, 50–55.

Hearn AB (1994). OZCOT: a simulation model for cotton crop management. *Agricultural Systems* **44**, 257–299.

Hearn AB, Bange MP (2002). SIRATAC and CottonLOGIC: persevering with DSSs in the Australian cotton industry. *Agricultural Systems* **74**, 27–56.

Hearn AB, Constable GA (1984). Cotton. In *The Physiology of Tropical Field Crops*. (Eds PR Goldsworthy, NM Fisher) pp. 495–527. Wiley, Chichester.

Hulugalle NR, Weaver TB, Scott F (2005). Continuous cotton and a cotton–wheat rotation effects on soil properties and profitability in an irrigated Vertisol. *Journal of Sustainable Agriculture* **27**, 5–24.

Hulugalle NR, Weaver TB, Finlay LA, Hare J, Entwistle PC (2007). Soil properties and crop yields in a dryland Vertisol sown with cotton-based crop rotations. *Soil and Tillage Research* **93**, 356–369.

IPCC (2007). *Climate Change 2007: Impacts, Adaptation and Vulnerability. Contribution of Working Group II to the Fourth Assessment Report of the Intergovernmental Panel on Climate Change*. (Eds

ML Parry, OF Canziani, JP Palutikof, PJ van der Linden, CE Hanson). Cambridge University Press, Cambridge.

Jordon WR, Ritchie JT (1971). Influence of soil water stress on evaporation, root absorption, and internal water status of cotton. *Plant Physiology* **48**, 783–788.

Mauney JR, Kimball BA, Pinter PJ, Lamorte RL, Lewin KF, Nagy J, Hendrey GR (1994). Growth and yield of cotton in response to free–air carbon–dioxide enrichment (FACE) environment. *Agricultural and Forest Meteorology* **70**, 49–67.

McRae D, Roth G, Bange M (2007). Climate Change in Cotton Catchment Communities – A Scoping Study Report for the Australian Greenhouse Office. Cotton Catchment Communities Cooperative Research Centre, Narrabri, NSW.

Milroy SP, Bange MP, Hearn AB (2004). Row configuration in rainfed cotton systems: modification of the OZCOT simulation model. *Agricultural Systems* **82**, 1–16.

Montgomery J, O'Halloran J (2008). A comparison of water use between solid plant and one–in–out skip. *The Australian Cottongrower* **29**, 21–25.

Morison K, Tomkins R (2008). 'Market opportunities for Australian long staple cotton'. Discussion paper prepared for Cotton Research and Development Corporation.

Neilsen J (2008). Evaporative effects on cotton water stress. In *Proceedings of the 14th Australian Cotton Conference*. 12–14 August, Broadbeach, Queensland. Australian Cotton Growers Research Association.

Nelson RA, Holzworth DP, Hammer GL, Hayman PT (2002). Infusing the use of seasonal climate forecasting into crop management practice in North East Australia using discussion support software. *Agricultural Systems* **74**, 393–414.

Pinter PJ, Kimball BA, Mauney JR, Hendrey GR, Lewin KF, Nagy J (1994). Effects of free-air carbon dioxide enrichment on PAR absorption and conversion efficiency by cotton. *Agricultural and Forest Meteorology* **70**, 209–230.

Rawson HM, Begg JE (1977). The effect of atmospheric humidity on photosynthesis, transpiration and water use of leaves of several plant species. *Planta* **134**, 5–10.

Richards D, Yeates S, Roberts J, Gregory R (2006). Does Bollgard II® cotton use more water? In *Proceedings of the 13th Australian Cotton Conference*. 8–10 August, Broadbeach, Queensland. Australian Cotton Growers Research Association.

Reddy KR, Hodges HF, McCarty WH, McKinion JM (1996). *Weather and Cotton Growth: Present and Future*. Office of Agricultural Communications, Division of Agriculture, Forestry, and Veterinary Medicine, Mississippi State University, Mississippi, pp. 1–23.

Reddy KR, Robana RR, Hodges HF, Liu XJ, and McKinion JM (1998). Interactions of CO_2 enrichment and temperature on cotton growth and leaf characteristics. *Environmental and Experimental Botany* **39**, 117–129.

Richards QD, Bange MP, Johnston SB (2008). HydroLOGIC: an irrigation management system for Australian cotton. *Agricultural Systems* **98**, 40–49.

Roberts GN, Constable GA (2003). Impact of crop management on cotton crop maturity and yield. *Proceedings of the 11th Australian Agronomy Conference*, Geelong, Vic. http://www.regional.org.au/au/asa/2003/

Rochester IJ (2007). Nutrient uptake and export from an Australian cotton field. *Nutrient Cycling in Agroecosystems* **77**, 213–223.

Rochester IJ, Peoples MB (2005). Growing vetches (*Vicia villosa* Roth) in irrigated cotton systems: inputs of fixed N, N fertiliser savings and cotton productivity. *Plant and Soil* **271**, 251–264.

Rochester IJ, Peoples MB, Constable GA (2001a). Estimation of the N fertiliser requirement of cotton grown after legume crops. *Field Crops Research* **70**, 43–53.

Rochester I, O'Halloran J, Mass S, Sands D, Brotherton E (2007). Monitoring nitrogen use efficiency in your region. *The Australian Cottongrower* **28**, 22–27.

Rochester IJ, Rea M, Dorahy C, Constable GA, Wright P, Deutscher S, Thongbai P, Larsen D (2001b). 'NUTRIpak – a practical guide to cotton nutrition'. Australian Cotton Cooperative Research Centre, Narrabri, NSW.

Roth G (2007). 'Towards Sustainable and Profitable Water Use in the Australian Cotton Industry'. http://www.cottoncrc.org.au/content/Industry/

Samarakoon AB, Gifford RM (1996). Water use and growth of cotton in response to elevated CO_2 in wet and dry soil. *Australian Journal of Plant Physiology* **23**, 63–74.

Sinclair TR, Purcell LC, Sneller CH (2004). Crop transformation and the challenge to increase yield potential. *Trends in Plant Science* **9**, 70–75.

Stiller WN, Read JJ, Constable GA, Reid PE (2005). Selection for water use efficiency traits in a cotton breeding program: cultivar differences. *Crop Science* **45**, 1107–1113.

Stockton JR, Walhood VT (1960). Effect of irrigation and temperature on fiber properties. In *Proceedings 14 Annual Beltwide Cotton Defoliation Physiology Conference*. pp. 11–14. National Cotton Council, Memphis, Tennessee.

Stubbs JD, Lux C, Powell R (2008). Measuring community wellbeing in cotton communities. In *Proceedings of the 14th Australian Cotton Conference*. 12–14 August, Broadbeach, Queensland. Australian Cotton Growers Research Association.

Tennakoon SB, Milroy SP (2003). Crop water use and water use efficiency on irrigated cotton farms in Australia. *Agricultural Water Management* **61**, 179–194.

Wight M, Laffan J (2008). 'Rising prices for agricultural commodities'. Department of Foreign Affairs and Trade, Trade and Economic Analysis Branch, Canberra.

Wilson LJ, Mensah RK, Fitt GP (2004). Implementing integrated pest management in Australian cotton. In *Novel Approaches to Insect Pest Management in Field and Protected Crops*. (Eds A Rami Horowitz, I Ishaaya) pp. 97–118. Springer-Verlag, Berlin.

Wu G, Chen F, Ge F, Sun Y (2006). Responses of multiple generations of cotton bollworm *Helicoverpa armigera* Hubner, feeding on spring wheat, to elevated CO_2. *Journal of Applied Entomology* **130**, 2–9.

Wu G, Chen F, Ge F, Sun Y (2007). Effects of elevated carbon dioxide on the growth and foliar chemistry of transgenic Bt cotton. *Journal of Integrative Plant Biology* **49**, 1361–1369.

Yeates S (2001). 'Cotton research and development issues for northern Australia – a review and scoping study'. Australian Cotton Cooperative Research Centre, Narrabri, NSW.

Yeates S, Strickland G, Moulden J, Davies A (2007). 'Cotton production and management guidelines for the Ord River Irrigation Area (ORIA)'. Cotton Catchment Communities Cooperative Research Centre, Narrabri, NSW.

Ziska LH, Teasdale JR, Bunce JA (1999). Future atmospheric carbon dioxide may increase tolerance to glyphosate. *Weed Science* **47**, 608–615.

5

RICE

DS Gaydon, HG Beecher, R Reinke, S Crimp and SM Howden

KEY MESSAGES:

- The rice industry has been highly successful in increasing water use efficiency over its history, and must continue to do so in adapting to climatic change. Rice farmers will need to consider a wide range of potential farming system changes (new varieties/crops, rotations, water priorities, irrigation methods, farm layouts, use of seasonal climate forecasts in management) to adapt to predicted changes in on-farm climate and water supply over the coming century. Research into the viability of new farming system ideas, in comparison with traditional systems, is urgently needed to allow for future farm planning.

- Projected declines in irrigation water supply under climate change are likely to have a significant negative impact on Australian rice production due to a total dependence on irrigation.

- In addition to reduced water supply, water demand may increase in response to greater rates of evapotranspiration during the rice-growing season. However the exact increase on water demand is still unclear due to potential offsets through faster development of the crop and a shorter growth period.

- The risk of low-temperature damage during the reproductive phase, one of the major historical limitations to rice production, is likely to be reduced under climate change. However, the net impact of increased temperatures and CO_2 on rice remains largely un-researched in Australia.

- There is some scope to adapt existing rice production in an attempt to reduce irrigation demand through reduction in the duration of ponding via operational (direct drilling) and breeding (yield/duration) means, as well as reduction in deep percolation losses through enhanced definition and regulation of rice-suitable soils.

- Significant improvements in water productivity will be difficult to achieve under existing production systems, and the immediate consequence of less water will be reduced rice production. However aerobic and alternate-wet-and-dry (AWD) rice may present the Australian rice industry with new options, and may allow increased water productivity (kg grain/ML) in a changing climate. The viability of these novel rice production systems for the Australian environment warrants immediate research.

- Potential new methods of rice production (aerobic culture) may allow expansion of rice growing to new areas or regions.

Introduction

Rice was first commercially grown in Australia in the early 1920s near the townships of Leeton and Yenda in the New South Wales Riverina. The current industry in Australia has a restricted geographical range, encompassing the irrigated regions of southern NSW and northern Victoria (see Figure 5.1). Recent years have seen a small number of farmers experimenting with rice in

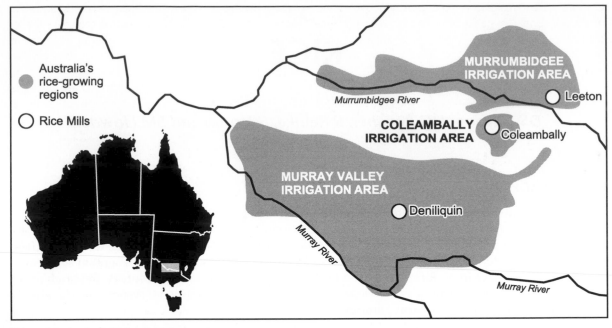

Figure 5.1: Australia's rice-growing regions. Courtesy of the Ricegrowers Association of Australia

parts of northern Australia (northern NSW, Queensland, Northern Territory, and Western Australia) due to enhanced water availability. Past efforts to establish rice production in these regions have had limited success and time will tell whether recent interest, together with potentially new cultivars from overseas, evolves into a genuine northern component to Australia's rice industry.

In a year without climate-induced restrictions the Riverina region produces around 1.3 M tonnes of rice. Of this, 85% is exported and 15% services the domestic market. In such a year, the industry earns around $800 million in revenue, which includes nearly $500 million from value-added exports. Rice is Australia's third largest cereal grain export, and the ninth largest agricultural export (all data from Sunrice website).

Australia's rice-growing region experiences evenly distributed annual rainfall, with the mean annual average ranging from 350–450 mm per annum. Rice is a summer-grown crop, sown in windows according to variety from mid-September to mid-November. The crop has a growing season of approximately six months, and due to the relatively low rainfall the crop is totally reliant on the supply of irrigation water during this period. Rice is grown in rotation with a range of other species including cereals, oilseeds, pulses and pastures, and is only one component in a diverse farming system. It is however the dominant broadacre crop,

in good seasons occupying 10–25% of the landscape in the major irrigation regions for about six months each year, and accounting for 50–70% of the total irrigation water use in any one year (Humphreys *et al.* 2006). Average Australian rice yields are among the highest in the world, averaging 10.4 tonnes grain yield per hectare in 2003 (Humphreys *et al.* 2004).

Surface water is the major source of irrigation water supply for rice growers in the Riverina (river water, pumped directly or diverted into canal systems), although some groundwater is also used (Humphreys *et al.* 2006). Farmers own licences entitling them to water allocations, however allocation amounts are regulated according to annual dam supply. The supply of water to rice growers is not affected by climatic factors alone. Environmental policies and the National Competition Policy have also resulted in decreased water availability to irrigators, and threaten increased water prices in the future (Humphreys and Robinson 2003).

Impacts of projected climate change

Irrigation water supply

Practically all of Australia's rice is grown in conditions of shallow (10 cm) ponded water.

Consequently, total rice production in Australia has a strong linear relationship with total irrigation water allocations in the rice-growing districts (Gaydon *et al.* 2008). Deviations around this linear relationship are likely to be due to a range of other less influential factors affecting rice production such as temperatures, disease, and market dynamics. The effect of projected climate change on irrigation water supply to the Riverina rice growing areas is therefore a dominating factor in assessing climate change impacts on rice production. Recent climate change projections suggest 16–25% reduction in average Murray–Darling streamflows by 2050 and a 16–48% reduction by 2100 (Pittock 2003; Christensen *et al.* 2007). This is likely to have dramatic implications for irrigation water allocations in the Riverina (Jones and Pittock 2003). The strong relationship between available irrigation water and production would suggest significant impacts in future production in response to likely declines in average streamflow, however there are a number of other confounding factors relating to the extent of changes in competing water sectors (i.e. domestic, environmental allocations, industrial, etc.) (Adamson *et al.* 2007; Humphreys and Robinson 2003).

Irrigation water demand

Seasonal irrigation water requirements in the Riverina have historically exhibited high levels of variability. A global modelling study on the impacts of climate change on irrigation water requirements found that significant variation

could be expected geographically (Döll 2002). For south-eastern Australia, this study found that the projected impact of climate changes on irrigation demand is smaller than the existing interannual climate variability. Figure 5.2 shows the comparison between historical pan evaporation at Griffith (1900–2006), and projected pan evaporation at Griffith using the HadleyCM2 GCM for low and high forcing climate change scenarios (IPCC B1 and A1FI respectively) for (a) 2030 and (b) 2070.

In summary we can expect that climate change will increase the irrigation water demand of rice crops. While the mean increases in water demand will most likely not exceed the range of historically experienced variability, individual years may exceed the range. Extending the projection to 2070 could see some large increases in water demand if a high forcing projection is considered. Also, projected decreases in regional rainfall (Christensen 2007) could increase the proportion of the total rice water requirement that has to be supplied by irrigation, thereby further adding to irrigation water demand.

Increased temperatures

There is little published work on impacts of increased maximum and minimum temperatures on Australian rice varieties, however there is substantial literature from international research. Flowering and booting (microsporogenesis) are the most susceptible stages of development to temperature in rice (Satake and Yoshida 1978; Farrell *et al.* 2006a). Studies detailed in Satake and

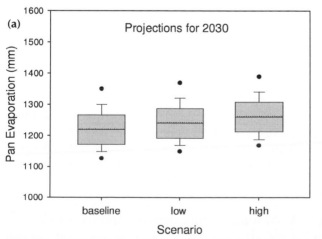
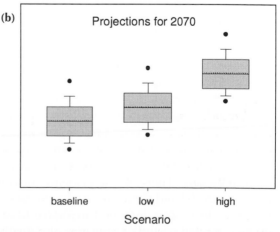

Figure 5.2: Historical and projected cumulative seasonal pan evaporation over the rice growing season (15 October to 15 April) for Griffith. Projections developed using the HadleyCM2 GCM for low and high forcing scenarios (IPCC B1 and A1FI respectively) for (a) 2030 and (b) 2070. 'Baseline' refers to historical records. Within the grey boxes, the dotted line is the mean, the solid line the median.

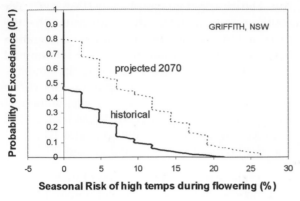

Figure 5.3: Effect of climate change scenarios on risk of experiencing dangerous high temperatures at rice flowering/booting for Griffith, NSW. Warming projection developed using the HadleyCM2 GCM for a high forcing scenario (IPCC A1FI).

Yoshida (1978) indicate that spikelets which are exposed to temperatures >35°C for about five days during the flowering period are sterilised and do not seed. There is generally significant cooling at the spikelet due to transpiration within the canopy, particularly in the low humidity rice growing environments of Australia. This has been measured at 4–6.8°C under conditions of 34.8°C ambient temperature and 20% relative humidity in Australian environments (Matsui *et al.* 2007). Australian rice crops regularly receive ambient temperatures in excess of 35°C during this period, hence if it were not for the low humidity and consequent evaporative canopy cooling, it appears that high temperature sterility issues could be a major limiting factor for rice production in Australia.

Flowering and booting stages last approximately five days for most rice genotypes, and in the Riverina occurs sometime between 5 January and 15 February (Farrell *et al.* 2006b). Figure 5.3 shows the significantly increased likelihood of dangerously high canopy temperatures (>35°C) during this period, assuming a 6°C cooling within the canopy at times of maximum temperature. The risk of exceeding this temperature threshold more than doubles for flowering rice crops at Griffith by 2070. There is also evidence that high temperature-related damage may be cumulative. Exposure to canopy temperatures close to 35°C for an extended period of time may also result in yield loss. This sensitivity to cumulative heat load may vary between cultivars (Jagadish *et al.* 2007),

however this has not been considered in the analysis below.

It is suggested that increases in average minimum temperatures may be more significant in overall terms than increases in maximum temperatures (Peng *et al.* 2004). At the International Rice Research Institute (IRRI) Farm, Los Banos, Philippines, between 1979 and 2003, rice grain yield declined by 10% for each 1°C increase in growing-season minimum temperature in the dry season, whereas the effect of maximum temperature on crop yield was insignificant. We can find no published evidence to show that trends observed by Peng *et al.* (2004) are mirrored in Australian rice yields, however similar trends have been observed elsewhere in the world (Pathak *et al.* 2003). There may be different factors at play in different environments, however the observed IRRI trend suggests that further investigation may be warranted for projected Australian climate scenarios.

There is general agreement that increased average temperatures will have a negative effect on rice yields internationally, with yield decreases due to temperatures alone (in isolation from CO_2 increases) estimated to range between 5 and 10% per 1°C rise in average season temperatures in some areas (Peng *et al.* 1995; Baker *et al.* 1992; Baker and Allen 1993). Peng *et al.* (1995) suggested a negative yield response to increasing average rice season temperatures above a threshold of 26°C. In Australia, the range of average temperatures during the rice growing season for historical, high warming and low warming scenarios do not exceed this threshold by 2030 (Gaydon *et al.* 2008), suggesting that a positive yield response to increasing mean temperatures may be possible in Australia in the near future as long as water is non-limiting. It is unknown whether the canopy cooling effect detailed by Matsui *et al.* (2007) would apply to aerobic and alternate-wet-and-dry (AWD) rice cultures, and the extent to which this would affect high temperature risk in these cultures. There is evidence of a likely strong interaction between temperature and CO_2 which is discussed in the 'Combined temperature and CO_2 effects' section below.

Cold temperature damage and frost risk

Under the present climate there is significant risk of low-temperature damage during the reproductive stage in rice under Australian conditions.

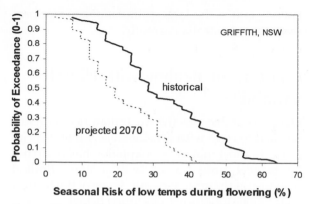

Figure 5.4: Risk of ambient temperatures below 15°C during rice flowering time under historical and projected climate for Griffith, NSW 2070 (general period of rice flowering is defined as 5 Jan–15 Feb (Farrell *et al.* 2006b), however in any given year occurs over only five days). Risk per season is defined as the number of days per 5 Jan–15 Feb period in which minimum temperature falls below 15°C. Warming projection developed using the HadleyCM2 GCM for a high forcing scenario (IPCC A1F1).

Low minimum temperatures can lead to pollen sterility and low yield in high N status crops. Minimum temperatures less than 17–19°C during panicle development are considered dangerous (Farrell *et al.* 2006b). The development of semi-dwarf varieties allowed the effective use of deep floodwater levels as a management strategy during this period to minimise risk by inundating the developing panicles (Williams and Angus 1994). At critical times, paddy water may be 5–7°C warmer than the ambient air conditions. The coldest temperature often occurs just before dawn. The temperature of the rice plant components exposed to the ambient air has been shown to be practically identical to ambient air temperatures at this time (Williams and Angus 1994). The probability of encountering such ambient air temperatures during flowering are around 20–25% in the Griffith area, for the most commonly planted cultivar Amaroo (Farrell *et al.* 2006b). The degree to which projected climate change scenarios might impact that risk is shown in Figure 5.4. Figure 5.4 shows how projected climate change (IPCC A1FI scenario, 2070) is likely to reduce the risk of low temperature occurrence at rice flowering by roughly one-third. Currently this risk is mediated by the use of increased water levels (20 cm) over this period, however adaptive options such as aerobic rice (see section below under 'Adaptation options') do not have this protection and the projection in

this risk is relevant to deliberations on this option. There is evidence that frost risk may increase with climate change in southern Australia (Rebbeck and Gnell 2007), hence further analysis of this matter is recommended.

CO_2 fertilisation

Several free air CO_2 enrichment (FACE) experiments have been conducted with rice, although none in Australia. Recent experiments in Japan and China using Japonica cultivars found that crop response to elevated CO_2 was associated with N uptake, and varied throughout the season (Kim *et al.* 2003; Yang *et al.* 2007). For CO_2 environments 200 ppm greater than ambient, Kim *et al.* (2003) recorded an average 15% increase in rice grain yield when N was in good supply, but lower responses for low N treatments. They found that green leaf area index response to increased CO_2 was positive during vegetative stages and negative after panicle initiation. This phenomenon was also observed in the Chinese experiments, with researchers suggesting that the recommended rates, proportions and timing of nitrogen application should be reconsidered under increased CO_2 to take full advantage of early uptake capacity and also facilitate subsequent N uptake (Yang *et al.* 2007). This study reported a rice grain yield increase of 13%. There is evidence from experiments in the Philippines that even higher CO_2 concentrations (300 ppm above ambient) can result in greater rice yield increases of (27%) (Ziska *et al.* 1997). Note that mid-range projections of CO_2 in 2100 are about 350 ppm higher than present. The implications of an increase in atmospheric CO_2 for rice production under Australian conditions are not known, however it would seem reasonable to assume that potential rice yield increases will be similar to those observed in the Japanese and Chinese studies.

Combined temperature and CO_2 effects

Several studies have shown that high air temperatures can reduce grain yield even under elevated CO_2 conditions (Baker and Allen 1993; Ziska *et al.* 1997). There is also evidence to indicate that the relative enhancement of rice yields due to CO_2 fertilisation is gradually reduced with increases in air temperatures (Matsui *et al.* 1997), as the critical temperature for spikelet sterility is reduced. It is difficult to make inferences on the implications for Australian rice production particularly as some studies have reported significant variance in the response of different cultivars to CO_2 fertilisation (Baker 2004; De Costa *et al.* 2007; Moya *et al.* 1998; Ziska *et al.* 1996). A modelling study covering a diverse range of agro-climatic zones in India, found that climate change outcomes on rice grain yield were positive regardless of the uncertainties and climate change forcings, with average rice grain yield increases ranging from 1 to 38% (Aggarwal and Mall 2002). In contrast, negative trends in rice yields were found in a modelling study in India due to increases in minimum temperature and decreases in incident radiation (Pathak *et al.* 2003). This result was supported by observed data, and the researchers also suggested that these declining trends should be taken as an indication of a future problem in food security. Given the disparity in existing international studies, and the lack of published Australian studies, it is difficult at this point to make any informed judgement on the combined effect of increased temperature and CO_2 concentrations on Australian rice production.

Summary of climate change impacts

The largest impact from projected climatic change on Australian rice production is likely to be from reduced supply of irrigation water. Projected reductions in Murray–Darling stream-flows of 16–25% by 2050 and 16–48% by 2100 are likely to result in similar levels of reduction in rice production, under current production and water use systems. The impacts on rice yields in the Australian situation from projected CO_2 fertilisation and changes in temperature are much more uncertain, and require further research. It would appear that yield reductions due to extreme high temperatures present a possible risk, given the current high ambient temperatures during rice microspore and the likelihood of increased temperature extremes in combination with potentially reduced evaporative cooling of the canopy under increased CO_2 scenarios. It is unlikely that increases in average temperatures over the growing period will present significant production risk, as a considerable buffer already exists between current average temperatures and the reported average temperature thresholds for incurring yield losses. Increased temperatures associated with climate change may reduce the risk of low temperature damage during micro-

spore – one of the largest limitations to rice production in Australia currently.

Options for dealing with climate variability

Over recent decades the Australian rice industry has had to cope with considerable climatic variability. In most instances adaptation has occurred on tactical time frames with a more limited response at the strategic scale. Some of the tactical adaptation options are described below.

Cropping area modification

Hot, dry growing seasons result in increased water demand, and historical climate variability in the rice-growing districts has been characterised by significant variance in seasonal water requirements to produce a rice crop. There are existing options for managing this variability in water demand. When calculating areas for sowing to rice, farmers are often conservative in their calculations to account for the possibility of encountering greater than average evaporative demand in the coming season. Alternatively, growers might adopt a strategy of sequential sowing, increasing over the planting window as progressive allocation increases are made and are anticipated into the future.

Purchase/sale of water on open market

Water can be bought and sold on the open market, and represents an option for managing situations in which a growers encounters water shortages due to climatic variability (e.g. http://www.waterfind.com.au/).

Flexible rotations

In managing variability in water supply, rice farmers have a high degree of inbuilt flexibility in their farming systems in comparison with operations based on permanent plantings such as grapes, fruit trees, etc. If water allocations are reduced, farmers can respond by growing reduced areas of rice in a given season and potentially more of less water-intensive crops.

Sowing winter crops directly after rice

During the rice phase of rotations, the soil is flooded until shortly before the crop matures in

April, and hence effectively has a full profile of stored moisture at harvest. Growing winter crops immediately after rice harvest (direct-drilling) minimises the impact of climatic variability on the subsequent crop by providing a soil water buffer against the likelihood of a dry winter. Also, if not followed by a winter crop, the rice field will usually lie fallow over the winter until the next rice crop (depending on the rotation). In addition to increasing the grower's production, the planting of a sod-sown winter crop creates the capacity in the profile to capture winter rainfall instead of losing it as runoff or deep percolation (Humphreys *et al.* 2006).

Nutrient management adjustment

The MaNage Rice decision-support software provides the farmer with guidance in determining nitrogen application rates on a seasonal basis (Angus *et al.* 1996; Williams *et al.* 1996) and is a key tool in managing climate variability for individual crops.

Water management

Low minimum (<15–18°C) temperatures during the reproductive stage in rice can cause catastrophic damage to crops, particularly those with high N status (Williams and Angus 1994). A management adaptation to this climate variability which allows growers to maintain high N–status (and hence potentially high yielding) rice crops is establishment and maintenance of increased water depth over the critical danger period. The normal ponded water depth is often approximately 50–100 mm deep, however the depth is increased to around 200 mm for two weeks, before returning to 50–100 mm for the remainder of the ponded period. The developing microspores are submerged or partially submerged to avoid low air minimum temperatures, using the thermal mass of the water as protection. High N applications have been shown to lead to high rice yields provided the microspores are protected from low temperatures in this manner (Williams and Angus 1994).

Stubble retention

As detailed in the grains discussion (Chapter 4), stubble retention for the purposes of soil moisture conservation is a widespread technique used to manage climate variability during non-rice phases in the rice-growing districts, however rice stubble is usually burnt prior to sowing of subsequent crops due to mechanical difficulties involved in sowing into the high-density biomass. Options are being researched to overcome this, such as the 'Happy Seeder' (Sidhu *et al.* 2007). It is likely that climate change will drive continued changes in residue management practices over Australia as growers seek to conserve soil moisture and modify greenhouse gas emissions (Howden and O'Leary 1997).

Options for dealing with projected climate change

Many of the adaptation options for dealing with climate variability will provide resilience against the potential negative impacts of climate change. Altering varieties, planting times, nitrogen management and irrigation management to better match the new environments experienced under climate change could mitigate potential yield losses and in some cases increase yields. When summarised over a large number of studies globally, the benefits of adaptation increase significantly with increasing levels of climate change (Figure 5.5). However, a similar analysis in wheat-based cropping systems suggests that the benefits of management adaptations will be limited if more

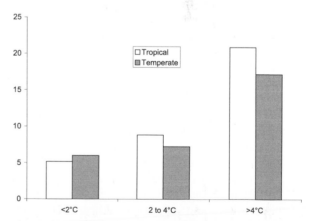

Figure 5.5: Mean benefit of adapting rice systems to impact of temperature and rainfall changes calculated as the difference between % yield changes with and without adaptation. Values are means for tropical and temperate systems. The mean benefit of adapting was not significantly different for temperate and tropical systems. Data sources are listed in Figure 5.2 of (Easterling *et al.* 2007). The temperature changes have associated changes in rainfall and CO_2 that vary between sites, scenarios and publications.

fundamental system changes are not made (Howden and Crimp 2005). In that case, the benefits were largely found with only a 1 to 2°C increase in temperature and the associated changes in rainfall and CO_2. Rice systems appear to have a greater range of adaptive response with positive benefits in yield simulated for a warming of up to 5°C. However, many of the studies on which this analysis is based assume continuing availability of irrigation water and this may not eventuate. When the negative impacts of climate change on rice production reach a point that they overwhelm all adaptation options, there will be a need to adopt more systemic change such as moving to intermittently-irrigated systems or dryland systems.

Improving water use efficiency

Reductions and increased variability in irrigation water supply are likely to represent the greatest challenge to the Australian rice industry from projected climate change. Options for increasing water use efficiency (WUE) are considered below:

(1) Better definition of rice-suitable soils: Historical methods of classifying rice-suitable soils depended on single-point paddock sampling. Until recently, a soil was deemed as suitable for rice production if one soil profile per 4 ha contained 2 metres or more of continuous medium or heavy clay textured material. However it has been demonstrated that clay content alone is a poor predictor of groundwater recharge (Beecher *et al.* 2002). Electromagnetic induction surveys of rice fields can show variation in soil physico-chemical properties across rice fields. This allows the delineation of distinctly different areas of a field so that soil sampling and measurements can be accurately targeted. Current rice soil suitability criteria incorporate electromagnetic surveying and measurement of soil exchangeable sodium percentage (a measure of soil sodicity) enabling soils with potentially high percolation characteristics to be removed from rice growing.

(2) Piping water on farm: Rice farms generally convey water around the property in open earth-lined channels and incur both seepage and evaporation losses of water as a result. Research into on-farm channel seepage losses has been performed on nine Riverina

rice farms previously identified with seepage problems (Akbar and Khan 2005; Khan *et al.* 2007; Khan *et al.* 2005). For a large rice farm with a licensed allocation of 1400 ML, it was shown that combined losses could be >60 ML per year, representing 4–5% of the farm's available water. Options to reduce these losses include lining or piping of channels, however there are significant practical limitations relating to the need to move large volumes of water and the energy costs of doing so.

(3) Piping water in the district, or lining supply channels: Over 42 GL per year of water is lost from 500 km of channels as seepage, in addition to 12.5 GL per year through evaporation from channel water (Akbar and Khan 2005). Seepage occurs primarily through 'hot spots', i.e. where channels/drains cross coarse-textured highly permeable materials. Potentially some 100 GL of water in the Murrumbidgee Irrigation Area and 105 GL in the Coleambally Irrigation Area (with potentially equivalent savings in the Murray Valley Irrigation Districts) can be saved each year through smarter and more integrated schemes for on-farm management. This represents roughly 16% of total water in these schemes so is a significant saving. However capital investment in improvements required to achieve these savings are large ($500–$4000 per ML water saved, depending on local conditions) (Akbar and Khan 2005; Khan *et al.* 2007; Khan *et al.* 2005).

(4) Whole farm planning: Through currently-available technology, electromagnetic surveys of whole-farm soil characteristics is possible. This has the potential to allow more efficient design of farm layouts, e.g. to avoid inclusion of old stream-beds in rice bays, and hence to reduce drainage losses in irrigated rice production.

(5) Raised beds in bays: Raised beds within irrigation bays have been suggested as a means of increasing water productivity in Riverina rice-based operations. High-yielding rice crops can be successfully grown on raised beds, but when beds were ponded after panicle initiation for the required cold-temperature protection, there

was no measured water saving compared with rice grown on a conventional flat layout (Beecher *et al.* 2006). The authors concluded that, through the use of beds on Riverina rice-suitable soils, there is little scope for saving water without proportionate levels of yield loss until ponded water is no longer required (for low temperature protection for current varieties, weed control, etc.). Raised beds may offer water productivity gains for aerobic rice cultivars, which may be more suited to higher permeability soils (see section below on aerobic rice). Beds within terraced, bankless channel systems are being adopted in the Murrumbidgee Irrigation, Coleambally Irrigation, Murray Valley and by riparian and groundwater irrigators, and allow a range of potential benefits including increased flexibility and WUE gains (Gaydon, Beecher *et al.* 2008).

(6) Irrigation scheduling: Water savings may be available from more efficient scheduling of irrigation for crops in rotation with rice. Irrigation scheduling aims to apply the required amount to the crop at critical times, and minimise losses through runoff and deep percolation, while maximising crop production. A study in California found that optimal irrigation required less (48–63%) water than required by what local growers referred to as 'full' irrigation. This also reduced both the deep percolation and runoff losses and caused a 31–43% increase in the application efficiency, and a 32–54% increase in net return (Raghuwanshi and Wallender 1998). There appears to be considerable scope within existing Australian irrigated rice-based farming systems to irrigate non-rice crops more efficiently and hence potentially divert more water for rice enterprises.

(7) Combine and sodsowing of rice: Combine and sodsowing of rice allows the possibility of delaying the introduction of standing water, with the aim of reducing water loss through evaporation and drainage (Tabbal *et al.* 2002). An ancillary benefit in Australian rice systems is a decreased risk of duck damage to the young crop compared with crops established by aerially sowing into standing water. Water savings have been reported, but further research is required to understand the process on a range of soil types. Drilling rice presents problems in the heavy clay sodic soils such as those in the Murray Valley where drilled rice has difficulty breaking through drying surface layers, due to the rice plant's weak emergence capacity.

(8) Breeding: The rice industry in Australia continues to invest in cultivar development to deliver increased yields from less water as a means of increasing water productivity. Also, it has been suggested that development of new varieties with optimum sowing dates which move ponded periods outside peak evaporation periods may represent another option for incremental gains in water productivity (Humphreys *et al.* 2005). Water use in ponded rice culture can also potentially be reduced by the development of varieties with shorter crop duration (up to 10%) (Reinke *et al.* 1994), however there is some evidence that shorter durations will also result in less yield and hence reduced WUE (Williams *et al.* 1999). In recent times, however, by focusing on seedling vigour and early growth, shorter duration varieties with high yield potential have been developed (Reinke *et al.* 2004). An associated benefit of shorter-season rice varieties from a farming systems perspective is that they may facilitate earlier establishment of following winter crops, hence leading to higher yields and better system WUE (Humphreys *et al.* 2005).

(9) Aerobic and alternate-wet-and-dry rice: Evidence from other countries shows that aerobic rice culture has the potential to increase water productivity by 32–88% over conventionally irrigated rice (Bouman *et al.* 2007; Bouman *et al.* 2005). Water losses are constrained by treating the rice like any other irrigated crop such as wheat or maize: once soil moisture has decreased to a specified level, irrigation is then applied to bring the soil moisture content in the root zone up to field capacity. Research in Australia on intermittent watering using existing rice cultivars showed that saturating the rice root zone with flooding every seven days throughout the season reduced water use by 60%, but the negative effect on yields was dramatic (1–2 tonnes per

hectare (t/ha) compared with 9 t/ha for conventional ponded culture) (Heenan and Thompson 1984). Grain quality was also negatively affected. Experiments were also performed in the Riverina with sprinkler irrigation (Humphreys *et al.* 1989) and although water use was reduced by 30–70%, yields declined by an even larger amount, resulting in decreased water productivity. These past experiences in Australia suggest a critical issue when considered alongside more positive reports (Bouman *et al.* 2005) is that cultivars should be selected for their aerobic potential. Ideal aerobic rice cultivars combine the drought-resistant characteristics of upland varieties with the high-yielding characteristics of irrigated varieties (Lafitte *et al.* 2002). Chinese breeders have produced aerobic rice varieties with an estimated yield potential of 6–7 t/ha. There are currently steps underway to trial some of these varieties in Australian conditions. AWD refers to 'alternate-wet-and-dry' culture, also aimed at reducing evaporation and deep drainage losses, but not to the extent of aerobic rice. For current Australian rice varieties, there is no yield loss with intermittent irrigation for the first 10 to 11 weeks, offering the opportunity to save irrigation water and increase water productivity immediately (Humphreys 2006). Following the intermittent irrigation phase, management proceeds as for normal irrigated rice, providing the low temperature protection with increased water depth at the critical stage. Water use savings of 10–15% may be possible for no appreciable loss in yield.

(10) Investment in more efficient irrigation technology: The majority of the other crops in the rice-growing districts are irrigated using surface irrigation methods (furrow, border-check bays) which may be inefficient on light-textured soils (Watson and Drysdale 2006), although growers are moving to use higher flows and modified layouts to improve efficiency. With increasing pressure on declining water supplies and the likelihood of increased variability in supply, some growers are now considering investment in more water-efficient irrigation methods (pressurised systems, lateral move,

subsurface drip) for non-rice crops on non-rice soils. For rice crops, shallow standing water is still necessary hence this potential adaptation does not apply, and improved surface application techniques/layouts are being pursued actively by growers (North 2007). However, savings on other parts of the farm have the potential to affect water available for rice. All of the more efficient irrigation technologies require considerable initial investment and the cost–benefits are largely unknown.

Use of seasonal climate forecasts

Seasonal climate forecasts may assist in early planning of farm operations from expectations of the coming season's water supply and demand. Limited research has been conducted into how this information should best be used in irrigated farming system management, yet potential benefits are large.

New crops, rotations and priorities for water

With climate change projections suggesting a future reduced and more variable water supply, some Riverina irrigators are questioning previous priorities for water and asking questions such as: What do optimal irrigation water priorities look like if my farm is no longer to be a fully irrigated enterprise? Is it better to partially irrigate everything or plan to intensively irrigate a small portion of the farm with the remainder dryland? How does a small intensively irrigated area of vegetables, maybe under a sub-surface drip irrigation system, compare with conventional rice production? How would aerobic rice (if found to be viable) affect water priorities on farm, and impact on other crops, rotations, weed control, diseases, etc? With reduced water, what is the optimal combination of livestock and cropping? Is water better diverted to winter crops as a first priority, due to their need for supplemental irrigation only, and should rice be considered an option only in years with high allocation? These are all very valid questions for which there is an absence of research answers at present, particularly under a changing climate.

Risks of maladaptation

The water-saving possibilities with aerobic rice culture could lead to increased problems with cold damage, nitrogen loss and weeds, with

Box 5.1: Example of the rice industry responding to water shortages

In Australia's Riverina, some rice growers have responded to water shortages by reverting to drill sowing in preference to aerial sowing. Aerial sowing has become more common in recent years, but requires that rice bays are flooded from the start in order to maximise establishment. If conventional sowing equipment is available, direct drilling allows the commencement of permanently flooded conditions to be delayed, thereby saving water from both evaporation and deep drainage (Tabbal *et al.* 2002). Water savings of 10–15% have been reported with negligible yield reductions, however further research is required to understand the process on a range of soils. An ancillary benefit in Australian rice systems is a decreased risk of duck damage to the young crop compared with crops established by aerially sowing into standing water. By the time a permanent flood is established (around panicle initiation), the plants are too large to be significantly damaged by ducks, which prefer to eat the young seedlings. On the downside, delaying permanent ponding may result in increased weed control requirements during the early period. Also, establishing rice in this fashion can present problems in heavy clay sodic soils, such as those in the Murray Valley, where drilled rice has difficulty breaking through drying surface layers due to the rice plant's weak emergence capacity. Hence this adaptive measure is geographically limited.

potentially more chemicals in the environment, and accumulation of salt in the root zone. Conventionally-irrigated rice results in a net movement of salt down through the profile (Humphreys *et al.* 1994), and is thus an important tool for managing salinity. Piping of water increases efficiency and reduces water loses, but could have significant negative effects on biodiversity and native vegetation, and could potentially increase water demand of other crops which benefited from the seepage. Reduced areas of irrigated rice culture may also have negative biodiversity consequences on farm (Doody *et al.* 2004).

Increased N fertilisation in response to higher CO_2 could have negative greenhouse gas implications. Life cycle analyses (i.e. breakdown of materials) of piping may also indicate increased associated greenhouse gases emissions. Furthermore, large investments by irrigated industries in infrastructure costs, new irrigation technologies and equipment are setting the scene for even greater losses if the current situation of no water continues. Increased water infrastructure costs may not offset water losses.

Costs and benefits

The costs of implementing several adaptive options (for example AWD culture) are minimal as no new infrastructure is required, and 10–15% savings are potentially achievable. However, for other options such as piping or lining channels, capital investments are large ($500–$4000 per ML water saved, depending on local conditions) (Khan *et al.* 2005; Khan *et al.* 2007). In view of potential earnings from irrigation water from rice and wheat (105 and 75 $/ML respectively), a pay-back period in the range of 6–53 years is estimated. In some cases, cost–benefit deliberations (for example, on investing in more efficient irrigation technology) are limited by a real and significant lack of understanding relating to the water productivity gains that could be achieved (lateral move, centre pivot, subsurface drip) under a changing climate. Lateral move systems cost roughly $1500 per ha, and centre-pivot systems $2000–$2500 per ha (Khan *et al.* 2005). Potential benefits are likely to depend heavily on soil type, enterprise mix, water and commodity prices. Water savings benefits from more efficient irrigation technology in the Riverina have been reported as ranging between 16–35% for cereals, 15% for soybeans, and 7% for maize (Khan *et al.* 2005). Centre-pivot and lateral move irrigation systems have low labour requirements but require increased skill levels (Raine and Foley 2002) which need to be accounted for in any economic analysis.

Discussion

The rice industry has been highly successful in increasing water productivity over its history, and must continue to do so in adapting to climatic change. In consideration of adaptive options, several categories exist as detailed in Table 5.1. Under 'Easily implemented', the table lists the technologies and options which are available now, and which offer demonstrable benefits in all or some situations. Combine or sodsowing of rice

Table 5.1: Summary of adaptation options for the rice industry. Priority 1 (high), 2 (medium) and 3 (low).

Adaptation option	Priority
Easily implemented	
Combine and sodsowing of rice	1
Alternate wet-and-dry rice culture	1
Better definition of rice-suitable soils	1
Irrigation scheduling	2
Potentially implementable after further R&D	
Aerobic rice evaluation	1
New AWD water management practices	1
Seasonal climate forecasts	1
Consideration of new crops, rotations, priorities for water	1
Conventional breeding for shorter season varieties, increased yields	2
Requiring major investment	
Investment in more efficient irrigation technology	2
Whole farm planning	2
Piping water on farm	3
Piping water in district	3
Raised beds in bays	2

represents an opportunity for water savings through delaying the application of permanent or ponded water. Its application is limited on the heavy sodic clay soils of the Murray Valley due to problems with seedling emergence, however in combination with delayed flooding (i.e. no permanent water for drill-sown crops until just prior to panicle initiation) it may present an option for 10–15% savings in rice water use for no appreciable loss of yield (Humphries *et al.* 2006). The cost of implementing this option is minimal for suitable soils, subject to the grower's access to suitable drilling equipment. Research and development into other different AWD water-management strategies (different wetting and drying regimes, and associated effect on practices such as fertilisation, weed control) is also considered high priority, specifically relating to costs (e.g. reduced production) and benefits (water savings).

The technology for enhanced identification of rice-suitable soils is currently available. There is significant potential to further increase water productivity through reduced water use in rice production by using the revised rice soil suitability criteria to more accurately determine the least permeable soils (Beecher *et al.* 2002). This technology has had varied adoption at this point; in some Riverina jurisdictions electromagnetic surveying

has been widespread while sodicity assessment is being adopted gradually. The approach has been adopted in principle by all authorities regulating rice land suitability, and it is suggested that this should progress as a high priority. Similarly, the technology for more effective irrigation scheduling (timing and quantity) for non-rice crops in rice-based farming systems is also available. However this has been listed as priority 2 due to the limited flexibility of many existing irrigation systems (e.g. flood irrigation in bays) in applying the exact amount of water required by the crop. This technology is most suited following investment in high efficiency irrigation systems such as drip, centre pivot and lateral move machines on suitable soils. High flow/bed layouts could present a better option for applying more targeted water amounts to non-rice crops on heavy rice-suitable soils.

Under the adaptation options (Table 5.1) which may be viable after further R&D, investigations into the potential for aerobic rice in Australian systems is considered a high priority, after demonstrated substantial increases in water productivity (32–88%) from overseas trials using specially adapted aerobic cultivars (Bouman *et al.* 2007; Bouman *et al.* 2005). Several subsequent years in the same experiment yielded variable results, indicating a need for research to observe longer-term effects in Australian conditions. Previous experiments in Australia have centred on growing traditional flooded varieties aerobically, and proved fruitless (Heenan and Thompson 1984; Humphreys *et al.* 1989). Australian trials with international aerobically adapted cultivars are considered a high priority for the Australian rice industry. A major issue which will affect success of new aerobic rice cultivars in Riverina growing conditions will be tolerance to low temperatures during microspore. It is possible that the best suited soils for aerobic rice production in Australia are the more freely draining soils currently not suitable for rice under ponded culture (due to high drainage losses). If this is the case, then another potential method of addressing the cold temperature constraints of the Riverina would be to relocate aerobic rice production further north to warmer climatic conditions hence mitigating low temperature risk. This would also move into summer-dominant rainfall locations meaning a higher proportion of in-season rainfall and decreased demand for irrigation water. Weed control costs in aerobic rice culture are also an unknown quantity for the Australian rice industry, as are issues relating to optimal fertilisation and irrigation regimes.

In summary, although various forms of AWD technology present the most immediate option for the Australian situation, it offers only modest increases in water productivity and limits rice production to existing rice-suitable soils. Aerobic culture offers the promise of significantly enhanced water productivity benefits, together with the potential for expansion in geographical range of production. Additionally, the removal of other constraints of a flooded system (e.g. vertebrate pests, mainly birds, which have caused problems in previous Northern Australian rice ventures), may also be an advantage with aerobic rice. It should be noted that issues related to disease risks in more northern locations would need to be targeted in research.

The use of seasonal climate forecasts in irrigated systems to aid in water and crop management decisions is listed as high priority due to the projected increases in future climatic variability, and the continuing effort being applied in scientific circles to improve climate/weather forecast skills

(Macintosh *et al.* 2007). Accurate forecasts would assist in early planning of farm operations from expectations of the coming season's water supply and demand. Limited research has been conducted into how this information should best be used in irrigated farming system management, yet preliminary studies suggest benefits may be substantial (Gaydon *et al.* 2006). Rice-growing irrigators in the Riverina are already experimenting with new crop options (for example, cotton) and priorities for water (for example, allocating available irrigation water to winter crops rather than rice), a process influenced by low allocations in recent years, but also by available returns from other crops during drought conditions. While this process will continue, it would benefit from dedicated R&D involving farmer participation and simulation modelling, which would allow consideration of various scenarios of water availability, climate change, water and commodity prices.

The rice industry in Australia continues to invest in fully ponded cultivar development to deliver increased yields from less water as a means of further increasing water productivity. At this point, potentially greater gains in water productivity are offered by AWD and aerobic rice, hence this ongoing development has been allocated a priority of 2 to recognise the greater urgency for R&D in the areas. Due to the current drought conditions, and consequent financial implications for many rice-growers, the adaptation options requiring significant financial investment have all been prioritised 2–3. These priorities may vary considerably between farmers, depending on both their financial position and potential benefits to their operations from these investments.

Ultimately, rice production in Australia under projected climate change scenarios will see significant changes in practice and potentially also in geographical region of production. For the irrigated Riverina, future water availability will determine whether rice maintains its position as a dominant crop in agricultural rotations, however the economic sustainability of local rice-related infrastructure (mills, etc.) may become a limiting factor if rice production continues to experience significant extended reductions, or becomes highly variable.

References

Adamson D, Mallawaarachchi T, Quiggin J (2007). Water use and salinity in the Murray–Darling Basin: a state-contingent model. *Australian Journal of Agricultural and Resource Economics* **51**, 263–281.

Aggarwal PK, Mall RK (2002). Climate change and rice yields in diverse agro-environments of India. II. Effect of uncertainties in scenarios and crop models on impact assessment. *Climatic Change* **52**, 331–343.

Akbar S, Khan S (2005). Saving losses from irrigation channels – technical possibilities vs common pool realities. In *Irrigation 2005 Restoring the Balance. Proceedings Irrigation Association of Australia, National Conference.* Townsville, Queensland.

Angus J, Williams R, Durkin C (1996). maNage rice: decision support for tactical crop management. In *8th Australian Agronomy Conference: Agronomy – Science with its sleeves rolled up.* Toowoomba, Queensland.

Baker JT (2004). Yield responses of southern US rice cultivars to CO_2 and temperature. *Agricultural and Forest Meteorology* **122**, 129–137.

Baker JT, Allen LH (1993). Contrasting crop species responses to CO_2 and temperature: rice, soybean and citrus. *Plant Ecology* **104–105**, 239–260.

Baker JT, Allen LH, Boote KJ (1992). Response of rice to carbon dioxide and temperature. *Agricultural and Forest Meteorology* **60**, 153–166.

Beecher HG, Dunn BW, Thompson JA, Humphreys E, Mathews SK, Timsina J (2006). Effect of raised beds, irrigation and nitrogen management on growth, water use and yield of rice in south-eastern Australia. *Australian Journal of Experimental Agriculture* **46**, 1363–1372.

Beecher HG, Hume IH, Dunn BW (2002). Improved method for assessing rice soil suitability to restrict recharge. *Australian Journal of Experimental Agriculture* **42**, 297–307.

Bouman B, Lampayan R, Tuong T (2007). *Water Management in Irrigated Rice – Coping with Water Scarcity.* International Rice Research Institute publication, Los Banos, Philippines.

Bouman BAM, Peng S, Castañeda AR, Visperas RM (2005). Yield and water use of irrigated tropical aerobic rice systems. *Agricultural Water Management* **74**, 87–105.

Bouman BAM, Tuong TP (2001). Field water management to save water and increase its productivity in irrigated lowland rice. *Agricultural Water Management* **49**, 11–30.

Christensen JH, Hewitson B, *et al.* (2007). IPCC WG1 Chapter 11. Regional climate projections. In *Climate Change 2007: The Physical Science Basis. Contribution of Working Group I to the Fourth Assessment Report of the Intergovernmental Panel on Climate Change.* (Eds S Solomon, D Qin, M Manning, Z Chen, M Marquis, K Averyt, M Tignor, H Miller) pp. 849–926. Cambridge University Press, Cambridge.

De Costa WAJM, Weerakoon WMW, Chinthaka KGR, Herath HMLK, Abeywardena RMI (2007). Genotypic variation in the response of rice (*Oryza sativa* L.) to increased atmospheric carbon dioxide and its physiological basis. *Journal of Agronomy and Crop Science* **193**, 117–130.

Döll P (2002). Impact of climate change and variability on irrigation requirements: a global perspective. *Climatic Change* **54**, 269–293.

Doody S, Osborne W, Bourne D (2004). Reconciling farming & wildlife – vertebrate biodiversity in Riverina rice fields. In *IREC Farmer's Newsletter Large Area.* pp. 21–23.

Easterling WE, Aggarwal P, *et al.* (2007). IPCC WG2 Chapter 5. Food, fibre and forest products. In *Climate Change 2007: Impacts, Adaptation and Vulnerability. Contribution of Working Group II to the Fourth Assessment Report of the Intergovernmental Panel on Climate Change.* (Eds M Parry, O Canziani, J Palutikof, P van der Linden, C Hanson) pp. 275–303. Cambridge University Press: Cambridge.

Farrell TC, Fox KM, Williams RL, Fukai S (2006a). Genotypic variation for cold tolerance during reproductive development in rice: Screening with cold air and cold water. *Field Crops Research* **98**, 178–194.

Farrell TC, Fukai S, Williams RL (2006b). Minimising cold damage during reproductive development among temperate rice genotypes. I. Avoiding low temperature with the use of appropriate sowing time and photoperiod-sensitive varieties. *Australian Journal of Agricultural Research* **57**, 75–88.

Gaydon D, Beecher HG, Reinke R, Crimp S, Howden SM (2008). Rice. In 'An overview of climate change adaptation in Australian primary industries – impacts, options and priorities'. Report prepared for the National Climate Change Research Strategy for Primary Industries, February 2008 (Eds CJ Stokes, SM Howden).

Gaydon DS, Lisson SN, Xevi E, Dassanayake D (2006). 'Value assessment of irrigation allocation forecasts for a rice-based operation in the Coleambally Irrigation District – in terms of water-use efficiency, production, and drainage – final report'. CSIRO Water for a Healthy Country Flagship, Project No. T4.S2.P1 Murray Region Irrigation.

Heenan DP, Thompson JA (1984). Growth, grain yield and water use of rice grown under restricted water supply in New South Wales. *Australian Journal of Experimental Agriculture and Animal Husbandry* **24**, 104–109.

Howden SM, Crimp S (2005). Assessing dangerous climate change impacts on Australia's wheat industry. *MODSIM 2005 International Congress on Modelling and Simulation.* Modelling and Simulation Society of Australia and New Zealand. December 2005. (Eds A Zerger, RM Argent) pp. 170–176.

Howden SM, O'Leary GJ (1997). Evaluating options to reduce greenhouse gas emissions from an Australian temperate wheat cropping system. *Environmental Modelling & Software* **12**, 169–176.

Howden SM, Soussana J–F, Tubiello FN, Chhetri N, Dunlop M, Meinke H (2007). Climate Change and Food Security Special Feature: Adapting agriculture to climate change. *Proceedings of the National Academy of Sciences* **104**, 19691–19696.

Humphreys E (1999). Rice crop water use efficiency in the southern Murray–Darling basin. In *Rice Water Use Efficiency Workshop Proceedings: Proceedings of a workshop at Griffith 12 March 1999.* (Ed DL Humphries) pp. 3–13. CRC for Sustainable Rice Production, Yanco, NSW.

Humphreys E, Lewin LG, Khan S, Beecher HG, Lacy JM, Thompson JA, Batten GD, Brown A, Russell CA, Christen EW, Dunn BW (2006). Integration of approaches to increasing water use efficiency in rice-based systems in southeast Australia. *Field Crops Research* **97**, 19–33.

Humphreys E, Meisner C, Gupta R, Timsina J, Beecher HG, Tang Yong Lu, Yadvinder-Singh, Gill MA, Masih I, Zheng Jia Guo, Thompson JA (2005). Water saving in rice–wheat systems. *Plant Production Science* **8**, 242–258.

Humphreys E, Meisner C, Gupta R, Timsina J, Beecher HG, Tang Yong Lu, Yadvinder-Singh, Gill MA, Masih I, Zheng Jia Guo, Thompson JA (2004). Water saving in rice–wheat systems. In *Proceedings 4th International Crop Science Congress.* Brisbane.

Humphreys E, Meyer WS (1996). How much water does rice really need? In *IREC Farmer's Newsletter Large Area.* pp. 39–41.

Humphreys E, Muirhead WA, Melhuish FM, White RJO, Blackwell J (1989). The growth and nitrogen economy of rice under sprinkler and flood irrigation in South East Australia. *Irrigation Science* **10**, 201–213.

Humphreys E, Robinson D (2003). Increasing water productivity in irrigated rice systems in Australia: institutions and policies. In *Rice Science: Innovations and Impact for Livelihood – Proceedings of the International Rice Research Conference.* IRRI, Chinese Academy of Engineering and Chinese Academy of Agricultural Sciences, Beijing, China.

Humphreys E, van der Lelij A, Muirhead WA, Hoey D (1994). The development of environmental restrictions for rice growing in New South Wales. *Australian Journal of Soil Water Conservation* **7**, 11–20.

Jagadish SVK, Craufurd PQ, Wheeler TR (2007). High temperature stress and spikelet fertility in rice (*Oryza sativa* L.). *Journal of Experimental Botany* **58**, 1627–1635.

Jones R, Pittock B (2003) *Climate Change and Water Resources in an Arid Continent: Managing Uncertainty and Risk in Australia.* Springer, The Netherlands.

Khan S, Abbas A, Gabriel H, Rana T, Robinson D (2007). Hydrologic and economic evaluation of water–saving options in irrigation systems. *Irrigation and Drainage* **57**, 1–14.

Khan S, Akbar S, Rana T, Abbas A, Robinson D, Paydar Z, Dassanayke D, Hirsi I, Blackwell J, Xevi E, Carmichael A (2005). Off-and-on farm savings of irrigation water. Murrumbidgee Valley water efficiency feasibility project. CSIRO, Canberra.

Khan S, Robinson D, Beddek R, Wang B, Dassanake D, Rana T (2004). Hydro-climatic and Economic Evaluation of Seasonal Climate Forecasts for Risk Based Irrigation Management. *CSIRO Land & Water Technical Report* **5/04**.

Kim HY, Lieffering M, Kobayashi K, Okada M, Miura S (2003). Seasonal changes in the effects of elevated CO_2 on rice at three levels of nitrogen supply: a free air CO_2 enrichment (FACE) experiment. *Global Change Biology* **9**, 826–837.

Lafitte HR, Courtois B, Arraudeau M (2002). Genetic improvement of rice in aerobic systems: progress from yield to genes. *Field Crops Research* **75**, 171–190.

Matsui T, Kobayasi K, Yoshimoto M, Hasegawa T (2007). Stability of rice pollination in the field under hot and dry conditions in the Riverina region of New South Wales, Australia. *Plant Production Science* **10**, 57–63.

Matsui T, Namuco OS, Ziska LH, Horie T (1997). Effects of high temperature and CO_2 concentration on spikelet sterility in indica rice. *Field Crops Research* **51**, 213–219.

McIntosh PC, Pook MJ, Risbey JS, Lisson SN, Rebbeck M (2007). Seasonal climate forecasts for agriculture: Towards better understanding and value. *Field Crops Research* **104** (1–3), 130–138.

Moya TB, Ziska LH, Namuco OS, Olszyk D (1998). Growth dynamics and genotypic variation in tropical, field-grown paddy rice (*Oryza sativa* L.) in response to increasing carbon dioxide and temperature. *Global Change Biology* **4**, 645–656.

Nicholls N (2004). The changing nature of Australian droughts. *Climatic Change* **63**, 323–336.

North S (2007). 'Improving the performance of basin (contour) irrigation systems in the southern Murray–Darling Basin – a scoping study'. Cooperative Research Centre for Irrigation Futures and NSW Department of Primary Industries, Deniliquin, NSW. April 2007. Land & Water Australia, Canberra.

Pathak H, Ladha JK, *et al.* (2003). Trends of climatic potential and on-farm yields of rice and wheat in the Indo-Gangetic Plains. *Field Crops Research* **80**, 223–234.

Peng S, Huang J, Sheehy JE, Laza RC, Visperas RM, Zhong X, Centeno GS, Khush GS, Cassman KG (2004). Rice yields decline with higher night temperature from global warming. *Proceedings of the National Academy of Sciences* **101**, 9971–9975.

Peng S, Ingram K, Neue H, Ziska LH (1995). *Climate Change and Rice*. Springer, Berlin and IRRI.

Pittock B (Ed.) (2003). *Climate Change – An Australian Guide to the Science and Potential Impacts*. Australian Greenhouse Office, Canberra.

Raghuwanshi NS, Wallender WW (1998). Optimal furrow irrigation scheduling under heterogeneous conditions. *Agricultural Systems* **58**, 39–55.

Raine SR, Foley JP (2002). Comparing application systems for cotton irrigation – what are the pros and cons? In *Field to Fashion, 11th Australian Cotton Conference*. Australian Cotton Growers Research Association Inc., Brisbane, Queensland.

Rebbeck MA, Gnell GR (2007). *Managing Frost Risk: A Guide for Southern Australian Grains*. South Australian Research and Development Institute. GRDC.

Reinke R, Snell P, Fitzgerald M (2004) Rice development and improvement. *IREC Farmer's Newsletter* **165**, 43–47.

Reinke RF, Lewin LG, Williams RL (1994). Effect of sowing time and nitrogen on rice cultivars of differing growth duration in New South Wales. 1. Yield and yield components. *Australian Journal of Experimental Agriculture* **34**, 933–938.

Satake T, Yoshida S (1978). High temperature induced sterility in indica rices at flowering. *Japanese Journal of Crop Science* **47**, 6–17.

Sidhu HS, Manpreet S, *et al.* (2007). The Happy Seeder enables direct drilling of wheat into rice stubble. *Australian Journal of Experimental Agriculture* **47**, 844–854.

Sunrice website – http://www.sunrice.com.au.

Tabbal DF, Bouman BAM, Bhuiyan SI, Sibayan EB, Sattar MA (2002). On farm strategies for reducing water input in irrigated rice; case studies in the Philippines. *Agricultural Water Management* **56**, 93–112.

Watson DJ, Drysdale G (2006). Irrigation practices on north-east Victorian dairy farms: a survey. *Australian Journal of Experimental Agriculture* **45**, 1539–1549.

Williams R, Angus J, Crispin C (1996). Progress with maNage Rice. *IREC Farmer's Newsletter* **148**, 42–43.

Williams R, Farrell T, Hope M, Reinke R, Snell P (1999). Short duration rice: implications for water use efficiency in the NSW rice industry. In *Rice Water Use Efficiency Workshop Proceedings: Proceedings of a workshop at Griffith 12 March 1999*. (Ed. DL Humphries) pp. 50–54. CRC for Sustainable Rice Production, Yanco, NSW.

Williams RL, Angus JF (1994). Deep floodwater protects high-nitrogen rice crops from low-temperature damage. *Australian Journal of Experimental Agriculture* **34**, 927–932.

Yang L, Huang J, Yang H, Dong G, Liu H, Liu G, Zhu J, Wang Y (2007). Seasonal changes in the effects of free-air CO_2 enrichment (FACE) on nitrogen (N) uptake and utilization of rice at three levels of N fertilization. *Field Crops Research* **100**, 189–199.

Ziska LH, Manalo PA, Ordonez RA (1996). Intraspecific variation in the response of rice (*Oryza sativa* L.) to increased CO_2 and temperature: growth and yield response of 17 cultivars. *Journal of Experimental Botany* **47**, 1353–1359.

Ziska LH, Namuco O, Moya T, Quilang J (1997). Growth and yield response of field-grown tropical rice to increasing carbon dioxide and air temperature. *Agronomy Journal* **89**(1), 45–53.

6

SUGARCANE

SE Park, S Crimp, NG Inman-Bamber and YL Everingham

KEY MESSAGES:

■ It is likely that the greatest direct climate change impact (and adaptation challenge) on Australian sugarcane production will be the projected change in the amount, frequency and intensity of future rainfall. In many of the sugarcane growing regions the amount of effective rainfall available to the crop will be reduced, while demand is likely to increase due to increased rates of evapotranspiration linked to atmospheric warming.

■ A range of adaptation strategies (both tactical and strategic) is needed across the entire sugar cane industry value chain if it is to remain sustainable under a changing climate. Strategies must be tailored to individual mill regions to take account of location-specific biophysical and logistical impacts.

■ Adaptation options available to the sugarcane industry include improvements to the management of limited water supplies; technological fixes based on reductionist analysis; engineering design principles, or computer-aided modelling; altered cropping system design and agronomic management; enhanced utilisation of decision-making tools, and effective institutional change (Park *et al.* 2007a).

■ Building capacity through targeted extension, improving skills and providing a more industry-wide knowledge base are all essential for future adaptation.

■ Many adaptation strategies involve an enhancement or extension of existing activities aimed at building resilience to climatic variability. Additional longer-term adaptation options will also need to be iteratively developed and evaluated in an adaptive management context if the industry is to remain sustainable into the future.

Introduction

Although sugarcane can be grown in a relatively wide range of climatic conditions, the time-critical nature of processing from crop harvest to juice extraction typically results in the evolution of discrete production regions containing a centralised mill, dedicated transport infrastructure and communities of farmers that all contribute to a tightly integrated supply chain. The Australian sugarcane industry reflects this, being made up of pockets of production spanning nearly 2100 km of eastern Australia, from Mossman in the Far North of Queensland (16° 27′ 37″ S, 145° 22′ 22″ E), to

Grafton in northern New South Wales (29° 41′ 28″ S, 152° 55′ 59″ E) (Fig. 6.1). Production occurs in four climatic zones, from the wet tropics in the north through to the dry tropics and humid subtropics. The majority of these regions are within 50 km of the coastline and in close proximity to tidal rivers and creeks. Approximately 94% of the country's raw sugar production occurs in Queensland, occupying approximately 380 000 ha of land (CANEGROWERS 2007; Australian Sugar Milling Council 2006). Northern New South Wales (NSW) accounts for around 4% of production while the remainder, prior to 2008, occurred in the Ord River Irrigation Area in Western Australia (but

Figure 6.1: The Australian sugarcane industry, from Mossman in the Far North of Queensland to Harwood in northern New South Wales. Source: Harman and Collins, modified from Geoscience Australia 2006 and NRM 1995

the area is not currently in production). Depending on the commodity price, the industry generates between $1.5 billion and $2 billion in direct revenue, with approximately $1.2 billion from export markets (CANEGROWERS 2007).

The production of sugarcane in Australia is strongly moderated by three climatic constraints. These are water availability, the amount of solar radiation received and minimum temperature at which growth starts and maximum temperature at which growth ceases. The production potential of the dry tropics, although providing an ideal temperature for the growth of sugarcane, is constrained by a limited supply of irrigation and rainwater. For this reason, almost 60% of Australian sugarcane production depends on some form of irrigation (Inman-Bamber 2007).

Impacts of projected climate change

The following section provides a broad overview of possible impacts of climate change on sugarcane production in Australia. This overview is based in part on modelling studies undertaken using climate change projections derived from those described in Chapter 2, as well as a synthesis of both national and international research drawing from physiological data for sugarcane and other crop species (with an emphasis on other C_4 grass species).

To date, much of the research undertaken to examine the potential impacts of climate change on the sugarcane industry has focused explicitly on primary production. However, to fully appreciate the net impact of climate change on the industry as a whole we must understand the integrated nature of the entire value chain. Any impacts experienced upstream in the value chain as a result of a change in climate variability, or the implementation of an adaptation strategy, will subsequently impact all sectors downstream. Results from a study in the Maryborough Mill region have been adapted from Park *et al.* (2007c) to demonstrate this point and highlight the potential impacts across all sectors of the sugarcane value chain.

Impacts on primary production

While it is generally considered that C_3 species will receive greater benefit from CO_2 fertilisation compared to C_4 species such as sugarcane, recent research using open top chambers shows substantial increases in the growth and productivity of sugarcane in response to higher CO_2 concentrations (Da Souza *et al.* 2008). When exposed to elevated levels of CO_2 in open top chambers, photosynthesis in sugarcane increased by approximately 30%, height by 17% and accumulated biomass by 40% over the course of 50 weeks. The plants also exhibited lower stomatal conductance and transpiration rates (-37 and -32%, respectively), and consequently 62% higher water use efficiency (WUE). These responses are notable given the lack of water stress experienced by the crop and further studies are required to address the limitations of this study. However, these findings suggest that increases in atmospheric CO_2 concentration will deliver more benefit to production than first thought, through increases in WUE and possibly biomass and sucrose content.

One possible consequence of increasing concentrations of atmospheric CO_2 is a modification of the carbon-to-nitrogen ratio (C:N) as more carbon is taken up in leaf and stem material, resulting in a slowing of residue decomposition (Ball 1997). This slowing of residue decomposition has been identified in many C_3 crop trials but has not yet been observed in sugarcane. The reduction of nitrogen in sugarcane leaf material has been shown to have a negative impact on radiation use efficiency and growth (Park *et al.* 2005) and although not yet tested in sugarcane, it is possible that the lower nutritional quality of the crop may result in increased insect herbivory as the sugar level in leaves increase, as seen in soybean (Hamilton *et al.* 2005).

Increases in temperature may result in a number of physiological impacts. These include:

- potential acceleration of crop phenology;

- potential shortening of time to crop maturity, and

- possible increases in maximum sucrose content, although higher temperatures may also favour vegetative growth and predicted effects on sucrose content are at this stage equivocal (Inman-Bamber 1994).

Higher temperatures will particularly benefit productivity in presently cool sugarcane growing locations around south-east Queensland and northern NSW. Rates of photosynthesis may start to decline some time towards the end of the century as maximum temperatures more regularly exceed the 34°C optimum (Kingston 2000). In the absence of water limitations and reduced frost risk due to warmer temperatures, sugarcane production could become commercially viable further south than at its current location.

The impact of a change in the amount and timing of rainfall events will be dependent on current production locations and rainfall patterns (Park *et al.* 2007a). In the central sugar growing region (encompassing Proserpine, Marian, Farleigh, Racecourse and Plane Creek mills) and other moisture limited areas, a decrease in the amount of effective rainfall (through a reduction in rainfall and/or an increase in rainfall intensity) will result in:

- heightened crop water stress (through decreased availability of soil moisture and supplementary water supply);

- potential decrease in the quality of supplementary water;

- increased solar radiation and evaporation with any reduction in cloud cover;

- reduced rate of early leaf area and canopy development through water stress, and

- reduced photosynthesis, decreased tillering and stalk length associated with changing irrigation water availability.

In sugar-producing regions currently experiencing seasonal waterlogging and flooding events, (e.g. the northern extent of the current region) any reduction in summer rainfall would likely increase yields as a result of:

- reduced soil anaerobic conditions and nutrient loss through less leaching and erosion due to lower annual rainfall (although a potential increase in the number of extreme events may increase the amount of nutrients lost in a single rainfall event);

- increased solar radiation and evaporation with any reduction in cloud cover;

- a potential increase in CCS (commercial cane sugar) through a more effective drying-off period, and

- increased trafficability for harvest machinery and the timeliness of operations.

Increased soil C decomposition and soil N mineralisation may occur under changed temperature and rainfall conditions (stimulating net primary production) resulting in a change in the C:N ratio of the soil. However, any benefits arising from increased temperatures may be negated by increased vapour pressure deficit reducing stomatal conductance and transpirational cooling, resulting in leaf damage. Further yield losses may result from a likely increase in the incidence of pests and diseases although little experimental research has been undertaken in this area (Sutherst *et al.* 1996).

The close proximity of much of the sugarcane industry to tidal creeks and rivers renders it vulnerable to rises in sea level as well as storm surge and riverine flood damage resulting from potential changes in the frequency and severity of extreme rainfall events. Increased intrusion of saltwater into coastal aquifers is also likely in response to sea level rises impacting on the quality and quantity of coastal fresh water supplies available for sugarcane production (Ghassemi *et al.* 1996; Murphy and Sorenson 2001). Strong winds have the potential to shred leaves and uproot individual sugarcane plants, as demonstrated by Cyclone Larry in 2006. Preliminary research has shown that for a 15% increase in the central pressure of a cyclone travelling on the same path as Cyclone Larry, an additional 125 km extent of damaging winds would occur (Crimp and Harper 2007), potentially impacting a wider distribution of sugarcane producers in the region.

Harvest sector

Any changes in yield or the volume of trash produced by the sugarcane crop will impact harvest capacity and management. Whether the impact is positive or negative will, in part, depend upon the end product (i.e. first or second generation biofuel or sugar). Similarly any change to the timing of crop maturity and harvesting will need to be accommodated by the harvesting sector. Benefits may arise from a reduction in rainfall in the Northern region and improved harvester efficiency as a result of reduced slippage of field machinery and less stool and paddock damage (Park *et al.* 2007a).

Transport sector

The transport sector will also need to manage climate change and variability impacts occurring earlier in the sugarcane value chain. In addition, sugarcane transport infrastructure (especially coastal highways and railways) will be vulnerable to an increase in the number and intensity of extreme events (McInnes *et al.* 2003).

Mill sector

Similarly, this sector will need to manage climate change and variability impacts occurring earlier in the sugarcane value chain. Key changes may include alteration in the timing and duration of the crushing season and production volume (Park *et al.* 2007a). Although important, of lesser concern is an increase in the volume of extraneous matter contained in the cane consignment due to increased lodging (Brotherton 1980), the suppression of sucrose extraction, increased wear and tear on milling infrastructure and increased volumes of mill mud requiring disposal. Any reduction in the speed of mill processing will incur penalties on productivity.

Regional impacts

In order to provide a more industry-wide impact assessment, present constraints to regional production were identified and likely impacts of climate and variability change on these were considered in collaboration with industry stakeholders (Park *et al.* 2007a) (Table 6.1).

Integrated climate change impacts

The preceding sections provided an indicative or qualitative assessment of the future challenges the sugarcane industry may face as a consequence of projected changes in individual climate drivers. However there is also a need to examine the

Table 6.1: Summary of potential climate change related impacts on sugarcane production regions on the eastern coast of Australia (adapted from Park *et al.* 2007a).

Present or potential future constraints	Likely impact of a change in climate and variability
Northern region	
Low radiation experienced when cloudy	May improve as rainfall declines
Damage to the crop during cyclone events	Increased with an increase in cyclone intensity
Excess rainfall during wet season results in waterlogging and limited access for operations	Improved with reduced rainfall and extent and frequency of waterlogging, enabling in-field access
Export of nutrients and sediment to Great Barrier Reef Marine Park (GBRMP)	Increased with increasing frequency of extreme events
Poor crop establishment	Exacerbated by reduced spring rainfall
Poor drainage	Exacerbated by sea level rise and rainfall intensity
Herbert/Burdekin regions	
Insecurity regarding Burdekin Dam water supply	Exacerbated by future competition from urban use and lower amounts of effective rainfall
Rising water table and salinity issues in the Burdekin River Irrigation Area	Exacerbated by increasing use of irrigation, especially in the absence of water use efficiency improvements
Rising saline groundwater table (Burdekin) and poor drainage (Herbert, lower Burdekin)	Exacerbated by sea level rise and potential intrusion into the ground water
Catchment hydrology and water availability (Burdekin floodplain)	Reduced amounts of effective rainfall will reduce recharge rate into the Burdekin aquifer
Poor harvesting efficiency	Reduced rainfall will increase harvesting efficiency, but increased cyclone intensity will reduce it
Export of nutrients and sediment to GBRMP	Increased with increasing frequency of extreme events
Tidal intrusion in Herbert and Burdekin deltas	Exacerbated by sea level rise
Central region	
Limited supply of irrigation water	Exacerbated by reduced effective rainfall and increased evapotranspiration
Limits to crop growth in frost–prone areas in the western districts	Reduced constraint due to increased minimum temperatures and decreased frost events
Export of nutrients, etc. to GBRMP	Increased with increasing frequency of extreme events
Variable annual yields	Absolute productivity may increase due to extended growing season (providing other inputs non-limiting)
Poor drainage/tidal intrusion in lower floodplains	Exacerbated by projected sea level rise
Southern region	
Limited supply of irrigation water	Exacerbated by reduced amount of effective rainfall
Crop growth limited by low winter temperatures and short growing season	Reduced constraint due to increased minimum temperatures, with the potential to increase productivity
Present competition for land-use from other crops *e.g.* horticulture and tree crops	Increased due to greater risk of 4–5 year sugarcane production compared with short-duration annual crops
New South Wales region	
Low levels of solar radiation constrain growth	Decreased due to reduced rainfall and cloud cover
Production prone to frost damage	Reduced due to increases in minimum temperatures
Growth of two-year crop necessary	Reduced due to increases in minimum temperatures
Careful management of drainage and watertable due to acid sulfate soils	Increased difficulty with a rise in sea level, and potential to reduce areas suitable for crop growth

integrated impacts of multiple climate variables across the entire sugarcane value chain. Efforts are already underway internationally to provide quantitative estimates of climate change impacts on sugarcane production using a range of biophysical modelling capabilities (Singh and

El Mayaar 1998; Cheeroo-Nayamuth and Nayamuth 2001). In the Australian context, a number of studies have been undertaken using the Agricultural Production Systems Simulator (APSIM) (Keating *et al.* 2003) together with historic climate data altered to reflect projections of future temperature, rainfall and CO_2. For example, simulations of sugarcane growth were produced for the Sunshine Coast (McDonald *et al.* 2006) and Rocky Point (Pearson *et al.* 2007) sugarcane regions around south-east Queensland. In the first study only the impacts of an annual change in temperature (+0 to 1.7°C) and rainfall (-13% to +7%) were considered, whereas in the second study monthly changes in temperature (+0.2°C to 1.6°C for spring; +0.2°C to 1.6°C for summer; +0.2°C to 1.3°C for autumn; +0.2°C to 1.6°C for winter) and rainfall (−20% to 0% for spring; -7% to +7% for summer; -13% to +7% for autumn; -13% to +7% for winter) were incorporated into the model simulations. An increased concentration of atmospheric CO_2 was also included in the latter study at a rate of 437 ppm. In both cases numerous combinations of agronomic management and soil type were considered in order to examine a range of external factors impacting crop growth and yield. Changes in sea level rise, rainfall intensity, number of dry days and average wind speed, while also likely to impact crop productivity, were not assessed.

For the Sunshine Coast study, sugar yields were projected to change by -2 to +7.4% by 2030 depending on soil type and crop management. Whereas at Rocky Point, greater variation in potential sugar yields were simulated (-19 to +12% by 2030). Comparison with a similar desktop study conducted on the most northerly (Mossman) mill region in Queensland showed likely greater losses in the cooler southern regions as opposed to the north as a result of reduced water availability and hence greater water stress (Park *et al.* 2007c).

The above results would suggest the greatest impacts of climate change are likely to be experienced by the grower sector of the Australian sugarcane industry (Park *et al.* 2007c). However, net productivity gains/losses will ultimately depend on the availability of a number of resources, particularly water, and the plant's physiological response to relative changes in all climate variables. While the capacity within the sugar industry value chain is considered sufficient to absorb an increase in yield, a decrease may financially challenge many mills.

Options for dealing with climate variability

A number of web-based sources of information and decision-support tools are presently available (or in design) to assist sugarcane industry stakeholders improve the responsiveness of their management to climate variability and change. Table 6.2 lists decision-support tools and forecasting systems specifically designed, applied and/or validated for the Australian sugarcane industry. These resources and tools aim to better inform decisions relating to irrigation scheduling, N fertiliser management, drying-off duration and the timing of harvest, yield forecasts, water storage investments and whole of industry planning. By utilising the seasonal forecasting prediction indices and statistical analyses of historic climate variables embedded in these resources and tools, operations are strategically adjusted to better manage the already highly variable climate as shown by the example in the box below.

Box 6.1: Example of managing climate change in the sugarcane industry

One strategy used by the Australian sugarcane industry to respond to inter-annual climate variability is increased use of climate forecasting systems to assist decision-making and risk management. For example the ENSO (El Niño-Southern Oscillation) signal is used early in the year in selected sugarcane-growing regions to predict likely climate, and hence sugarcane productivity, around the time of harvest. Having this information seven months prior to the harvest period enables the scheduling of the harvest start date to be optimised for productivity. Such information is also useful to marketers planning customer allocations, shipping schedules and storage requirements for the following season. This information has resulted in financial benefits for the sugarcane industry (Anthony *et al.* 2002).

A similar forecasting approach is currently being used to aid the rostering of harvests and labour and equipment resources during the harvest period (Everingham *et al.* 2001b). By better managing increasing variability in the climate, industry stakeholders are also automatically tracking long-term changes in mean climate.

Table 6.2: Decision-support tools and sources of information for managing climate variability developed for, applied and/or validated for the Australian sugarcane industry. Sectors utilising the tools or information sources are indicated as: 1 whole of the value chain; 2 grower; 3 harvest and transport; 4 milling, and 5 marketing.

Tool/information source/use	Reference	Details
RainForecaster [2,3,4]	Everingham et al. (2006); Everingham (2007a)	Produces 3-monthly forecasts of rainfall and wetdays for a limited number of sugarcane locations
WaterSense [2]	Inman-Bamber et al. (2005, 2006)	Web-based irrigation water optimisation tool
Irrigation scheduling [2]	Inman-Bamber et al. (2001)	Crop simulations and climate data combined to determine El Niño/La Niña impacts on irrigation management
Nitrogen management [2]	Everingham et al. (2006)	Crop simulations and climate data combined to determine optimum nitrogen fertiliser management given Southern Oscillation Index (SOI) phase
Long lead forecasting for estimated yield [5]	Everingham et al. (2007)	Bayesian discriminant analysis procedure for determining scale of crop production
Long lead rainfall forecasts [1]	Everingham et al. (2008b)	Long-lead statistical ENSO prediction model for estimating likelihood of disruption to harvest by rainfall
El Niño-Southern Oscillation (ENSO) signal [5]	Everingham et al. (2003; 2001a)	Five phases of SOI used to indicate sugarcane yield anomalies seven months prior to harvest
Decadal forecasting system [2,3,4]	Jaffres and Everingham (2005)	Identified relationship between variations in sea surface temperature (SST) and mean sea level pressure (MSSLP)
Irrigation optimisation [1]	Park (2006); Everingham et al. (2006)	Three- and five-phase seasonal climate forecasting used to optimise irrigation in La Niña and El Niño years
Relationship between ENSO and decadal systems [2,3]	Jones and Everingham (2005)	Demonstrates ENSO signals (e.g. SOI and equatorial SST) may provide insights into coming season climate
Comparison of phase-based seasonal climate forecasting systems [2,3]	Everingham (2007b); Jones and Everingham (2006); Everingham et al. (2006)	Statistically compares a number of climate forecasting systems against the five-phase SOI system (Stone et al. 1996)
Estimated risk associated with drying-off strategies [2]	Robertson et al. (1999)	Relationships between crop attributes, simulations and climate data used to determine optimum drying-off time
Value chain based decision-making [1]	Everingham et al. (2002)	Climate forecasting used to improve risk management in value chain during yield forecast, harvest and irrigation management
DamEa$y [2]	Lisson et al. (2003)	Estimates economic and environmental implications of building a dam on a sugarcane farm
Forecasting water allocations [2]	Everingham et al. (2008a)	Forecasts water allocations using phases of the SOI and links these to an irrigation scheduling system

Options for dealing with projected climate change

The sugarcane industry in Australia has a long history of managing the impacts of climate on crop production. Nevertheless, additional adaptation measures will be required to reduce the adverse impacts of projected climate change and variability, regardless of the scale of mitigation undertaken over the coming decades (IPCC 2007). Similar to other cropping systems, many of the management-level adaptation options suitable to the sugarcane production system are largely extensions or intensifications of existing climate risk management or production enhancement activities (Howden et al. 2007). Adaptation strategies can be categorised into the approaches detailed in the following five areas (Park et al. 2007a).

Improved management of limited water supplies

Water management could be improved through more efficient use of water supplies (e.g. use of improved irrigation water delivery technologies such as trickle tape, land grading, modified furrow irrigation techniques and schedules aimed at maximising production and limiting expansive growth in favour of sucrose accumulation); improved soil structure for increased infiltration and moisture conservation, trash blanketing, cover crops, and minimum tillage for reduced evaporation); improved capture and storage of water (e.g. harvesting rainfall and excess surface water in on-farm storage facilities, laser levelling and the re-use of tail water), and the maintenance of below-ground water sources (e.g. restrictions on groundwater pumping and the construction of new bores, abandonment of saline bores, and ongoing monitoring of water quality). Integration of streamflow forecasting, storage options and crop water demand could also improve the use of catchment runoff for crop production. A change in pricing and regulations, such as the dates of the water year (i.e. the year within which water allocations are assigned), may also create incentives for more efficient use of catchment runoff.

Technological fixes

Technological advances could contribute in several areas including improved crop varieties that express desirable traits consistent with prevailing climate conditions (e.g. greater drought resistance, water use efficiency, tolerance of increased temperatures, reduced lodging, low vegetative growth and desirable stalk fibre content), or machinery technologies (e.g. wet-weather harvesters and machinery able to effectively harvest lodged cane, switch between rail and road transport infrastructure, and improved designs for key elements of the sugar processing system, such as improved mill clarifier design and mud scrapers). These technical fix adaptive options are generally based on reductionist analysis, engineering design principles, and computer-aided modelling.

Management strategies

Cropping system design and agronomic management strategies include farm-scale planning and design (e.g. tree planting for shelter and soil protection, introduction of precision agriculture, laser levelling, and diversification into alternative/ additional crop species), or improved and more flexible agronomic management (e.g. adjustment of planting dates and crop varieties, revision of best management practice for erosion, pest, disease and weed control, trash blanketing, nutrient management (Brouder and Volenec 2008) and improved soil structure, and the optimisation of resources to achieve maximum yield in the prevailing climate). These adaptive options require a greater element of attitudinal change than the technical fixes detailed above.

Decision-support tools for managing climate variability

Enhanced use of existing decision-support tools for managing climate variability may improve resilience to further changes in variability brought about by anthropogenic climate change (e.g. Table 6.2). Such tools may also track developing climate change trends, encouraging autonomous incremental year-to-year adjustments in management practices. Clearly defined pathways to adoption may increase the presently limited use of these tools.

Policy and institutional change

Broader scale changes in policy and institutions may manifest in changes in physical infrastructure (e.g. construction of seawalls and storm surge barriers, dune reinforcement, land acquisition and the creation of marshland/wetlands as buffer zones against sea level rise and flooding, and the protection of existing natural barriers). It may also be realised through a switch between sugar production and biofuel (depending on market conditions (Kagatsume 2006)), and/or industry reform, such as revision of quarantine boundaries, relocation of the sugarcane production to more southerly areas to track poleward shifts of climate zones (taking into account present milling capacity and location, supporting infrastructure and competing land uses), increased flexibility in capacity and operations in the value chain to track changes in the quality and quantity of throughput to maintain optimal efficiency, or greater diversification into alternative rural enterprises and off-farm income.

Risks of maladaptation

No research has yet been undertaken explicitly defining the risks of maladaptation to climate change for the sugar industry. However, the approach taken in Park *et al.* (2007b) which expresses impacts and adaptation strategies in a value chain context, highlights a number of possible conflicts and maladaptations.

(1) An extended harvest season may have benefits for the harvesting, transport and milling sectors since it would reduce the requirement for capital stock (e.g. harvesters and trucks), but may be negative for the growing sector since a greater proportion of the crop will be harvested at a time of year when sucrose levels are suboptimal.

(2) An increased use of irrigation water in some areas may result in an increased risk of salinity or exacerbate rising water tables.

(3) The use of increasing volumes of trash on the soil surface may immobilise soil N stores to levels below crop demand, particularly in the period shortly after application when crop N demand is high. This may necessitate using increased amounts of N fertiliser, resulting in increased greenhouse gas emissions and the risk of runoff into waterways.

(4) If trash were removed from the paddock for co-generation (the generation of electricity for sale by the mill) as a mitigation strategy, this might reduce crop yields (primarily through a reduction in the input of C and N to the soil) and increase the cost of alternative weed control measures (Thorburn *et al.* 2006). It would also be likely to negatively impact erosion control and soil moisture retention.

Adaptation strategies must also be regionally specific in design, as exemplified by the localised impacts identified in the example of altered planting dates. Similarly, the use of trash blanketing for reduced evaporation and increased water infiltration is not suitable for those regions where furrow irrigation is used (e.g. the Burdekin). It will also be necessary to iteratively develop and apply adaptation strategies in line with progressive changes in climate to avoid maladaptation.

Costs and benefits

The adaptation strategies identified above are generic in nature and will result in different productivity and environmental responses in different regions. In order to formally evaluate these adaptation options in terms of their productivity and environmental outcomes, an ecophysiological modelling approach is required. Models such as APSIM, can not only quantify the potential impacts of climate change in terms of crop yield, but also assess the efficacy of adaptation strategies. For example, simulations have shown that when sugarcane is planted at Rocky Point 60 days earlier than is currently practised, median cane yield may increase by up to 5% on 1990 levels for 2030, while delaying planting by up to 30 days has little effect (Park *et al.* 2007c). In contrast, simulated planting 60 days earlier in the Mossman region showed a decrease in yield by nearly 10%, while planting up to 30 days later in the season may increase yields marginally.

These very different regional responses highlight the importance of accounting for localised biophysical, operational and institutional differences across the Australian sugarcane industry when considering both the impacts of climate change and variability, and adaptation strategies on crop production. Interestingly, the simulation study also suggested that a change in planting date of up to 60 days earlier and 30 days later than is presently practised is unlikely to fully ameliorate yield losses resulting from changes in rainfall and temperature by 2030 in the Rocky Point region. A later planting date in Mossman may offset much of the potential yield losses predicted for the region. This method can be extended to a range of adaptation strategies including changing the time and duration of the harvest season, variety selection and irrigation and N management.

APSIM has also been used to estimate gross margin impacts of the interactions between climate change, land use and management options (Roebeling *et al.* 2007). In this Tully-Murray catchment study in tropical north Queensland, the interaction of climate change with different tillage or fallow management options resulted in gross margin changes between -20 and +10%.

A limited number of studies have moved beyond simple econometrics (i.e. gross margins) to examine the economic impacts on a range of sectors in the value chain. One such study examined the incremental impact of changes in sugarcane production on the harvesting and transport sectors of the value chain in the Maryborough region (Park *et al.* 2007a). The analysis was conducted using a value chain model (Archer *et al.* 2004; Thorburn *et al.* 2006) and yield projection output from APSIM. The value chain model was run to

Table 6.3: Adaptation options identified for the Australian sugarcane industry prioritised in terms of their potential for implementation (1 immediately, 2 may offer potential adaptive capacity in the short term, 3 requires further development in order to provide adaptive potential).

Adaptation options	Priority
Value chain integration	
Australian industry futures	1
Biosecurity risk assessment	2
Global industry futures	3
Alternative products (e.g. biofuels)	3
Farming and harvesting	
Best management practices (aimed at improving resilience through, for example, soil health and water use efficiency)	1
Precision agriculture	2
Transport, milling and marketing	
Transport and milling system analysis to identify product and efficiency opportunities	3
Derivatives for risk management	3
Genetics and breeding	
New varieties developed with desirable traits	2
Improved breeding systems	2
Improved knowledge of sugarcane physiology	3
Individual and community capacity	
Climate knowledge and external collaborative experience	1

produce estimates of the percentage change in costs (relative to the year 2003). The model was parameterised with the worst-case scenario estimates of yield change of -4% by 2030 (consistent with a warming of 1°C and annual rainfall decline of 14%) and -47% by 2070 (consistent with a warming of 4°C and decline in rainfall of 42%) respectively, for the Maryborough region. Total harvesting costs across the region were projected to decrease by between 3 and 27%, but the cost of production per tonne of cane is likely to increase by between 2 and 34% due to reduced harvesting efficiency and lower returns on capital. Total transport costs for the region similarly reduced under lower sugarcane yields, but increased per tonne of cane transported by up to 13%. Production in the Maryborough region relies primarily on road transport as opposed to rail facilities used in other regions and hence can be more readily altered without incurring significant costs. In such circumstances, the impact on costs per tonne of cane in the transport sector is likely to be less than that for harvesting in this region. This is unlikely to be the case in regions where rail transport is used since it involves a large amount of capital investment in rail track, locomotives and wagons, which are generally owned by the mill.

Discussion

In this chapter we have reviewed the specific actions and issues needing to be addressed by the Australian sugarcane industry in dealing with climate change, with some priorities summarised in Table 6.3. Further climate change adaptation issues common across primary industries in Australia are provided in Table 16.1. The industry-specific actions and activities (Table 6.3) have been drawn from detailed feedback provided by sugarcane industry stakeholders attending workshops in 2006 where they were asked to consider opportunities for improving the profitability and sustainability of their industry in the face of a changing climate (Park *et al.* 2007a). From this we have identified gaps in knowledge that present barriers to the implementation of adaptive management strategies to deal with climate change. We supplement this information by prioritising the issues and actions to indicate those that could be promoted immediately and those that may have longer-term potential after further evaluation or development (Table 6.3).

The range of sugarcane industry-specific ideas collated from stakeholders regarding opportunities for climate-related adaptation options can be grouped into value chain integration; farming and harvesting; transport, milling and market-ing; genetics and breeding, and individual and community capacity. The following discussion of these five priority action areas for adaptation is based on that by Park *et al.* (2007a).

(1) To enhance industry preparedness it will be necessary to have a greater understanding of the implications of climate change on the specific value chain interactions operating in each mill region, as adaptation strategies identified for one mill region may not be equally applicable to other regions (Park *et al.* 2007c). Recognising these differences within adaptation response strategies is vital to whole-of-industry sustainability. Given the integrated nature of mill region value chains, issues such as biosecurity, or changes in production focus (e.g. biofuel), will be more effectively considered at this scale.

In preparing adaptation strategies it will also be important to consider how other sugar-producing nations will be affected by and respond to climate change, and how this will influence Australia's competitiveness in global markets.

(2) The challenges for farming and harvesting relate to whether the tactical and strategic responses highlighted above will be sufficient to maintain a resilient sugarcane industry in Australia. Further evaluation is required across the value chain in order to identify region-specific demands and appropriate adaption responses (Park and Attard 2005). Some immediate opportunities where adaptation strategies could be implemented include more extensive use and improvement of seasonal forecasting and decision-support tools aimed at increasing system resilience, reducing exposure to climate risks and mitigating off-farm impacts. Greater use of these tools offers immediate potential to substantially improve climate-related farm management and harvesting practices, while at the same time addressing the growing pressure for the sugarcane industry to improve its environmental performance, e.g. in compliance with the Queensland and Commonwealth Governments' Reef Water Quality Protection Plan. Areas likely to benefit from improvements in climate-related best-management practices include soil health and water use efficiency. Precision

agriculture may offer substantial benefits in both improved environmental performance and economic and environmental resilience (Esquivel *et al.* 2007; Halpin *et al.* 2008).

(3) Opportunities for improved adaptive capacity may be obtained from greater integration of the transport, milling and marketing sectors of the industry. Using systems-based operations analysis and seasonal forecasting decision tools will enable opportunities for increased profitability to be explored and assist in evaluating alternative products, such as biofuels, and marketing opportunities, such as sugar derivatives. The uncertain nature of the changing climate means that adaptation strategies and options must be considered in a flexible, adaptive management context if the challenges related to climate and other external pressures are to be effectively addressed.

(4) As with other industries, genetics and breeding programs in sugarcane should aim to deliver varieties that are better suited to a changing climate. This area of adaptation research and development is already active within the sugarcane industry, where research is aimed at better identifying desirable plant traits such as drought tolerance and water use efficiency. This may result in reduced time from parent selection and crossing to the commercialisation of a cultivar, and improved cultivar adoption programs. Additional long-term benefits for adaptation may be gained by improving the presently limited knowledge of sugarcane physiology, particularly related to the process of sucrose accumulation, the relationship between CO_2 and crop growth, and other climate-related interactions (Inman-Bamber *et al.* 2008).

(5) Opportunities exist to enhance the capacity of the sugarcane industry to better manage climate change and variability. This will require developing sufficient knowledge and expertise among farmers for them to understand and assess the impacts, risks, and adaptation options related to their individual circumstances. At the community level, opportunities for collaboration within the value chain and with other agribusinesses will offer the greatest value.

References

Anthony G, Everingham YL, Smith DM (2002). Financial benefits from using climate forecasting – a case study. *Proceedings of the Australian Society of Sugar Cane Technologists* **24**, 153–159.

Archer AA, Higgins AJ, Thorburn PT, Hobson P, Antony B, Andrew B (2004). Employing participatory methods and agent-based systems modelling to implement agricultural supply chain systems. In *Modern Supply Chain Management: From Theory to Practice. Second Annual Symposium on Supply Chain Management*. DeGroot School of Business, McMaster University, Toronto.

Australian Sugar Milling Council (2006). 'Annual review 2006 season'. Australian Sugar Milling Council Pty Ltd. http://www.asmc.com.au/content/

Ball AS (1997). Microbial decomposition at elevated CO_2 levels: effect of litter quality. *Global Change Biology* **3**, 379–386.

Brotherton GA (1980). The influence of extraneous matter on CCS. *Proceedings of the Australian Society of Sugar Cane Technologists* **2**, 7–12.

Brouder SM, Volenec JJ (2008). Impact of climate change on crop nutrient and water use efficiencies. *Physiologia Plantarum* **133**, 705–724.

CANEGROWERS (2007). http://www.canegrowers.com.au/information-centre/about-the-industry/index.aspx

Cheeroo-Nayamuth FB, Nayamuth ARH (2001). Climate change and sucrose production in Mauritius. *Proceedings of the Australian Society of Sugar Cane Technologists* **24**, 107–112.

Crimp SJ, Harper B (2007). Impacts of climate change on tropical crop production. *Proceedings of the Seventh Australian Banana Industry Congress*. Gold Coast September 2007.

De Souza AP, Gaspar M, Da Silva EA, Ulian EC, Waclawovsky AJ, Nishiyama NY Jr., Dos Santos RV, Teixeira MM, Souza GM, Buckeridge MS (2008). Elevated CO_2 increases photosynthesis, biomass and productivity, and modifies gene expression in sugarcane. *Plant, Cell and Environment* **31**, 1116–1127.

Esquivel M, Cabezas L, Hernàndez B, Marrero S, Fernàndez F, Ponce E, Quintana L, Gonzàlez L (2007). Harvester automation and precision agriculture trials in Queensland 2006. *Proceedings of the Australian Society of Sugar Cane Technologists* **29**, 352–360.

Everingham Y (2007a). 'Moving from case studies to whole of industry: implementing methods for wider industry adoption'. SRDC Final Report CSE009.

Everingham Y (2007b). Comparing phase based climate forecasting systems. *Proceedings of the International Congress on Modelling and Simulation* **17**, 574–581.

Everingham YL, Muchow RC, Stone RC (2001a). Forecasting Australian sugar yields using phases of the Southern Oscillation Index. *Proceedings of the International Congress on Modelling and Simulation* **4**, 1781–1786.

Everingham YL, Muchow RC, Stone RC (2001b). An assessment of the 5 phase SOI climate forecasting system to improve harvest management decisions. *Proceedings of the Australian Society of Sugar Cane Technologists* **23**, 44–50.

Everingham YL, Muchow RC, Stone RC, Inman-Bamber NG, Singels A, Bezuidenhout CN (2002). Enhanced risk management and decision-making capability across the sugarcane industry value chain based on seasonal climate forecasts. *Agricultural Systems* **74**, 459–477.

Everingham YL, Muchow RC, Stone RC, Coomans DH (2003). Using southern oscillation index phases to forecast sugarcane yields: a case study for north eastern Australia. *International Journal of Climatology* **23**, 1211–1218.

Everingham YL, Jakku E, Inman-Bamber G, Thorburn P, Webster T, Attard S, Antony G (2006). Understanding the adoption of knowledge intensive technologies in the Australian sugar industry – a pilot study. *Proceedings of the Australian Society of Sugar Cane Technologists* **28**, 76–85.

Everingham YL, Inman-Bamber NG, Thorburn PJ, McNeill TJ (2007). A Bayesian modelling approach for long lead sugarcane yield forecasts for the Australian sugar industry. *Australian Journal of Agricultural Research* **58**, 87–94.

Everingham YL Baillie B, Inman-Bamber NG, Baillie J (2008a). Forecasting water allocations for canefarmers. *Climate Research* **36**, 231–239.

Everingham YL, Clarke AJ, Van Gorder S (2008b). Long lead rainfall forecasts for the Australian Sugar Industry. *International Journal of Climatology* **28**, 111–117.

Ghassemi F, Howard KWF, Jakeman AJ (1996). Seawater intrusion in coastal aquifers and its numerical modelling. *Environmental Modelling* **3**, 299–328.

Halpin NV, Cameron T, Russo, PF (2008). Economic evaluation of precision controlled traffic farming in the Australian sugar industry. A case study of an early adopter. *Proceedings of the Australian Society of Sugar Cane Technologists* **30**, 34–42.

Hamilton JG, Dermody O, Aldea M, Zangerl AR, Rogers A, Berenbaum MR, DeLucia EH (2005). Anthropogenic changes in tropospheric composition increase susceptibility of soybean to insect herbivory. *Environmental Entomology* **34**, 479–485.

Howden SM, Soussana J–F, Tubiello FN, Chhetri N, Dunlop M, Meinke H (2007). Climate change and food security special feature: adapting agriculture to climate change. *Proceedings of the National Academy of Sciences of the United States of America (PNAS)* **104** (50), 19691–19696.

Inman-Bamber NG (1994). Temperature and seasonal effects on canopy development and light interception of sugarcane. *Field Crops Research* **36**, 41–51.

Inman-Bamber NG (2007). Economic impact of water stress on sugar production in Australia. *Proceedings of the Australian Society of Sugar Cane Technologists* **29**, 167–175.

Inman-Bamber NG, Everingham YL, Muchow RC (2001). Modelling water stress response in sugarcane: validation and application of the APSIM–Sugarcane model. *Proceedings of the 10th Australian Agronomy Conference*. http://www.regional.org.au/au/asa/2001/

Inman-Bamber NG, Attard SJ, Baillie C, Lawson D, Simpson L (2005). A web-based system for planning use of limited irrigation water in sugarcane. *Proceedings of the Australian Society of Sugar Cane Technologists* **27**, 170–181.

Inman-Bamber NG, Webb WA, Verrall SA (2006). Participatory irrigation research and scheduling in the Ord 1: R&D. *Proceedings of the Australian Society of Sugar Cane Technologists* **28**, 155–163.

Inman-Bamber NG, Bonnett GD, Spillman MF, Hewitt ML, Jackson J (2008). Increasing sucrose accumulation in sugarcane by manipulating leaf extension and photosynthesis with irrigation. *Australian Journal of Agricultural Research* **59**, 13–26.

IPCC (2007). 'Summary for policymakers of the synthesis report of the IPCC Fourth Assessment Report'. Draft copy 16 November 2007.

Jaffres J, Everingham YL (2005). An exploratory investigation on the relationship between decadal rainfall and climate indices. *Proceedings of the Australian Society of Sugar Cane Technologists* **27**, 96–108.

Jones K, Everingham Y (2005). Can ENSO combined with Low-Frequency SST signals enhance or suppress rainfall in Australian sugar-growing regions. *Proceedings of the International Congress on Modelling and Simulation* **16**, 1660–1666.

Jones K, Everingham Y (2006). Assessing statistical climate forecasting methods. In 'Advanced climate forecasting – helping industry make better decisions more often'. Final report (Technical Report 1) L&W Australia project JCU20. (Ed. Y Everingham). Townsville.

Kagatsume M (2006). Biofuel production in Australia and its implications – towards resource recycled farming through renewable petro-substitute fuels. *Natural Resource Economics Review, Kyoto University (1341–8947)* **12**, 31–50.

Keating BA, Carberry PS, Hammer GL, Probert ME, Robertson MJ, Holzworth D, Huth NI, Hargreaves JNG, Meinke H, Hochman Z, McLean G, Verburg K, Snow V, Dimes JP, Silburn M, Wang E, Brown S, Bristow KL, Asseng S, Chapman S, McCown RL, Freebairn DM, Smith CJ (2003). An overview of APSIM, a model designed for farming systems simulation. *European Journal of Agronomy* **18**, 267–288.

Kingston, G (2000). Climate and management of sugarcane. In *Manual of Cane Growing*. (Eds M Hogarth, P Allsopp) pp. 7–26. BSES, Brisbane.

Lisson SN, Brennan LE, Bristow KL, Keating BA, Hughes DA (2003). DAM EA$Y – software for assessing the costs and benefits of on-farm water storage based production systems. *Agricultural Systems* **76**, 19–38.

McDonald G, Park S, Antony G, Thorburn P, Dawson S, Harman B (2006). 'Future use of Sunshine Coast cane landscapes'. CSIRO Sustainable Ecosystems, St Lucia.

McInnes KL, Walsh KJE, Hubbert GD, Beer T (2003) Impact of sea-level rise and storm surges on a coastal community. *Natural Hazards* **30**, 187–207.

Murphy SF, Sorenson RC (2001). Saltwater intrusion in the Mackay coastal plains aquifer. *Proceedings of the Australian Society of Sugar Cane Technologists* **23**, 70–76.

Park S (2006). Linking climate forecasts with irrigation decision support systems. In 'Advanced climate forecasting – helping industry make better decisions more often'. Technical report No. 2. L&W Australia project ref JCU20.

Park S, Robertson M, Inman-Bamber G (2005). Decline in the growth of a sugarcane crop with age under high input conditions. *Field Crops Research* **92**, 305–320.

Park S, Creighton C, Howden M (2007a). 'Climate change and the Australian sugarcane industry: impacts, adaptation and R&D opportunities'. SRDC Technical Report.

Park S, Howden M, Higgins A, McRae D, Horan H, Hennessy K (2007b). Climate change: informing the Australian sugar industry of potential impacts, possible strategies for adaptation and best-bet options for future R&D. In the final report of projects CSE019 and SRD011 by S Park, C Creighton, M Howden. SRDC.

Park S, Howden M, Horan H (2007c). Evaluating the impact of and capacity for adaptation to climate change on sectors in the sugar industry value chain in Australia. *Proceedings of the International Society of Sugar Cane Technologists* **26**, 312–324.

Park SE, Attard SJ (2005). Potential impacts of climate change on the Queensland sugar industry and the capacity for adaptation. *Proceedings of the Australian Society of Sugar Cane Technologists* **27**, 61–74.

Pearson L, McDonald G, Park S, Harman B, Heyenga S, Horan H (2007). 'Future use of the Rocky Point Cane landscapes, Gold Coast'. CSIRO Sustainable Ecosystems, St Lucia.

Robertson MJ, Muchow RC, Donaldson RA, Inman-Bamber NG, Wood AW (1999). Estimating the risk associated with drying-off strategies for irrigated sugarcane before harvest. *Australian Journal of Agricultural Research* **50**, 65–77.

Roebeling PC, Webster AJ, Biggs J, Thorburn P (2007). *Financial-economic analysis of current best-management practices for sugarcane, horticulture, grazing and forestry industries in the Tully-Murray catchment.* CSIRO Sustainable Ecosystems, Canberra.

Singh B, El Maayar M (1998). Potential impacts of greenhouse gas climate change scenarios on sugar cane yields in Trinidad. *Tropical Agriculture (Trinidad)* **75**, 348–338.

Stone RC, Hammer GL, Marcussen R (1996). Prediction of global rainfall probabilities using phases of the Southern Oscillation index. *Nature* **384**, 252–255.

Sutherst RW, Yonow T, Chakraboty S, O'Donnell C, White N (1996). A generic approach to defining impacts of climate change on pests, weeds and diseases in Australasia. In *Greenhouse: Coping with Climate Change.* (Eds WJ Bouma, GI Pearman, MR Manning) pp. 190–204. CSIRO Publishing, Melbourne.

Thorburn PJ, Archer AA, Hobson PA, Higgins AJ, Sandel GR, Prestwidge DB, Andrew B, Antony G, McDonald LM, Downs P, Juffs R (2006). Value chain analyses of whole crop harvesting to maximise co-generation. *Proceedings of the Australian Society of Sugar Cane Technologists* **28**, 37–48.

7
WINEGRAPES

L Webb, GM Dunn and EWR Barlow

KEY MESSAGES:

- A warmer climate will hasten the progression of phenological stages of the vine (e.g. budburst, flowering and veraison) so that ripening will occur earlier in the season.

- In most cases, quality of existing mainstream winegrape varieties will be reduced if no adaptation measures are implemented.

- Water requirements for grapevines are likely to increase while at the same time rainfall and associated runoff to water storages is likely to decrease.

- Vintage, the period when grapes are harvested and processed in wineries, is likely to become more compressed requiring possible changes to winery infrastructure and vintage staffing levels.

- Yield can affect wine quality, so effects of elevated CO_2 and increased temperature on yield and its components will need to be closely monitored and, if necessary, managed.

- Budburst in some of the more maritime climates may become uneven due to less chilling during the winter dormancy period.

- Shifting to cooler sites will alleviate some warming impacts. As vineyard blocks have an average life of 30+ years, this option will need to be considered with some urgency.

- Within regions, existing varieties can be replaced with 'later season' varieties to compensate for the warmer temperatures and compressed phenology.

- Consumer education relating to new wine styles and varieties will be important, e.g. the typical style for any given region is likely to change.

Introduction

The Australian wine industry is an important contributor to the Australian economy with wine exports in 2006–07 being second only to beef exports as the largest valued agricultural export commodity (ABARE 2006). Exports for the year ending May 2008 were reported to be $2.8 billion (Wine Australia 2008), with vineyard-bearing hectares expected to increase to an estimated 171 000 ha by 2009–10 (Jackson *et al.* 2008). In 2008, the Australian wine industry's wine grape intake

increased by almost 31% over 2007 to 1.83 million tonnes (Mt) (AWBC, 2008). This was just 4% less than the 2005–06 record intake of 1.9 Mt.

Winegrapes are planted in diverse climatic regions in Australia (Smart *et al.* 1980) and mainly between the latitudes of 30°S–40°S (Figure 7.1). These regions range in climate type from some of the warmest wine growing regions in the world to cool climate regions capable of producing more delicate wine styles (Johnson 1989). The 'Mediterranean' and 'cool-climate temperate' climate

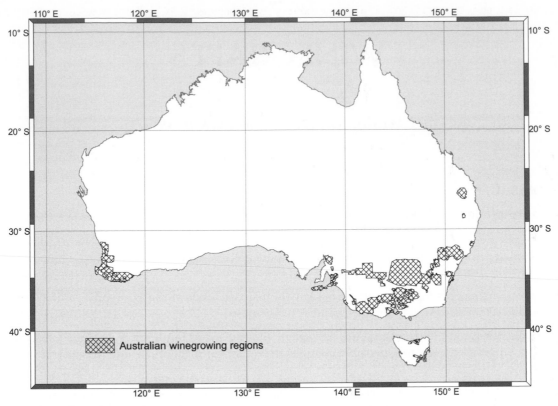

Figure 7.1: Australian winegrowing regions.

zones of the 10 climatic zones described by Hobbs and McIntyre (2005) contain the majority of the wine regions of Australia (Figure 16.1).

Wine production is intimately wedded to the concept of *terroir:* matching premium grape varieties to particular combinations of climate and soils to produce unique wines of particular styles (Seguin 1986). Climate changes will alter these *terroirs* and challenge the adaptive capacity of the industry (Seguin and de Cortazar 2005). Investors in the wine industry, as well as consumers of its products, are therefore alert to the prospect of such consequences resulting from climate variability and change and are increasingly exploring the available options to enable the industry to adapt to change.

Impacts of projected climate change

Temperature increases

Winegrapes *(Vitis vinifera* L.) have four main developmental stages: budburst, flowering, veraison (berry softening and colour change, signalling the onset of rapid sugar importation), and maturity (when ripe grapes are harvested) (McIntyre *et al.* 1982). The timing and duration of these stages (phenology) varies with grapevine variety and climate (Jones and Davis 2000). Matching developmental phases to climate is an important consideration in vineyard planning. As climates warm, phenology will progress more swiftly and grapes will ripen earlier (Webb 2007b). This will have both positive and negative effects on the Australian wine industry, depending on the current climate regime, and being able to maintain consistent quality against this background will be critical.

Grapes are monitored for sugar and acidity and often undergo berry sensory assessment during the latter stages of ripening. Fruit is harvested when the balance of these is expected to produce the desired wine style. The rate of change in fruit composition is strongly influenced by temperature, with higher temperatures increasing the speed of sugar development, hastening acid degradation, and altering flavour compounds (Coombe and Iland 2004; Lund and Bohlmann 2006). If an environment is too warm, grape flavour compounds may develop more slowly than other compounds in the berry and while the winemaker waits for certain flavours to either appear or disappear, acidity is lost and sugar content increases. This can lead to high alcohol wines with low acidity (Jones 2007). Generally, the sugar content of the berries determines alcohol content of the wine and higher alcohol wines have been observed in response to a warming climate (Duchene and Schneider 2005; Godden and Gishen 2005; Petrie and Sadras 2008).

Aside from sugars and acids, other compositional characters are likely to be affected. Higher temperatures during ripening can reduce the colour of red wine grapes (Haselgrove *et al.* 2000; Mori *et al.* 2007) and alter aroma profiles of both white and red grapes (Marais 2001; Marais *et al.* 2001). Studies modelling the projected impact of climate change on grape quality (Webb *et al.* 2008a,b) and wine quality (Jones *et al.* 2005) indicate that for winegrapes that are currently able to ripen in a given climate, warming will reduce compositional quality if no adaptive measures are taken.

Temperature bands within which particular wine grape varieties are grown currently will shift, leading to a severe reduction in the potential area available in Australia for growing varieties and making wine styles suited to cooler regions (Webb *et al.* 2007a). Based on current winegrape plantings and mainstream varieties, if the overall quality of grapes is to remain equivalent to that of the present day the area suitable for viticultural production may be reduced by up to 40% by the year 2050 (see Figure 7.2, page 46). The area of land suitable for viticulture in the United States is also likely to lessen with future warming (White *et al.* 2006).

Projected phenological shifts are likely to lead to shorter harvest periods (vintages) and varieties that are currently harvested sequentially will tend to reach maturity at a similar date (Webb *et al.* 2007b). This will create problems for intake scheduling and make it more difficult to process each and every batch of fruit at the time when grape quality in the vineyard is deemed 'optimal'. This will have important implications for planning of infrastructure and staffing during the vintage period.

Winter temperatures are likely to increase by 0.3°C–1.5°C by 2030 in most winegrape growing areas (CSIRO and Australian Bureau of Meteorology 2007). Grapevines, like many other perennial horticultural plants, have a physiological requirement to accumulate a quota of cold temperatures before dormancy can be broken (Lavee and May 1997). This chilling requirement is known as vernalisation. The mean July temperature in the Margaret River, Western Australia, (13.2°C), which is relatively high in comparison to many other wine grape regions, has already been associated with non-uniform (patchy) budburst related to insufficient chilling (Dry 1988). A recent study has highlighted the trend towards reduced hours of chilling in this region (Lyons and Considine 2007), which is likely to be further exacerbated by warming.

An increasing frequency of extreme events has the potential to affect crops more than higher temperatures *per se*. For instance, a 15 day 'hot spell' in South Australia in March 2008 (Australian Bureau of Meteorology 2008) resulted in large quality reductions and fermentation problems (Hook 2008). Furthermore, many ripe grapes were unable to be processed by wineries which were already at full capacity (AWBC 2008). Grapevine studies on the effects of extreme temperatures and the determination of 'damage thresholds' are yet to be undertaken (Soar *et al.* 2007). Although anecdotal reports describe vine 'shut-down' and sunburn impacts, these require validation.

Australian viticulture, which is concentrated in southern regions of Australia, is affected by indigenous insect pests, especially light brown apple moth (LBAM), and fungal diseases like downy mildew, powdery mildew, black spot, phomopsis and botrytis (exacerbated by LBAM infestation). The life cycles of all of these are strongly influenced by climate. The current trend towards increased night-time temperatures is likely to result in an increased risk of downy mildew and related diseases. Currently, temperatures below 10°C at night preclude the formation of many primary downy mildew infections (Magarey *et al.* 1994). This effect may be moderated by lower projected rainfall during the growing season, as leaf wetness is also a required precursor for downy mildew infection.

Carbon dioxide concentration increases

Increased growth under higher atmospheric carbon dioxide (CO_2) concentrations may lead to excessive vegetative vigour and within-canopy shading of winegrapes (Dry 1988). Heavy shading over a prolonged period around flowering (four weeks) has been found to reduce fruitfulness (May *et al.* 1976). It is unlikely fruitfulness will be affected in this way as any negative effect is likely to be more than offset by increased temperatures around flowering favouring fruit bud initiation and differentiation (Buttrose 1974) and, in some climates, fruit set (Carmona *et al.* 2008). The effect of increased CO_2 on the growth of vines *in situ* has been modelled for Europe (Bindi *et al.* 1996). Bindi and his co-authors predicted a 35% increase in fruit yield if CO_2 was increased from 350 ppm to 700 ppm without any corresponding temperature change, though which component of yield was affected was not reported. In a related study on sour orange, yield increase was attributed to increased fruit number with fruit size decreasing slightly (Idso *et al.* 2002).

The effect of both CO_2 and temperature increases together may increase year-to-year variability in yield, thus increasing economic risk for growers (Bindi *et al.* 1996). However, most of the costs related to fluctuating yields are incurred by wineries (e.g. tank space, ordering oak and chemicals, matching supply to demand; Dunn and Martin 2003). These costs and revenue losses are substantial and likely to increase in the face of increased yield variability.

The effect of increasing atmospheric CO_2 concentrations on the quality of wines in Europe revealed some enhancement of sugar and acid concentrations through the ripening process but this effect did not persist through to maturity (Bindi *et al.* 2001).

Plants grown in elevated atmospheric CO_2 typically have lower protein and nitrogen concentrations (Drake *et al.* 1997; Morison and Lawlor 1999) and higher C:N ratios (Drake *et al.* 1997). However, a recent meta-analysis of free air CO_2 enrichment (FACE) studies showed that C:N ratios in enriched CO_2 environments may not be as large as initially anticipated (Ainsworth and Long 2005) (see Chapter 1). What may be of interest to winemakers is that the reduction of the availability of nitrogen for yeast nutrition during fermentation may increase the risk of 'stuck' fermentations (Bell and Henschke 2005).

Water balance changes

Most of Australia's vineyards are irrigated and rely on a secure irrigation supply (McCarthy *et al.* 1992), therefore the largest impact of climate change is likely to be reduced water availability due to lower rainfall in catchments supplying irrigation. Lower rainfalls are associated with proportionally larger reductions in streamflow that will substantially reduce available water in public storages, rivers and on-farm dams (Cai and Cowan 2008b; Potter *et al.* 2008). When water is not available, yields can decline (Jackson *et al.* 2008) and vines can suffer long-term damage (Hardie and Considine 1976).

Growers accessing state or privately-run irrigation schemes purchase access to a share of the available water, or water right. In drought years only a proportion of a water right may be available to the growers and this is known as an allocation (usually described as a percentage of the water right). Trading of water normally occurs throughout the growing season whereby some water licence owners may sell (or lease for one season) part, or all, of their share to growers requiring more water than their allocation allows. The cost of water increases in drier low allocation years (Bjornlund and Rossini 2005) and these 'lower-allocation' years are projected to increase (Hennessy *et al.* 2008), so vineyard profitability may be adversely affected in future. Increased pressure on water supplies will mitigate against the licensing of more farm dams or bores. Regulations and restrictions already exist on water drawn from water sources in most areas (e.g. The State of Victoria Department of Sustainability and Environment 2007).

Annual rainfall totals for most of the grape growing regions are likely to decrease by 2–10% by 2030 and 5–20% by 2070 (CSIRO and Australian Bureau of Meteorology 2007). As variation in annual rainfall differs between winegrowing regions, impacts on viticulture will need to be assessed region by region. For most regions, the greatest percentage projected reduction in rainfall is in spring, with little change in summer. The Hunter Valley in New South Wales is one of the few winegrowing regions likely to experience an increase in summer rainfall.

As well as the projected decline in rainfall, potential evapotranspiration (PEt) will increase as a function of projected increased temperature in all seasons and in all regions (CSIRO and Australian Bureau of Meteorology 2007). Water budgeting will need to be revised to take this into account. The effect of increased PEt on seasonal vine water use may be moderated by compressed phenology

(Webb *et al.* 2007b) and reduced stomatal conductance in response to increased vapour pressure deficit (Berry and Bjorkman 1980).

Reduced in-season rainfall in most winegrowing regions is likely to decrease the risk of fungal disease. However, in some regions (e.g. the Hunter Valley), the probability of increased summer rainfall may increase fungal pressure (Magarey *et al.* 1994). This leaf wetness may be counteracted by increased evaporation within canopies due to higher temperatures.

Although average rainfall is projected to decrease, the frequency of extreme rainfall events is likely to increase by up to 4% in south-eastern Australia during the harvest period in autumn (CSIRO and Australian Bureau of Meteorology 2007). Along with the risk of berry splitting and berry drop due to abscission, *Botrytis cinerea* (grey mould) develops in wet and humid conditions, and can devastate winegrape crops close to and during harvest. Therefore, any heavy rainfall at this time represents a serious risk (Magarey *et al.* 1994).

Salinisation of arable land is a significant problem affecting winegrape growers (Walker *et al.* 2002). Recent studies suggest that the amount of groundwater recharge may decline significantly across southern Australia reducing the longer-term risk of salinisation (CSIRO 2008) though the relative impact on recharge and aquifer productivity will vary by region, by soil type and with management. However, reduced rainfall and water allocations in the warm irrigated regions may heighten the risk of salinisation by reducing the capacity for flushing any salt build-up through the profile in the winter period (Clark 2004).

Frost and fire

Climate change projections indicate fewer frost days so it might be expected that in a warmer climate growers would experience less frost damage. However, as grapevine phenology is strongly controlled by temperature, budburst in grapevines will occur earlier, and frost risk may not be reduced (Nemani *et al.* 2001; Hanninen 1991). Projections of lower rainfall in spring and associated drier soils, fewer clouds and lower dew points may even increase frost risk (Snyder and Paulo de Melo-Abreu 2005; Jones 1991). Furthermore, an increase in day-to-day climate variability may lessen any 'positive' impact of mean warming on reducing frost frequency (Rigby and Porporato 2008). It is noteworthy that in 2006, the warmest spring on record (dataset from 1950 to 2007) in southern Australia (Australian Bureau of Meteorology 2008), widespread frost damage decimated vineyards in many of Australia's wine regions. Widespread frost damage also occurred in 2003, another hot and very dry year.

Smoke from bushfires and controlled burning can reduce the sensory characteristics of wine, leading to 'smoke taint' (Kennison *et al.* 2007) and unsaleable wine. Smoke taint due to wildfires cost Australian grape growers more than $7.5 million during the 2003 and 2004 vintages. In 2003, Victorian growers suffered at least $4 million losses due to fires in adjoining national parks (Whiting and Krstic 2007). With the incidence of bushfires projected to increase (Lucas *et al.* 2007), the risk of smoke taint is likely to increase.

Options for dealing with climate variability

Variability in temperature

With the exception of ensuring that grapevines are well watered, there are few practical options available for managing high temperatures (>40°C) in vineyards. To protect grapevines from heat stress it may be necessary to water them up to three days in advance of heat stress events. Fortunately, the accuracy of short-term temperature prediction, even at a local level, is very good. However, any given vineyard contains many blocks of different varieties that might be at different stages of development and water status and thus at different levels of heat stress risk. To manage in this complex environment, the alert grower will need to understand the level of risk for each block, have an accurate and timely temperature prediction and be able to manage within the vineyard's irrigation capacity and infrastructure to ensure that priority blocks are irrigated. Adding to this complexity is the need not to over-water vines, especially post-veraison, which could cause excessive berry swelling and reduced grape and wine quality. Excessive water between flowering and veraison can also reduce berry quality (McCarthy *et al.* 1992).

Other management strategies for reducing the risk of heat damage include creating a protective leaf-canopy cover for fruit, avoiding north-south row orientation (where the west-aspected side of the canopy can become over-exposed in the afternoon), wider trellises and retention of inter row cover crops to minimise reflection from the soil and potentially, the use of particle film technology (PFT) to reduce leaf and fruit temperature (Cooley *et al.* 2008).

High light intensity and high temperature exposure can result in sunburn on berry skins (Greer *et al.* 2006) and growers can be penalised for this berry fault. The impact of sunburn is most severe after leaf plucking, a practice sometimes used to improve fruit exposure. Again, reliable weather forecasting will be needed to inform the timing of leaf plucking.

Windbreaks can protect the outside rows of a vineyard from hot, dry, northerly winds. However, trees can impact negatively by housing birds (a pest problem) competing with vines for nutrients and water, and shading significant portions of vineyards. Their position relative to slopes needs to be considered in order not to interfere with any cold air drainage that can potentially increase frost risk (Freeman *et al.* 1982).

Variability in the ripening of grapes due to cool or hot summers has been managed through adjusting harvest dates and, in some instances, by crop thinning. Cooler than average seasons can delay ripening and lead to associated problems. On the other hand, a potential reduction of aroma compounds in warmer seasons can lead to less desirable wine styles (Jackson and Lombard 1993; Lund and Bohlmann 2006). For some, one of the interesting and intrinsic values of wine is the season-to-season variation that results in 'better vintages' and 'poorer vintages'. This acceptance of variation by consumers represents an inbuilt flexibility to adapt to climate variability. However, Australia is known for its consistency of product, especially in the United Kingdom (Toni 2005) and thus will need to manage the impacts of climate change to retain consistency. Large wine companies have the scope to do this through varying the regional sources of winegrape supply from year to year.

Some grape varieties can have a range of possible end uses, which can facilitate adaptation to inter-annual temperature variability. For instance Chardonnay and Pinot Noir grapes can be used for making sparkling wine or to produce more full-bodied table wine depending on the temperature of the growing season. Furthermore, winemakers can blend wine from different regions, or different varieties, to take advantage of the complementary flavour profiles developed in different climates.

Frost is presently managed by avoiding risky sites and matching varieties to sites phenologically. Large wind fans or helicopters (to create air movement and mixing), overhead sprinklers, maintaining soil moisture through irrigation, increasing the height of the canopy, keeping inter-rows free of mulch and vegetation, and foggers can also be used to ameliorate frost risk (Snyder and Paulo de Melo-Abreu 2005).

Variability in rainfall

In Australia most grape production occurs in regions where plant water requirements during the growing season are far higher than rainfall, thereby avoiding disease pressure due to humidity and leaf wetness. For this reason, irrigation has been widely adopted in Australia to maintain commercial yields. Most vineyards are equipped with soil moisture monitoring devices to help schedule irrigation. Furthermore, the timely withholding of water to expose vines to mild stress can be a strong lever for improving fruit and wine quality.

Improved watering strategies have recently been applied or trialled in some regions, such as regulated deficit irrigation (RDI) (Goodwin and Jerie 1992), partial root-zone drying (PRD) (Dry and Loveys 1998; Possingham 2002), and sustained deficit irrigation (Chalmers *et al.* 2008). While most of these strategies were introduced to improve winegrape quality, with the recent reductions in water availability (MDBC 2008), these strategies have proven beneficial in improving water use efficiency. Drought-tolerant rootstocks (e.g. Ramsay, 140 Ruggeri) have provided another strategy for growers to exploit when building vineyard resilience to drier conditions (Dry 2007). In areas with salinity issues, computer-controlled irrigation systems are used to increase efficiency while reducing the impact on the environment (Bramley and Lanyon 2002). There is still considerable scope within the industry to further improve water use efficiency.

It has become increasingly important to understand the needs of other water users, and relationships between irrigation allocation and price (Bjornlund and Rossini 2005). The pasture sector, for instance, used about 27% of the water from the MDB in 2005–06, while grapevines used about 7% (Australian Bureau of Statistics 2008) despite the value of produce per volume of water ($ per ML) being at least three times greater for viticulture (NLWRA 2001). This demonstrates that, as a proportion of the value of the crop, the cost of water for viticulture is not as great as it is for many other industries so purchasing water is an option, provided the water is available to purchase. Various tools are available to assist managers in making informed decisions about water budgeting during low irrigation allocation years, e.g. CCW Co-operative Limited (2008). Many of the water adaptation measures that are discussed in the following chapter are also relevant for viticulture, e.g. increasing efficiency of irrigation delivery systems and increasing on farm water use efficiency.

Rainfall, and corresponding humidity and leaf wetness, affects disease incidence (Magarey et al. 1994). Grape purchase contracts have inbuilt penalties linked to various levels of disease, whereby disease must be avoided to reduce these penalties and the potential rejection of fruit. Pest and disease management in the current climate is dependent on a limited understanding of the ecology of light brown apple moth (LBAM) and long experience with the fungi that affect grapevines. LBAM larvae physically damage the developing grape berries predisposing them to botrytis fungal infection which can cause the berries to rot and become unsuitable for wine making. Releases of moth parasites, or hormonal disruption strategies, can be used to control LBAM. Even without LBAM, rainfall up to and during harvest can still cause botrytis infection. Chemical control is expensive and does not always provide effective control of botrytis (Magarey et al. 1994). Downy mildew and powdery mildew, two other important fungal diseases, are effectively controlled with chemical spray applications in most climates.

Options for dealing with projected climate change

Adaptation to temperature increases by adjusting vineyard situation, variety selection and/or management strategies

To maintain consistency in wine styles, the industry will need to consider altering the balance of varieties growing in different areas to better match varieties to growing season conditions. Ripening can then coincide with the best possible climate conditions (Schultz 2000). It is only in the hotter winegrape growing regions

that currently-grown varieties may become redundant, or new varieties may be required (Webb et al. 2007a). Moving to new varieties may be difficult as markets are often based on traditional grape varieties. Some breeding programs aim to breed wine grape varieties that ripen later in the season and are able to maintain a good sugar to acid balance (Clingeleffer 1985). However, recent experience indicates that it takes 20–30 years to introduce new varieties (Rose 2007).

French, Italian and German legislation allows for only certain grape varieties to be grown in certain regions for wines produced to be awarded the regional quality classification (Johnson 1989), for example, the Appellations Contrôlées system (France), and the Denominazione di Origine Controllata (Italy). These may restrict adaptation potential in Europe but there are no such legislative restrictions in Australia.

New regions that historically have been too cool for wine grapes are likely to become more suitable (Hood et al. 2002). Infrastructure, water availability and quality (McCarthy et al. 1992), labour availability, along with soil type (Maschmedt 2004) will all need to be considered when selecting new regions and sites within a region. The spatial analysis methodology of Kenny and Harrison (1992) can be used to identify unacceptably high frequencies of temperature extremes and may be expanded to also consider water issues (droughts, floods) and frost risk. The nature of vineyards, with their underlying infrastructure (e.g. trellising), wine processing facilities and established wine tourism cannot be moved on a year-to-year basis, and this geographical inflexibility means that these decisions have long-term implications.

If bud-break becomes uneven or protracted due to the chilling requirement of vines not being met, use of chemical dormancy breakers (Shulman et al. 1983) may offer some adaptive measures. Certain varieties like Cabernet Sauvignon and Sauvignon Blanc are currently cane-pruned in cooler climates due to low basal bud fertility. Warming in these regions will improve basal bud fertility allowing less expensive spur pruning to be practised (Tassie et al. 1992).

Modifying phenology will be difficult to influence in situ, however there is some scope to delay budburst by delaying pruning to reduce frost risk. For instance, Dunn and Martin (2000) were able to delay budburst by four days by postponing pruning by six weeks. However, labour availability and the requirement to prune large areas may preclude lengthy delays in pruning. Selection of rootstock material can affect timing of budburst (Dry 2007), but this effect is limited.

Because the opportunities to manipulate phenology by viticultural management (e.g. pruning) or chemical means are limited, in cases where demand for varieties has changed, top grafting of different varieties has been practised (Nicholas *et al.* 1992). The rate of climate change will determine the rate of variety change and the lifespan of vineyard blocks may become less than the accepted 30 or so years.

Adaptation to temperature increases through winemaking and winery infrastructure changes

The problem of higher than desirable alcohol in wine is widespread and techniques to manage this are currently under investigation. Alcohol removal through reverse osmosis or spinning cone techniques is costly (De Barros Lopes *et al.* 2003). The use of yeast strains that may produce lower alcohol wines is being studied (Bartowsky *et al.* 2007) but these are unreliable and difficult to control. Options for management in the vineyard to reduce alcohol content while maintaining or enhancing key quality attributes include the adoption of lighter pruning, deficit irrigation, low-to-moderate vigour rootstocks and crop thinning (Clingeleffer 2008). In some countries (but not Australia) it is permitted to add water or low sugar grape juice during the winemaking process (Galpin 2006).

It will be important to utilise forecasting to improve winery scheduling. Winery capacity is built around expected winegrape intake. Winery infrastructure has some flexibility in that more (or temporary) processing vats can be introduced. With more fruit coming into the winery over a shorter time frame, there will be added pressure on crushing/pressing operations. Again, more, or larger capacity, units can be installed. These adaptations are not, however, without major cost penalties, especially as a large proportion of this infrastructure is only used for short periods each year.

Adaptation to carbon dioxide enrichment

The effects of elevated CO_2 will need to be monitored closely with particular attention being paid to yield and its components. Although the relationship between yield and wine quality is complex, major growers and purchasers of winegrapes stipulate that particular yield targets be met, in the belief that this will improve and maintain wine quality and to prevent winery capacity from being exceeded. It will be important to monitor the effects of climate change on yield and, if required, manage any variability, particularly when water availability is uncertain. Coefficients of variation for year-to-year yields typically vary from 30–50% (Dunn *et al.* 2004). Better yield regulation is possible, but this relies on accurate yield forecasts to begin with; a quantitative understanding of yield compensation in response to regulation at different times throughout the season, and accurate crop thinning methods (Dunn *et al.* 2004). Again, flexibility in the winery with regard to processing capacity (tonnage) will be required if the projected increasing yield variability (with increasing temperatures and CO_2 enrichment) is realised (Bindi *et al.* 1996).

Supply of unwanted varieties and substandard quality winegrapes is currently being managed by a prohibitive pricing policy. If the supply became more variable the question is: Who bears the risk of extreme climate-induced reductions in winegrape quality, the winegrape grower or the winery? An essential adaptation to climate change may include establishment of fair policy to distribute the risk appropriately to each partner.

The impact of an increase in within-canopy berry shading associated with enriched CO_2 conditions may necessitate the need for an increased number of passes of vine hedging equipment post-veraison. RDI and PRD are water-saving management practices that could be utilised to regulate vigour in lower rainfall/warmer regions (Goodwin and Jerie 1992) to offset any potential growth enhancement from elevated CO_2. In most environments the aim is to increase water deficits post-flowering and pre-veraison to stop growth. Coordinated adjustment of irrigation scheduling and leaf area may be needed in response to CO_2 changes.

There are many techniques available in the vineyard to manage C:N ratios including fertilisation (Monk *et al.* 1986) and rootstock selection (Treeby *et al.* 1998) and thus mitigate against any deleterious effects during fermentation. Also, the importance of nitrogen on wine-yeast nutrition is well understood (Bell and Henschke 2005) and standard management techniques such as addition of diammonium phosphate (DAP) can address this issue.

Adaptation to rainfall changes

The nature of impacts due to lower rainfall will depend on existing water sources, evaporative demand, soil type and competition for water from other users. These factors can be incorporated into models that evaluate water balance and irrigation needs (CSIRO 2008). If there is a need to allocate water more efficiently, then the seasonal timing of water supply, which has a large bearing on the yield, will need to be carefully monitored in the context of future reduced rainfall. Night-time watering may result in reduced evapotranspiration losses. Applying saline water strategically may assist some growers to cope with severely reduced access to water in drought situations, though effects on soil structure will need to be considered (Clark 2004). Cost and efficiency of distribution, holding dams and water quality maintenance will also need to be reviewed (Van Dijk *et al.* 2006).

Rain, or the threat of rain, may cause growers to pick early (to reduce risk of botrytis and juice dilution) and deliver immature fruit for processing (Jackson and Lombard, 1993). Short-term forecasting of extreme rainfall events close to harvest will become crucial given the projected increase in these events during vintage.

In some regions soil salinity is managed in the root zone by applying a 'leaching fraction' during winter. However, reduced access to reliable water may jeopardise this management option in some very dry periods. Use of salt-tolerant rootstocks (Dry 2007) and better irrigation management may only partly overcome this problem (Stevens *et al.* 1999; Walker 2004).

Waste water from wineries, or large population centres, can be used on vineyards and this option warrants further investigation, including an evaluation of the effects of elevated levels of dissolved mineral salts in the water (Ewert 1993; McCarthy and Downton 1981).

Schultz (2000) explains that in Europe, shifts in precipitation patterns may necessitate

Box 7.1: Example of using recycled water to irrigate grape vines

Due to increasing water use in recent decades, and record low inflows over the past decade, storages in the Murray–Darling Basin are well below long-term averages (Potter *et al.* 2008). Consequently, allocations of water to irrigators have been very low in recent years (MDBC 2008). While water has been available to purchase from other users, viability of some farms has been adversely affected depending on the price of water. The Willunga Basin Pipeline (WBWC) (http://www.wbwc.com.au/) began operating in 1999, and treated wastewater from its plant is used for drip irrigation of vines, fruit trees, nut crops and flowers. The use of this water on grapevines had no negative impact on yield or quality of the resulting wine, and some improvements, compared to vines watered with mains water, were noted (Rawnsley 2007). This scheme provides an alternative water source for vineyard owners in this region where other options for sourcing irrigation water can be problematic.

introduction of cover crops between vine rows over winter in order to minimise soil erosion and to maximise water and nutrient storage. Modelled future rainfall patterns indicate longer dry spells interrupted by heavier rainfall, increasing the need for both water conservation practices and also erosion control measures. Such measures could include the planting of inter-row groundcover using 'drought-tolerant' grass and legume species (Pardini *et al.* 2002).

Whether elevated atmospheric CO_2 concentration reduces vine transpiration depends on the effects on leaf area index (LAI) as well as stomatal conductance (Drake *et al.* 1997), interactions between stomatal conductance and increased VPD and choice of rootstock (Dry 2007). While increases in plant water use may result from possible increases in LAI, the balance of this and possible reduced water use due to decreased stomatal conductance will need to be better understood to determine future water requirements for grapevines.

Pest and disease risk management

Grapevine disease is currently managed by canopy manipulation and with use of chemical pesticides/fungicides. Better targeted application methods for pesticides, increased knowledge of vine and pest dynamics, and technological advances in machinery are continually being developed. Successful adaptation of viticulture pest management to climate change will rely on having quality decision support systems, based on a quantitative understanding of the ecology of each pest (Aurambout *et al.* 2006). The effect of increased temperature and CO_2 enrichment on plant tissues may change disease dynamics by altering the suitability of host plants for pests. Host–pathogen interactions have been found to change in high CO_2 environments (Coakley *et al.* 1999). Adaptation will need to account for this.

Risks of maladaptation

Winegrape varieties that ripen later than 'ideal' under current climate conditions, but with the overall 'best' phenological suitability for future climate conditions may prove advantageous in the long term. However, this may incur opportunity costs earlier on as the variety will not be optimal for current conditions. In some regions early varieties may have some advantages in that the crop is harvested sooner, potentially avoiding late summer heat waves and exhibiting improved water use efficiencies. Hence, some mix of strategies may be needed. A cost–benefit analysis for planting longer season varieties that incorporates ripening dates and projections of regional climate variability (Timbal and McAvaney 2001) could help inform these decisions.

One measure that could be used to reduce temperatures in the vineyard would be to select sites at higher elevations. Many of the more elevated sites may have been used for forestry or remain uncleared. These sites may therefore have higher risks of bushfires and subsequent risk of exposure to smoke.

Costs and benefits

Some projected climate change risks for the industry with regard to wine grape quality have been evaluated (Webb *et al.* 2008a) suggesting potential future suitable winegrape varieties for each growing region and potential future suitable sites for growing winegrape vines may need to be considered. With careful planning, matching the variety to the climate to achieve the best quality wine over the life of a vineyard should be achievable. An alternative to this option is to

bear the cost of replanting vineyards more frequently or top-working with more suitable varieties if trellis and rootstocks are still amenable to this. Evaluation of threshold events to determine the risk of climate extremes is being actively researched with consideration of different macroclimates and mesoclimates (Soar *et al.* 2007).

It will be necessary to determine both future water requirements and also water availability. Climate models, scaled down to a regional level, can be analysed and impact assessments made of the effect of climate change on water budgets in present and potential future vineyard sites. Continued improvement of irrigation technology is essential. The effect of enhanced CO_2 concentrations on water requirements will need to be better understood.

Effects of elevated CO_2 and temperature on the risk of pest and diseases will need to be modelled using regional scale rainfall, temperature and humidity projections.

Discussion

The Australian wine industry is currently engaged in assessing the potential impacts of climate change on a national and regional basis. The industry is also investing in research to adapt to both short-term and long-term climate change and variability. Adaptation options have been described and ranked in order of urgency in Table 7.1.

Potential new regions suitable for development of viticulture in a warmer climate will be evaluated. The existing regions that are most vulnerable to the warming climate and reduced water availability will need to be identified and adaptation options assessed so that future risks to continued viticulture are understood. This will include calculating the risk of increased exposure to hot spells and also reduced exposure to chilling. New management techniques to ameliorate some of the projected temperature extremes are being explored in current research projects.

For all regions the suitability of winegrape varieties should be re-examined in light of projected warming so that ripening can occur when conditions are suited to producing a desired quality of wine grapes. In some cases, it may be possible to

Table 7.1: Adaptation options for the Australian viticulture industry. The most urgent options are given the highest priority ranking (1).

Adaptation options	Priority
Temperature increase	
Assess new and existing sites for future climate-related risks	1
Vineyard design strategies to ameliorate climate impacts	1
Harvest logistics: infrastructure capacity modelling	1
Evaluate varieties to best match changing climate	1
Determine extent of consumer and product flexibility	2
CO_2 enrichment	
Determine the effect of CO_2/temperature on vine–water interactions	1
Determine changes in viticultural management required to deal with possible increased growth	2
Economic and legal adaptations to manage the risk of yield variation	2
Manage vine nutrition to address possible imbalance in C:N ratios	3
Rainfall changes	
Water demand and supply predictions: vineyard and regional scale	1
Irrigation management to increase efficiency	1
Alternative sources of water e.g. water recycling	1
Irrigation and other viticultural management strategies (e.g. rootstock selection) to address salinity	1

continue growing the same grape variety in the same location, even after the climate becomes warmer, by adjusting the style of wine produced, provided consumers accept the change in style. Consumers and marketers can explore these alternatives to their advantage.

The effects of rising CO_2 and its interaction with increasing temperatures on vine water use, growth and yield formation will need to be further investigated since little is known about these interactions for this crop.

The wine industry has already shown some vulnerability to recent water shortages and the issue is now a major concern for growers. Short- and mid-term forecasting is being undertaken but the solutions are not apparent if recent rainfall trends continue (Cai and Cowan 2008a; Murphy and Timbal 2008). Water trading/pricing may benefit winegrowers over some other irrigators but the increasing cost of water will have to be passed on to the consumer eventually. Furthermore, given persistent structural change in the medium to longer term, where water is permanently traded out from lower value ($/ML) sectors there will remain little of the current flexibility of demand for water. Water recycling offers a high security source of water that is currently underutilised in

many regions but will play some part in the future of the industry.

Research aimed at identifying and quantifying the responses of winegrape vines to climate, assessment of regional exposure (both climatically and socially) and planning of future national strategies will all serve to inform the industry of its current exposure and adaptive potential. Frameworks can be established for assessing regional risk and assisting with the decision processes. As thresholds are identified, these can be fed into these risk management frameworks with outcomes dependent upon the backdrop of the current climate, the rate of change projected, and how these affect the intended end product.

Acknowledgements

The authors wish to thank Ian Macadam, Peter Clingleleffer and Rob Walker (CSIRO) for their thorough editorial comments and Paul Petrie (Fosters Group Ltd) for his comments and assistance in providing an example of adaptive measures currently being undertaken by the industry. Also, the authors are grateful to Mark Walpole and Kelly Drysdale (Brown Brothers, Milawa) for comments on an earlier draft.

References

ABARE (2006). 'Australian commodity statistics 2007'. Australian Bureau of Agricultural and Resource Economics, Canberra. http://www.abareconomics.com/publications_html/acs/acs_07/acs_07.pdf

Ainsworth EA, Long SP (2005). What have we learned from 15 years of free-air CO_2 enrichment (FACE)? A meta-analytic review of the responses of photosynthesis, canopy properties and plant production to rising CO_2. *New Phytologist* **165**, 351–371.

Aurambout J, Finlay K, Constable F, Rowles-van Rijswijk B, Luck J (2006). 'A review of the impacts of climate change on plant biosecurity'. Report to the CRC for National Plant Biosecurity. Victorian Government Department of Primary Industries Landscape Systems Science, Victoria.

Australian Bureau of Meteorology (2008). 'Special climate statement 15: an exceptional and prolonged heatwave in Southern Australia.' Issued 20th March 2008 – updated 3rd April 2008. National Climate Centre, Melbourne, Australia. http://www.bom.gov.au/climate/current/statements/scs15b.pdf

Australian Bureau of Statistics (2008). 'Water use on Australian farms, 2005–06.' Report No. 46180DO012, Canberra. http://www.abs.gov.au/AUSSTATS/abs@.nsfDetailsPage/4618.02005-06?OpenDocument

AWBC (2008). '2008 vintage report.' Winemakers Federation of Australia, National Wine Centre, Adelaide, Australia. https://www.awbc.com.au/winefacts/data/free.asp?subcatid=235

Bartowsky EJ, Bellon JR, Borneman AR, Chambers PJ, Cordente AG, Costello P, Curtin C, Forgan A, Henschke PA, Kutyna D, McCarthy J, Macintyre OJ, Schmidt SA, Tran T, Swiegers JH, Ugliano M, Varela C, Willmott R, Pretorius IS (2007). Not all wine yeasts are equal. *Microbiology Australia* **28**, 55–58.

Bell SJ, Henschke P (2005). Implications of nitrogen nutrition for grapes, fermentation and wine. *Australian Journal of Grape and Wine Research* **11**, 242–295.

Berry J, Bjorkman O (1980). Photosynthetic response and adaptation to temperature in higher plants. *Annual Review of Plant Physiology* **31**, 491–543.

Bindi M, Fibbi L, Gozzini B, Orlandini S, Miglietta F (1996). Modelling the impact of future climate scenarios on yield and yield variability on grapevine. *Climate Research* **7**, 213–224.

Bindi M, Fibbi L, Miglietta F (2001). Free Air CO_2 Enrichment (FACE) of grapevine (*Vitis vinifera* L.) II. Growth and quality of grape and wine in response to elevated CO_2 concentrations. *European Journal of Agronomy* **14**, 145–155.

Bjornlund H, Rossini P (2005). Fundamentals determining prices and activities in the market for water allocations. *International Journal of Water Resources Development* **21**, 355–369.

Bramley RGV, Lanyon DM (2002). 'Vineyard leakiness.' CSIRO Land and Water Report to GWRDC, GWR01/04, Waite Campus, Adelaide. http://www.gwrdc.com.au/downloads/ResearchTopics/GWR%2001-04%20Vineyard%20leakiness.pdf

Buttrose MS (1974). Climatic factors and fruitfulness in grapevines. *Horticultural Abstracts* **44**, 319–326.

Cai W, Cowan T (2008a). Dynamics of late autumn rainfall reduction over southeastern Australia. *Geophysical Research Letters* **35**, L09708.

Cai W, Cowan T (2008b). Evidence of impacts from rising temperature on inflows to the Murray-Darling Basin. *Geophysical Research Letters* **35**, L07701.

Carmona MJ, Chaib J, Martinez-Zapater JM and Thomas MR (2008). A molecular genetic perspective of reproductive development in grapevine. *Journal of Experimental Botany* **59**, 2579–2596.

CCW Co-operative Limited (2008). 'CCW Water budgeting tools.' http://www.ccwcoop.com.au/

Chalmers YM, Krstic MP, Downey MO, Dry PR, Loveys BR (2008). Impacts of sustained deficit irrigation on quality attributes and flavonoid composition of Shiraz grapes and wine. *Acta Horticulturae* **792**, 163–169.

Clark LJ (2004). Impact of drip irrigation on the properties of red brown earths following changing management practices in vineyards. PhD thesis, University of Adelaide, Waite Campus.

Clingeleffer PR (1985). Breeding grapevines for hot climates. *The Australian Grapegrower and Winemaker* **256**, 99, 101–2, 104.

Clingeleffer PR (2008). Viticultural practices to moderate wine alcohol content. *Proceedings of the ASVO Seminar Series: Toward best practice through innovation in winery processing*, Tanunda, Barossa Valley, South Australia, October 2007. pp. 37–39.

Coakley SM, Scherm H, Chakraborty S (1999). Climate change and plant disease management. *Annual Review of Phytopathology* **37**, 399–426.

Coombe BG, Iland P (2004). Grape berry development and winegrape quality. In *Viticulture Volume 1 -Resources*. 2nd edn. (Eds PR Dry, BG Coombe) pp. 210–248. Winetitles, Adelaide, South Australia.

Cooley NM, Glenn DM, Clingeleffer PR, Walker RR (2008). The effects of water deficit and particle film technology interactions on Cabernet Sauvignon grape composition. *Acta Horticulturae* **792**, 193–200.

CSIRO (2008). Murray–Darling Basin sustainable yields project, a report to the Australian government from the CSIRO. http://www.csiro.au/partnerships/MDBSY.html

CSIRO and Australian Bureau of Meteorology (2007). 'Climate change in Australia.' CSIRO and Bureau of Meteorology through the Australian climate change science program: Melbourne, Australia. http://www.climatechangeinaustralia.com.au/resources.php

De Barros Lopes M, Eglinton JM, Henschke PA, Høj PB, Pretorius IS (2003). The connection between yeast and alcohol: Managing the double-edged sword of bottled sunshine. *Australian and New Zealand Wine Industry Journal* **18**, 17–22.

Drake BG, Gonzalez-Meler MA, Long SP (1997). More efficient plants: a consequence of rising atmospheric CO_2. *Annual Review of Plant Physiology and Molecular Biology* **48**, 609–639.

Dry N (2007). 'Grapevine rootstocks. Selection and management for South Australian vineyards.' Lythrum Press in association with Phylloxera and Grape Industry Board of South Australia, Adelaide.

Dry PR (1988). Climate change and the Australian grape and wine industry. *The Australian Grapegrower and Winemaker* **300**, 14–15.

Dry PR, Loveys BR (1998). Factors influencing grapevine vigour and the potential for control with partial rootzone drying. *Australian Journal of Grape and Wine Research* **4**, 140–148.

Duchene E, Schneider C (2005). Grapevine and climatic changes: a glance at the situation in Alsace. *Agronomy for Sustainable Development* **24**, 93–99.

Dunn GM, Martin SR (2000). Do temperature conditions at budburst affect flower number in *Vitis vinifera* L. cv. Cabernet Sauvignon? *Australian Journal of Grape and Wine Research* **6**, 116–124.

Dunn GM, Martin SR (2003).The current status of crop forecasting in the Australian wine Industry. *Proceedings of the ASVO Seminar Series: Grapegrowing at the Edge*, Tanunda, Barossa Valley, South Australia, July. pp. 4–8.

Dunn GM, Martin S, Petrie P (2004). Managing yield variation in vineyards. In *Proceedings of the 12th Australian Wine Industry Technical Conference*, 24th – 29th July. Melbourne. (Eds Blair R, Williams P, Pretorius S) pp. 51–56.

Ewert AJW (1993). 'The effect of saline irrigation water on wine quality.' Report to the Grape and Wine Research and Development Corporation. http://www.gwrdc.com.au/downloads/ResearchTopics/UA%207.pdf

Freeman BM, Kliewer WM, Stern P (1982). Research note: influence of windbreaks and climatic region on diurnal fluctuation of leaf water potential, stomatal conductance, and leaf temperature of grapevines. *American Journal for Enology and Viticulture* **33**, 233–236.

Galpin VC (2006). A comparison of legislation about winemaking additives and processes. Assignment submitted in partial requirement for the Cape Wine Master Diploma. http://www.cs.wits.ac.za/~vashti/ps/vgalpin-cwm-print.pdf

Godden P, Gishen M (2005). Trends in the composition of Australian wine. *Australian and New Zealand Wine Industry Journal* **20**, 21–46.

Goodwin I, Jerie P (1992). Regulated deficit irrigation: concept to practice. Advances in vineyard irrigation. *Australian and New Zealand Wine Industry Journal* **7**, 258–261.

Greer DH, Rogiers SY, Steel CC (2006). Susceptibility of Chardonnay grapes to sunburn. *Vitis* **45**, 147–148.

Hanninen H (1991). Does climatic warming increase the risk of frost damage in northern trees? *Plant Cell and Environment* **14**, 449–454.

Hardie W, Considine J (1976). Response of grapes to water-deficit stress in particular stages of development. *American Journal of Enology and Viticulture* **27**, 55–61.

Haselgrove L, Botting D, Van Heeswijck R, Høj PB, Dry PR, Ford C, Iland PG (2000). Canopy microclimate and berry composition: the effect of bunch exposure on the phenolic composition of *Vitis vinifera* L cv. Shiraz grape berries. *Australian Journal of Grape and Wine Research* **6**, 141–149.

Hennessy K, Fawcett R, Kirono D, Mpelasoka F, Jones D, Bathols J, Whetton P, Stafford Smith M, Howden M, Mitchell C, Plummer N (2008). An assessment of the impact of climate change on the nature and frequency of exceptional climatic events. CSIRO and BOM, Melbourne. http://www.daff.gov.au/__data/assets/pdf_file/0007/721285/csiro-bom-report-future-droughts.pdf

Hobbs RJ, McIntyre S (2005). Categorizing Australian landscapes as an aid to assessing the generality of landscape management guidelines. *Global Ecology and Biogeography* **14**, 1–15.

Hood A, Hossain H, Sposito V, Tiller L, Cook S, Jayawardana C, Ryan S, Skelton A, Whetton P, Cechet B, Hennessy K, Page C (2002). 'Options for Victorian agriculture in a "new" climate – a pilot study linking climate change scenario modelling and land suitability modelling'. Volume One: Concepts and Analysis; Volume Two: Modelling Outputs. Department of Natural Resources and Environment: Victoria.

Hook J (2008). Heatwave effects on South Australian vineyards – observations in 2008. *Australian and New Zealand Grapegrower and Winemaker* **533**, 25–26.

Idso SB, Kimball BA, Shaw PE, Widmer W, Vanderslice JT, Higgs DJ, Montanari J, Clark WD (2002). The effect of elevated atmospheric CO_2 on the vitamin C concentration of (sour) orange juice. *Agriculture, Ecosystems and Environment* **90**, 1–7.

Jackson DI, Lombard PB (1993). Environmental and management practices affecting grape composition and wine quality – a review. *American Journal for Enology and Viticulture* **44**, 409–430.

Jackson T, Shaw I, Dyack B (2008). 'Australian wine grape production projections to 2009–10.' Grape and Wine Research and Development Corporation, ABARE research report 08.5, Canberra. http://www.abareconomics.com/publications_html/crops/crops_08/winegrapes. pdf

Johnson H (1989). *The World Atlas of Wine*. Mitchell Beazley Publishers, London.

Jones GV (2007). Climate change and the global wine industry. In *Thirteenth Australian wine industry technical conference*. Adelaide, South Australia. . (Eds R Blair, P Williams, S Pretorius). pp. 91–98. Australian wine industry technical conference Inc.

Jones GV, Davis RE (2000). Climate influences on grapevine phenology, grape composition, and wine production and quality for Bordeaux, France. *American Journal for Enology and Viticulture* **51**, 249–261.

Jones GV, White MA, Cooper OR, Storchmann KH (2005). Climate change and global wine quality. *Climatic Change* **73**, 319–343.

Jones PA (1991). Historical records of cloud cover and climate for Australia. *Australian Meteorological Magazine* **39**, 181–189.

Kennison KR, Wilkinson KL, Williams HG, Smith JH, Gibberd MR (2007). Smoke-derived taint in wine: effect of postharvest smoke exposure of grapes on the chemical composition and sensory characteristics of wine. *Journal of Agriculture and Food Chemistry* **55**, 10897–10901.

Kenny GJ, Harrison PA (1992). The effects of climate variability and change on grape suitability in Europe. *Journal of Wine Research* **3**, 163–183.

Lavee S, May P (1997). Dormancy of grapevine buds – facts and speculation. *Australian Journal of Grape and Wine Research* **3**, 31–46.

Long SP, Ainsworth EA, Rogers A, Ort DR (2004). Rising atmospheric carbon dioxide: Plants FACE the future. *Annual Review of Plant Biology* **55**, 591–628.

Lucas C, Hennessy KJ, Mills GA, Bathols J (2007). 'Bushfire weather in Southeast Australia recent trends and projected climate change impacts.' Consultancy report prepared for the Climate Institute of Australia. Melbourne: Bushfire Cooperative Research Centre, Australian Bureau of Meteorology and CSIRO Marine and Atmospheric Research. http://www.cmar.csiro.au/e-print/open/2007/hennesseykj_c.pdf

Lund ST, Bohlmann J (2006). The molecular basis for wine grape quality – a volatile subject. *Science* **311**, 804–805.

Lyons TJ, Considine JA (2007). Modelling meso-climate in Margaret River. *Australian and New Zealand Grapegrower and Winemaker* **524,** 65–69.

Magarey PA, Wachtel MF, Nicholas PR (1994). Diseases. In *Diseases and Pests. Grape Production Series. Number 1*. (Eds P Nicholas, P Magarey, M Wachtel) pp. 2–44. Winetitles, Adelaide, South Australia.

Marais J (2001). Effect of grape temperature and yeast strain on Sauvignon Blanc wine aroma composition and quality. *South African Journal for Enology and Viticulture* **22**, 47–51.

Marais J, Calitz F, Haasbroek PD (2001). Relationship between microclimatic data, aroma component concentrations and wine quality parameters in the prediction of Sauvignon Blanc wine quality. *South African Journal for Enology and Viticulture* **22**, 22–26.

Maschmedt DJ (2004). Soils and Australian viticulture. In *Viticulture Volume 1 – Resources*. 2nd edn. (Eds PR Dry, BG Coombe) pp. 56–89. Winetitles, Adelaide, South Australia.

May P, Clingeleffer PR, Brien CJ (1976). Sultana (*Vitis vinifera* L.) canes and their exposure to light. *Vitis* **14**, 278–288.

McCarthy MG, Downton WJS (1981). Irrigation of grapevines with sewerage effluent. II Effects on wine composition and quality. *American Journal of Enology and Viticulture* **32**, 197–199.

McCarthy MG, Jones LD, Due G (1992). Irrigation – principles and practices. In *Viticulture. Volume 2: Practices*. (Eds BG Coombe, PR Dry) pp. 104–128. Winetitles, Adelaide, South Australia.

McIntyre GN, Lider LA, Ferrari NL (1982). The chronological classification of grapevine phenology. *American Journal for Enology and Viticulture* **33**, 80–85.

MDBC (2008). 'River Murray system drought update No. 14 July 2008.' Murray Darling Basin Commission. http://www.mdbc.gov.au/__data/page/1366/Drought_Update_Issue_14_-_July_2008.pdf

Monk PR, Hook D, Freeman BM (1986). Amino acid metabolism by yeasts. In *Proceedings of the Sixth Australian Wine Industry Technical Conference*. (Ed. T Lee) 14–17 July, Adelaide, South Australia.

Mori K, Goto-Yamamoto N, Kitayama M, Hashizume K (2007). Loss of anthocyanins in red-wine grape under high temperature. *Journal of Experimental Botany* **58**, 1935–1945.

Morison JL, Lawlor DW (1999). Interactions between increasing CO_2 concentration and temperature on plant growth. *Plant, Cell and Environment* **22**, 659–682.

Murphy BF, Timbal B (2008). A review of recent climate variability and climate change in south-eastern Australia. *International Journal of Climatology* **28**, 859–879.

Nemani RR, White MA, Cayan DR, Jones GV, Running SW, Coughlan JC, Peterson DL (2001). Asymmetric warming over coastal California and its impact on the premium wine industry. *Climate Research* **19**, 25–34.

Nicholas PR, Chapman AP, Cirami RM (1992). Grapevine propagation. In *Viticulture. Volume 2: Practices*. (Eds BG Coombe, PR Dry) pp. 1–22. Winetitles, Adelaide, South Australia.

NLWRA (2001). 'Australian water resources assessment 2000.' National Land and Water Resources Audit, Land and Water Australia. http://www.anra.gov.au/topics/irrigation/production/index.html (Accessed 15 July 2008).

Pardini A, Faiello C, Longhi F, Mancuso S, Snowball R (2002). Cover crop species and their management in vineyards and olive groves: review paper. *Advances in Horticultural Science* **16**, 225–234.

Petrie PR, Sadras VO (2008). Advancement of grapevine maturity in Australia between 1993 and 2006: putative causes, magnitude of trends and viticultural consequences. *Australian Journal of Grape and Wine Research* **14**, 33–45.

Possingham JV (2002). The influence of controlled water inputs on grape quality in regions of Australia with hot Mediterranean climates. *Acta Horticulturae* **582**, 101–107.

Potter NJ, Chiew FHS, Frost AJ, Srikanthan R, McMahon TA, Peel MC, Austin JM (2008). 'Characterisation of recent rainfall and runoff in the Murray-Darling Basin.' A report to the Australian Government from the CSIRO Murray-Darling Basin Sustainable Yields Project.

Water for a Healthy Country Flagship. http://www.csiro.au/resources RecentRainfallAndRunoffMDBSY.html

Rawnsley B (2007). Irrigating with reclaimed water and the effect on soil microbes. In 'Water, Friend or Foe?' Mildura Arts Centre, Mildura, Victoria. (Eds W Cameron, K DeGaris, C Dundon, G McCorkelle, P Pattison) p. 56. Australian Society of Viticulture and Oenology.

Rigby JR, Porporato A (2008). Spring frost risk in a changing climate. *Geophysical Research Letters* **35**, L12703.

Rose L (2007). Creating our future – new varieties and styles. In *Thirteenth Australian Wine Industry Technical Conference*. (Eds R Blair, P Williams, S Pretorius) pp. 135–141. Australian wine industry technical conference Inc., Adelaide.

Schultz HR (2000). Climate change and viticulture: a European perspective on climatology, carbon dioxide and UV-B effects. *Australian Journal of Grape and Wine Research* **6**, 2–12.

Seguin B, de Cortazar IG (2005). Climate warming: consequences for viticulture and the notion of 'terroirs' in Europe. *Acta Horticulturae* **689**, 61–71.

Seguin G (1986). 'Terroirs' and pedology of wine growing. *Cellular and Molecular Life Sciences* **42**, 861–873.

Shulman Y, Nir G, Fanberstein, Lavee S (1983). The effect of cyanamide on the release from dormancy of grapevine buds. *Scientia Horticulturae* **19**, 97–104.

Smart RE, Alcorso C, Hornsby DA (1980). A comparison of winegrape performance at the present limits of Australian viticultural climates – Alice Springs and Hobart. *The Australian Grapegrower and Winemaker* **184**, 28 and 30.

Snyder RL, Paulo de Melo-Abreu J (2005). 'Frost protection: fundamentals, practice, and economics.' FAO Environment and Natural Resources Series 10, Rome. http://www.fao.org/docrep/008/y7223e/y7223e00.htm

Soar C, Collins M, Sadras VO (2007). A comparison of experimental systems for increasing canopy and bunch temperature. In *Thirteenth Australian Wine Industry Technical Conference*. (Eds R Blair, P Williams, S Pretorius) p. 297. Australian wine industry technical conference Inc., Adelaide.

Stevens RM, Harvey G, Partington DL, Coombe BG (1999). Irrigation of grapevines with saline water at different growth stages 1. Effects on soil, vegetative growth, and yield. *Australian Journal of Agricultural Research* **50**, 343–355.

Tassie E, Freeman BM (1992). Pruning. In *Viticulture. Volume 2: Practices.* (Eds BG Coombe, PR Dry). Winetitles, South Australia.

The State of Victoria Department of Sustainability and Environment (2007). 'Your dam your responsibility: a guide to managing the safety of farm dams.' Published by the Victorian Government Department of Sustainability and Environment, Melbourne. http://www.dse.vic.gov.au/CA256F310024B628/0/F629C733D187D6F0CA25735C0027B054/$File/Your+Dam+Your+Responsibility.pdf

Timbal B, McAvaney BJ (2001). An analogue based method to downscale surface air temperature: application for Australia. *Climate Dynamics* **17**, 947–963.

Toni MW (2005). Australian Chardonnay: past, present and future. *Journal of Wine Research* **15**, 135–169.

Treeby MT, Holzapfel BP, Walker RR and Nicholas PR (1998). Profiles of free amino acids in grapes of grafted Chardonnay grapevines. *Australian Journal of Grape and Wine Research* **4**, 121–126.

Van Dijk A, Evans R, Hairsine P, Khan S, Nathan R, Paydar Z, Viney NR, Zhang L (2006). 'Risks to the shared water resources of the Murray-Darling Basin.' Report No. MDBC Publication No. 22/06, Murray–Darling Basin Commission, Canberra. http://www.mdbc.gov.au/__data/page/1131/CSIRO_Part_2_risks_to_shared_water_resources.pdf

Walker R, Read PE, Blackmore DH (2000). Rootstock and salinity effects on rates of berry maturation, ion accumulation and colour development in Shiraz grapes. *Australian Journal of Grape and Wine Research* **6**, 227–239.

Walker R, Blackmore DH, Clingeleffer PR, Correll RL (2004). Rootstock effects on salt tolerance of irrigated field-grown grapevines (*Vitus vinifera* L. cv. Sultana) 2. Ion concentrations in leaves and juice. *Australian Journal of Grape and Wine Research* **10**, 90–99.

Walker R, Blackmore DH, Clingeleffer PR, Godden P, Valente L, Robinson E (2002). The effects of salinity on vines and wines. *Australian Viticulture* **6**, 11–21.

Webb L, Whetton P, Barlow EWR (2007a). Climate change impacts on Australian viticulture. In *Thirteenth Australian Wine Industry Technical Conference Proceedings*. (Eds R Blair, P Williams, S Pretorius) pp. 99–105. Australian wine industry technical conference Inc., Adelaide.

Webb L, Whetton P, Barlow EWR (2007b). Modelled impact of future climate change on phenology of wine grapes in Australia. *Australian Journal of Grape and Wine Research* **13**, 165–175.

Webb L, Whetton P, Barlow EWR (2008a). Climate change and wine grape quality in Australia. *Climate Research* **36**, 99–111.

Webb L, Whetton P, Barlow EWR (2008b). Modeling the relationship between climate, winegrape price, and winegrape quality in Australia. *Climate Research* **36**, 89–98.

White MA, Diffenbaugh NS, Jones GV, Pal JS, Giorgi F (2006). Extreme heat reduces and shifts United States premium wine production in the 21st Century. *Proceedings of the National Academy of Sciences* **103**, 11217–11222.

Whiting JR, Krstic M (2007). 'Understanding the sensitivity to timing and management options to mitigate the negative impacts of bush fire smoke on grape and wine quality – scoping study.' Department of Primary Industries, Primary Industries research, Knoxfield, Victoria. http://www.gwrdc.com.au/downloads/ResearchTopics/GWR%2006-03%20final%20report.pdf

Wine Australia (2008). 'Wine export approval report – May 2008.' Australian Wine and Brandy Corporation, Adelaide, South Australia. http://www.wineaustralia.com/australia/News/Reports/WineExportApprovalReports/tabid/204/Default.aspx

8
HORTICULTURE

L Webb and PH Whetton

KEY MESSAGES:

- Site suitability may change for some horticultural crops as a result of climate change. In particular, there may be a reduction in areas suitable for growing stone- and pome-fruit varieties that require chilling, and an expansion in areas suitable for growing subtropical crops.

- Changes in rainfall and evaporation are likely to reduce soil moisture and runoff. Increased crop water demand combined with reduced water supply poses significant challenges. Efficient water use will become paramount. Lower winter and spring rainfall may indirectly cause increased frost risk.

- Increased temperatures will advance phenology (timing of crop developmental stages) with likely effects on flowering, pollination and harvest dates. Warmer temperatures may also increase sunburn incidence and reduce colour development. Also of concern for vegetable growers is a potential increase in premature flowering (bolting).

- For some perennial horticultural crops increased night-time temperatures will lead to increased respiration and thus affect the distribution of assimilates to reproductive sinks. It may be more difficult to obtain desired fruit size classes.

- Varietal selection can be used to match crops to new climate regimes. Utilising existing varieties or breeding new varieties can facilitate adaptation. For example, drought-tolerant plants for amenity horticulture (parks and gardens) will be favoured in a drier climate.

- The net effect of increasing atmospheric carbon dioxide concentrations is crop-specific. Elevated concentrations can enhance photosynthesis and water use efficiency in some plants. There are likely to be changes in the nitrogen status of horticultural crops as CO_2 levels increase, affecting crop management and quality.

- Decreasing rainfall and humidity may reduce fungal pressure in some regions. In other regions increased summer rain may favour fungal growth as will increases in extreme rainfall. Cold-season suppression of some pest species may be reduced. Efficacy of parasites and beneficial organisms may change in a future climate.

- Consumers may require assistance in accepting some changes in the availability, cost and quality of produce. For example, the cost of some produce tends to rise during droughts, which are likely to occur more often.

Introduction

Horticulture in Australia is botanically diverse and includes fruit (annual and perennials), perennial berry and tree crops, vegetables, tree nut crops, nursery, extractive crops, cut flowers and turf. In contrast to many other agricultural industries, high value products are grown on relatively small areas of land (Australian Bureau of Statistics 2005, 2007). Horticulture consequently has a

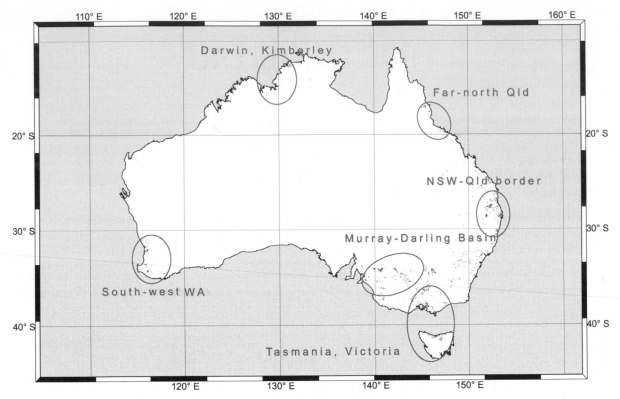

Figure 8.1: Major horticultural production regions in Australia 2001–02. ANRDL 2007

high level of management input, often aimed at ameliorating climate risks (e.g. via irrigation) or maximising 'quality'. Despite this intense management, many activities retain considerable exposure to various climate-related risks.

In 2005, the gross value of horticultural production in Australia was $2.55 billion for fruits and nuts (excluding grapes), $2.13 billion for vegetables and $768 million for nursery produce (Australian Bureau of Statistics 2005). Together these accounted for about 15% of the total gross value of Australia's agricultural production. Key fruit species (in order of production value in 2005) include grapes (winegrapes are discussed

in Chapter 7), apples, oranges, bananas, strawberries, mangoes, macadamias, melons, mandarins, peaches, pears, nectarines, avocados and cherries. Key vegetables (in order of production value in 2005) include potatoes, mushrooms, onions, tomatoes, carrots, lettuce, capsicum, broccoli, beans (French and runner), sweet corn and asparagus (Australian Bureau of Statistics 2005). Other significant contributors to the horticultural sector include the nursery, floriculture and turf industries (Australian Bureau of Statistics 2007).

Horticultural crops are grown in three distinct latitudinal zones (1) temperate and Mediterranean

Table 8.1: Groupings of major horticultural activities in Australia.

Horticulture grouping	Major vegetable crops grown	Other major horticultural crops grown
Temperate and Mediterranean	Potatoes, onions, brassica, lettuce, tomatoes, carrots and pumpkins	Stone fruit, pome fruit (apples, pears), berries, citrus (oranges), table grapes, and nuts (almonds)
Subtropical	Tomatoes, potatoes, capsicum, brassica, beans, lettuce, pumpkin	Bananas, pineapples, avocados, citrus (mandarins), nuts, avocados, some stone fruit and strawberries in the elevated areas
Tropical	Cucumbers, melons, chillies, capsicum and tomatoes	Mangoes, pineapples, bananas, avocados, coffee

(30°S–40°S), (2) subtropical (20°S–30°S), and (3) tropical (10°S–20°S) (Figure 8.1 and Table 8.1). These align with the five separate agro-climatic zones described by Hobbs and McIntyre (2005) that are mentioned in the summary chapter (Figure 16.1).

Impacts of projected climate change

Impact of temperature on plant growth, yield and crop quality

Higher temperatures tend to shorten the period of growth of individual crops. The opportunity to plant earlier in the season, or harvest later, will effectively extend the potential length of the growing season for annual crops such as lettuce (Pearson et al. 1997), French bean (Wurr et al. 2000), and tomato (Maltby 1995). With shorter phenological cycles, planting two crops per season rather than one may become possible.

In the case of perennial crops, timing of phenological stages will change with warmer temperatures being linked to accelerated phenological development in most species studied (Estrella et al. 2007). Apple trees, for example, flower and reach maturity earlier if grown at higher temperatures, with possible increased fruit size (Austin et al. 2000) if access to water during fruit expansion is not limiting. Some perennial tree crops require matching of cross-pollinating varieties for fertilisation (e.g. pears; Baxter 1997). If climate change has different effects on the timing of flowering of the different cultivars this may lead to problems with fertilisation and hence fruit production. As well as tree cultivars losing synchronicity, pollinators themselves may become climatically isolated from their food source (e.g. butterflies and Prunus spp.; Doi et al. 2008).

Most deciduous fruit and nut trees need sufficient accumulated chilling, or vernalisation, to break winter dormancy. Inadequate chilling due to higher temperatures results in prolonged or uneven dormancy break (Lavee and May 1997), leading to reduced fruit quality and yield. A warmer climate will reduce the suitability of certain regions for growing crops with these requirements (Carbone and Schwartz 1993; Hennessy and Clayton-Greene 1995).

High spring temperatures can cause reductions to peach yields (Lopez et al. 2007). Models like 'Peach' can be used to investigate yield sensitivity to temperature (DeJong et al. 1996). Model simulations of crop yields of almonds, table grapes, oranges, walnuts and avocados suggest reductions in future yields due to climate change (Lobell et al. 2006). Berry collapse in table grapes (Tilbrook and Tyerman 2008) has been associated with high temperature post-berry set, where giberrellic acid was applied, in three out of the last 10 seasons.

For citrus fruit, winter temperatures between 0 and 14°C are required during the 'resting period' for optimum production, so some warmer sites may lose suitability. Excessively warm temperatures during the bloom or early fruit set period are known to induce abscission (premature fruit fall) in citrus fruit (Rosenzweig et al. 1996) and adversely affect pollination of avocado (Schaffer et al. 2002) and some tomato varieties (Sato et al. 2000). Temperatures above 30°C in capsicum and chillies can result in flower buds falling off and yield being affected (Murison 1995).

Higher temperature can inhibit the formation of anthocyanin, the pigment causing colouration of apples (Ewa Ubia et al. 2006) and increase sunburn damage (Piskolczi et al. 2004). Citrus fruit quality, with respect to both development of sugars and colour, is also influenced by warmer temperatures, with a decrease in tree storage time and rind re-greening increasing as temperatures rise (Rosenzweig et al. 1996). The red colour of ripening capsicums develops between 18°C and 25°C but if temperatures rise above 27°C during the ripening a yellowish colour results (Murison 1995). Yellowing can also occur in tomatoes experiencing high temperatures when ripening (Maltby 1995).

Warmer-than-average growing seasons can result in 'less than desirable' ripening of some fruit. Reduced sugar content in fruit such as pea, strawberry and melon produced under warmer nights is often attributed to increased night-time respiration. However, this effect may also be caused by high temperatures reducing the period over which fruit develops (Wien 1997). Elevated temperatures have been demonstrated to reduce vitamin C levels in fruit production in woody perennials, the effect being influenced by the timing of heat exposure (Richardson et al. 2004). An increase in maximum temperatures may

Table 8.2: Examples of some temperature threshold effects that may occur more frequently in a warmer climate.

Crop affected	Temperature threshold exceedance effects
Lettuce	Lettuce tipburn, a browning of the inner leaf margins, occurs due to inadequate calcium distribution combined with temperatures in excess of 30°C (Dioguardi 1995). Lettuce quality and yield is adversely affected if maximum temperature exceeds 24°C (Titley 2000)
Onion	Cannot 'cure' (dry for the purposes of storage) onions in the field if temperatures exceed 30°C (Salvestrin 1995)
Tomato	Poor pollen germination with temperatures above 27°C (Maltby 1995). Heat stress is displayed in tomato plants grown at 35°C (Riveroa et al. 2003)
Strawberry	Strawberry plants cease to fruit and commence runner production in late spring and summer as day length increases and temperatures exceed 28–30°C on a regular basis (Morrison 1995)
Apples, table grapes	Sunburn of some vegetable and fruit crops will result from extremely hot days (Piskolczi et al. 2004; Wand 2007)

negatively impact yield and quality of leafy crops such as lettuce and spinach (Titley 2000). Bolting (premature formation of the seed head/flowers) is an example of a negative impact resulting from growing these crops in high temperature regimes (Dioguardi 1995).

If mangoes are grown in the subtropics of Australia, where winter temperatures regularly fall below 10°C, leaf yellowing occurs due to photoinhibition of the photosynthetic apparatus (Sukhvibul et al. 2000). This low temperature photoinhibition has also been reported for lychee and rambutan (Diczbalis and Menzel 1998) and banana (Damasco et al. 1997). In a warmer climate, cold-induced photoinhibition will be reduced and greater agronomic potential may exist in some areas currently considered to be marginal.

After harvest, many goods, including beans, melons, cherries and strawberries, require cooling so as to remove field heat quickly (Coombs 1995; Alique et al. 2006). With projected temperature rises, the costs and benefits of shifting harvest time to a cooler part of the day, or increasing refrigeration, will need to be assessed. Higher growing temperatures may reduce the shelf-life of lettuce (Rogers, 2007). For production of dried fruit, such as Thompson Seedless (sultana) and raisins, there could be significant benefits to solar drying as temperatures rise and humidity and rainfall decrease in some of the regions of Australia with Mediterranean climates (Possingham 2008).

While the different species of fruit and vegetables are suited to different climate regimes, it is where temperatures exceed a threshold that many negative impacts will be experienced, and this will vary for each crop. Examples of some temperature threshold exceedance impacts are listed in Table 8.2.

Impact of water balance changes

For many of the horticultural production regions of Australia, rainfall is projected to decline. Lower rainfall combined with increased evaporation under climate change will result in decreased inflow into catchments (Potter et al. 2008) and this means less water will be available for irrigation. At the same time water demand for crops is likely to increase in non-water stressed plants (to an upper limit of about 35°C) (Berry and Bjorkman 1980). Overall vulnerability to water shortages is greater for perennial crops than for annual crops since the perennial crops need to be watered every year to survive. On the other hand, a more opportunistic approach can be taken with annual crops that need not be grown in years of extremely low water availability.

Extremes: extreme rainfall, hail, frost, drought, cyclones

Consequences of increased intensity of extreme rainfall events include flood damage, erosion damage and increased disease pressure after such events. Splitting of berries and berry drop due to abscission is a major problem in Sultana grown for wine, dried fruit or table grapes after extreme rainfall events (Tilbrook and Tyerman 2008; Possingham 2008). A doubling of the risk of large hail frequency to between four and six days per year is projected for the end of the century in eastern Victoria and along the coast of NSW (CSIRO and Australian Bureau of Meteorology 2007). Hail damage can result in serious financial losses in perennial tree crops such as apples and cherries (Coombs 1995).

Frost, a major threat to horticultural production, is expected to decline in the future, so for annual crops, earlier planting (e.g. broccoli; Deuter 1995), or expansion of the region suitable for planting (e.g. pineapple; Scott 1995), may become viable. If perennial crops experience earlier budburst with the warmer climate, the frost risk may stay the same or increase (Hanninen 1991). Solanaceous vegetable crops (e.g. tomatoes, capsicum), currently seeded in heated glasshouses and transplanted into the field when danger of frost is past (Peet and Wolfe 2000), may be able to be direct seeded if frost risk reduces, and soil temperatures increase.

It is worth noting that reduction in rainfall may counteract some of the warming trend with regard to risk of frost in the future. Drier soil (Trought *et al.* 1999) and less cloud cover (Jones *et al.* 1991) are conducive to the occurrence of frost. In 2006 (a drought year) frosts decimated orchards in the Goulburn Valley even though the mean temperature during that spring was one of the highest on record (Australian Bureau of Meteorology 2008). In this case, increases in daily temperature variance (which increase frost risk) counteracted the increases in the mean temperature (which decreases frost risk) (Rigby and Porporato 2008).

Drought intensity and duration is projected to increase in the future (Chapter 2). Some examples of the major impacts of the extended 2001–08 drought (Australian Bureau of Meteorology 2008) on horticulture in south-eastern Australia were the threatened survival of perennial vine and tree crops in the Murray–Darling Basin (MDBC 2007) and reduced access to sporting fields as turf survival was impacted (Sport and Recreation Victoria 2007).

Studies in the Australian region indicate a likely increase in the proportion of tropical cyclones in the more intense categories but a possible decrease in the overall frequency of cyclones (CSIRO and Australian Bureau of Meteorology 2007). The magnitude of the impact cyclones can have is demonstrated by the devastating $300 million crop loss to the banana industry caused by Severe Tropical Cyclone Larry in 2006 (Australian Bureau of Meteorology 2006), with a further $150 million being required for replanting or renovating after the cyclone (ABGC 2006).

Pests, diseases and weeds

Elevated temperatures may result in increased pest and disease incidence and severity due to increased populations of pests and pathogens, caused by shorter life cycles and decreases in generation times (Aurambout *et al.* 2006), while other factors such as reductions in rainfall and humidity may reduce disease pressure (Coakley *et al.* 1999). Increased survival of pests that normally do not withstand cold winters, e.g. the cabbage moth (*Phitella xylostella*) (Deuter 1989), is of some concern in a warmer climate (Coakley *et al.* 1999). Disease threat may also increase, as in the case of citrus canker, where warmer winter temperatures would support a continuous life cycle in subtropical and temperate regions, and potentially favour establishment in regions outside its historic range (Aurambout *et al.* 2006). A 1.0°C rise in temperature is expected to increase the cost of controlling the Queensland Fruit Fly, estimated to be $28.5 million per year ($25.7–49.9 million), by 38%, without accounting for any increased risk to current exclusion zones (Sutherst *et al.* 2000).

Severe storms were reported as precursors to the two incursions of citrus canker in Emerald, Queensland, in 2004–05 (Aurambout *et al.* 2006) and these types of climatic events are projected to increase. An increase in the frequency of extreme rainfall events may also lead to conditions favouring root-invading fungus like *Phytophthora cinnamomi*, which affects avocado (Howden *et al.* 2005).

Temperature changes, altered wind patterns, or flooding, may allow tropical and frost-tolerant weed species to move further south. Competitiveness both between different weed species and between weeds and the crop may be altered (McFadyen 2008). Prolonged survival of weed hosts for crop pests through winter increases the ability of pests, e.g. the Silverleaf Whitefly, to overwinter (known as a 'green bridge') (Aurambout *et al.* 2006). Many agricultural pesticides and herbicides cannot be used in very hot conditions; there may be a reduction in the number of days or hours when these chemicals can be applied (McFadyen 2008).

Carbon dioxide enrichment effects

Recent studies confirm that the effects of elevated carbon dioxide (CO_2) on plant growth and yield will depend on photosynthetic pathway, species, growth stage and management regime and its implications for water and nitrogen applications (Easterling *et al.* 2007; and discussed in Chapter 1 in this volume). Horticultural crop responses vary. Wurr *et al.* (2000) found a null response of French bean to CO_2 enrichment in contrast to positive effects on onion (Daymond *et al.* 1997), beetroot and carrots (Wurr *et al.* 1998), avocado (Schaffer *et al.* 2002), citrus (Rosenzweig *et al.* 1996) and banana (Schaffer *et al.* 1996). Tree crops are shown to respond more than herbaceous crops in a CO_2-enriched environment (Ainsworth and Long 2005).

As well as assessing the implications of these crop-specific CO_2 enrichment effects, consideration of the likely related temperature and water supply changes will be imperative. These changes will modify, and often limit, direct CO_2 effects on plants (Easterling *et al.* 2007). For instance, with lettuce, increasing CO_2 should increase yield, but this will be partially offset by warmer temperatures (Pearson *et al.* 1997). Positive effects of CO_2 on potato crop growth counteracted the negative effects of warmer temperatures depending on the initial temperature regime (Miglietta *et al.* 2000). In contrast, Rosenzweig *et al.* (1996) found minimal compensating effect of CO_2 on potato yields. Wurr *et al.* (1998) found carrots had a temperature optimum of about 15.8°C for maximum responsiveness to CO_2 enrichment.

Plant responses to elevated atmospheric CO_2 include increases in the carbon to nitrogen ratios (C:N) in plant tissue (Drake *et al.* 1997; Morison and Lawlor 1999), which may reduce the nutritional value of some vegetables to both human consumers and pests. In stating this however, a recent review of many free air CO_2 enrichment (FACE) studies has reported this effect to be small (Ainsworth and Long 2005).

High CO_2 concentrations change host morphology and phenology, pest fecundity and life cycle, and plant disease epidemiology (Chakroborty 2004). These effects are often associated with altered plant physiology, canopy size and composition (including increased C:N), but may also be associated with changes in the geographical range of native and cultivated host plant species (Aurambout *et al.* 2006).

Ziska and Teasdale (2000) have shown that sustained stimulation of photosynthesis and growth of perennial weeds could occur as atmospheric CO_2 concentrations increase. This may result in a reduction in the effectiveness of some chemical controls and increases in weed/crop competition.

Increasing concentrations of CO_2 may also improve water use efficiency in some instances (Grant *et al.* 2004; Chapter 1 in this volume).

Options for dealing with climate variability

Site selection

Site selection to avoid unsuitable climate factors is crucial in horticulture. Temperature is the dominant factor that determines where and when horticultural crops are grown and has a significant influence on crop performance (i.e. on time to harvest, product quality and, to a lesser extent, yield). For example, warmer winters in the period 1994–2000 led to a halving of the area planted with kiwifruit in New Zealand due to chilling requirements not being met adequately (Salinger and Kenny 1995).

Low rainfall during harvest (and, to some extent, during other development phases of many horticultural crops) is often advantageous for managing disease pressure and product quality so sites are also selected on this basis. Floods and frosts are avoided by selecting less risky sites or ensuring adequate water or cold air drainage is provided.

Crop management

Variability in the ripening of fruit and vegetables due to seasonal and annual temperature variability has long been monitored and, where possible, managed as a matter of course. The timing of management activities, such as planting, fertilising and irrigation is adjusted in response to climate variability (Krug 1997).

Reduced fruit sunburn, improved colour and skin finish, and elimination of bird damage are all benefits of hail netting, found to be particularly suited to high-yielding, intensive apple orchard systems in Australia (Middleton and McWaters 2002). Plastic shelters are used to minimise damage from untimely or extreme rainfall events reducing berry splitting and berry drop in table grape production (Possingham 2008). Commercial kaolin-based coatings (also used as a pest repellent) can reduce sunburn (Thomas *et al.* 2004), and misting can also be used to lower

Box 8.1: Example of developing stone fruit varieties for warmer climates

Stone fruit varieties are defined according to their chill hours, that is, by the number of hours below 12°C that they need to accumulate in winter to trigger flowering (Saure 1985). Low-chill varieties need as little as 100 to 200 chill hours a year while high-chill types may need up to 1200 hours. Crop losses can occur due to poor fruit set if inadequate chill hours are obtained so inter-varietal variation in chilling requirement is utilised when selecting stone and pome fruit for particular sites (Carbone and Schwartz 1993).

Low chill stone fruit, peaches, nectarines and plums, are being bred for subtropical areas where very low winter chilling is experienced (Topp and Sherman 2000). A variety evaluation trial established in 1999 in North NSW has assessed quality and chill requirements of over 600 varieties (Wilk 2005). Over 18 000 hybrid seedlings have been created and these have provided the segregating populations from which to select superior genotypes (Topp *et al.* 2008).

These crops have been developed to take advantage of a marketing niche. Becoming available in September, they are the first stone fruit sold on Australian markets enabling growers in subtropical regions to receive premium prices for their fruit.

Though the motivation for exploiting the varietal differentiation was not the warming climate, this example serves to illustrate a mode of autonomous adaptation already underway in this industry.

the seedling is big enough to support formation of a large flower head (Coombs 1995). Crops with extended potential flowering periods like peas and pumpkin are less sensitive to periods of heat stress compared with those that have more tightly determined flowering times, e.g. cauliflower and broccoli (Peet and Wolfe 2000). Varietal selection is practised to match harvest timing and day length requirements to particular sites, e.g. strawberries (Morrison 1995) and selecting varieties of table grapes with later budburst, e.g. Merbein Seedless, can reduce exposure to frost risk (Uhlig and Clingeleffer 1998). Breeding varieties more adapted to high temperatures can reduce quality concerns with regard to lettuce (Dioguardi 1995; Rogers 2007) and yield concerns with peas (Olesen and Grevsen 1993) and potatoes (Manrique and Bartholomew 1991).

Water management

In Australia, most horticultural production occurs where the water requirement of crops is far higher than that provided by effective rainfall. Irrigation has been widely adopted to maximise yield and quality, giving more certainty about harvest dates and also facilitating marketing plans of growers. Many vegetable growers use drip irrigation technologies, which provide a significant improvement to water use efficiencies when compared to furrow or sprinkler irrigation systems (Fereres *et al.* 2003). The best irrigation practice to implement, however, can vary with soil type and crop morphology (Al-Jamal *et al.* 2001).

Diverting irrigation water away from broadacre activities has increased overall water security for the horticulture sector in periods of drought where access to water has been limited (Bjornlund and Rossini 2005). This is due to the value of produce per volume of water used ($ per ML) being at least three times greater for horticultural crops than broadacre uses like pasture production (NLWRA 2001).

Amenity horticulture and parks and gardens management will be required to adapt to drier conditions. In many areas in Australia (e.g. in Perth, WA) watering of gardens is restricted to twice per week, once only on each allocated day (Waterwise Ways for WA 2007). The current drought in south-eastern Australia has resulted in reduced access to sports grounds as turf cover declines. Funding for water recycling/saving

temperature and improve colour in apple (Thomaia *et al.* 1998).

Application of hydrogen cyanamide, as a way to promote budburst and adapt to inadequate chilling, is becoming more common in perennial fruit growing operations, e.g. table grapes (George and Nissen 1990). Evaporative cooling by water sprinkling (Nir *et al.* 1988) and timing of winter pruning (Hatch and Ruiz 1987) can also be used.

Varietal selection

Cultivar selection and planting dates can be changed to either suppress flower initiation, in the case of celery, onion or cabbage, or delay flowering, in the case of broccoli and cauliflower, until initiatives, drought-resistant turf installation, more efficient irrigation systems, and improved water harvesting and storage has been provided in many regions to address this problem (Sport and Recreation Victoria 2007).

Pests and diseases

Management of horticultural pests using pesticides, cultural practices and biological control methods in relation to current climatic variability depends on effective monitoring and predictive systems (e.g. Hetherington 2005), varies with crop type, and is undertaken as a matter of course by horticulturists. Predictive models can be employed to assist in control of insects and plant diseases (Deuter 2006).

Market behaviour

Availability of most horticultural crops (e.g. strawberries) tends to vary throughout the year and may be reduced in some seasons due to production difficulties caused by effects of higher temperatures, droughts or other climatic events. In these cases consumers can pay higher prices to Australian growers or for an imported product (if available). Banana prices increased dramatically after Severe Tropical Cyclone Larry devastated the crop in 2006 (ABGC 2006). Ongoing marketing programs are increasing consumer awareness of new and varied alternative produce.

Seasonal forecasts

Seasonal forecasts can be accessed to reduce exposure to unsuitable weather, and while such linkages have not been developed broadly for the fruit and vegetable growing industries (Deuter 2006), peanut processing and marketing bodies profitably use forecasts of likely production to adjust their operations strategically (Meinke and Hammer 1997). Incorporating crop-specific information can make existing climate forecasting tools more horticulturally relevant (Deuter 2006).

Options for dealing with projected climate change

Site selection

In the future, the practice of carefully selecting appropriate sites for different annual crops will be maintained. However, when selecting sites for perennial varieties it will become increasingly important and urgent to consider the changing climate. Mapping of threshold temperatures and other relevant aspects of the climate for a range of future climate regimes can expose risky, or less risky, areas in which to plant crops (e.g. Hood et al. 2002). The amount of damage suffered due to exposure to climatic extremes often depends on the stage of development when the extreme conditions occur and thus phenological information needs to be incorporated into these analyses.

Under climate change, the area suitable for growing tropical and subtropical crops is projected to expand, such as for citrus, avocados and bananas in the Mediterranean region of Europe (Houerou et al. 1992) and for avocados and pecan nuts in Southern Africa (Schulze and Kunz 1995). A similar trend is likely for Australia, though this has yet to be investigated rigorously.

Crop management

For annual crops, the timing of planting may be changed. In some areas, autumn soil temperatures could become too high for good germination of celery (Peet and Wolfe 2000). For potatoes, planting later in the season to avoid the very hot temperatures may be compromised by the shorter day lengths later in the year, which have a negative impact on yield (Rosenzweig et al. 1996). Thus, any planned change in sowing needs to be accompanied by considerations of day length requirements. In perennial crops, adaptation will be required to manage the variability and protracted full bloom of pome- and stone-fruit and nut trees if bud dormancy is affected in a way that leads to non-uniform budburst (Saure 1985). For example, pesticides with longer residual action may be required to protect the crop.

Because climate change is expected to reduce the duration from sowing to harvest for some crops there can be consequent reductions in yield (Olesen et al. 1993). These yield decreases may be compensated by sowing annual crops earlier. Where crops mature more rapidly, it may be necessary to plant smaller areas of crop more frequently in an attempt to reduce market supply fluctuations, e.g. for cauliflower (Olesen and Grevsen 1993).

Canopy management of some crops in an enriched CO_2 environment may need to be addressed. If yields rise with increasing CO_2 concentrations there is scope for improvements in the efficiency of water use (Grant et al. 2004). In the case of perennial fruit crops for instance, consideration of canopy structure could be scored for useful traits such as natural self-shading ability, and these could be included for consideration in breeding programs aimed at avoiding sunburn.

Varietal selection

Breeding varieties of various horticultural crops that may be more suited to new climatic regimes will assist in avoiding many of the impacts mentioned above. Product quality under enhanced CO_2 and elevated temperatures will need to be evaluated and negative impacts compensated for in breeding programs. In some cases, long season varieties may be more suited to a future climate than short season varieties (e.g. onion) (Daymond et al. 1997) as the hastened progression through phenological stages can be somewhat negated. One advantage of this is that ripening in a later and possibly cooler part of the season could reduce the risk of potential curing problems (Table 8.2) (Salvestrin 1995).

Water

Increasing irrigation efficiency by adopting the most suitable water saving techniques (Fereres et al. 2003) and benchmarking against other producers and sectors (e.g. Giddings et al. 2002), will assist growers to cope with the potentially limiting supply of water. Partial rootzone drying (Zegbe–Domínguez et al. 2003) is one system that could be explored more rigorously in some horticultural cropping regimes. Mulching the soil surface with organic matter for example will reduce the loss of water through evaporation (Tindall et al. 1991). Kaolin clay applied to plants, usually for pest and disease control, has been demonstrated to reduce the effects of water and heat stress (Thomas et al. 2004) though results have been variable so further studies are needed (Glenn et al. 2001).

Initiatives undertaken to address the reliability of the supply of water from public and private irrigation schemes are described by the Victorian Department of Sustainability and Environment (2008). Increased use of reclaimed wastewater for agricultural/horticultural enterprises needs to be considered. A summary of the major impediments to the use of reclaimed water by the Australian horticultural industry has been compiled (Hamilton et al. 2005). Insufficient knowledge of impacts on market acceptance, insufficient knowledge of food safety issues, inadequate understanding of consumer perceptions and uncertainty about pricing of reclaimed water were raised as some of the potential issues that need to be resolved.

Adaptation to increases in atmospheric carbon dioxide concentrations

Investment into research to identify plant characteristics that will be responsive or non-responsive to projected increases in atmospheric CO_2 concentrations will be beneficial. These could be incorporated into breeding programs so that genetic variation in these characteristics is maximised and maintained and a suite of genetic responses can be sourced in the future (Richards 2002).

Research into the effect of increasing atmospheric CO_2 on the nutritional quality of horticultural crops can inform crop management issues. Additional fertiliser applications may be required to maintain product quality as atmospheric CO_2 concentrations increase (Monk et al. 1986). These may, however, increase greenhouse gas emissions.

Certain weeds are likely to benefit from higher levels of carbon dioxide, thus necessitating increased application of herbicides. This suggests it would be worth examining how rising atmospheric CO_2 influences the tolerance of weeds to various methods of herbicide application. The possible increasing costs of fertilisers, pesticides and herbicides will need to be considered.

Pest and disease management and risk

Under climate change, disease assessment models should account for projected changes to overwintering of a wide range of insect pests and plant diseases and changes in the timing and severity of pest outbreaks. To date, the most common bioclimatic models used to predict potential range shift of species under climate change are CLIMEX and BIOCLIM (Aurambout et al. 2006). Geographically sensitive models that use projected climate information focussed on the use of bio-pesticides and natural enemies can be constructed to improve pest control. CO_2 effects on disease and hosts have to be considered when addressing future disease risks.

Impact on consumers and marketers

Traditional vegetables such as turnip or swede may suffer from consumer neglect as more exotic vegetables or fruits become available and/or affordable. Prices paid for produce may change. Whether prices increase due to the impact of reduced water supply or extreme events, or decrease due to enhanced growth and supply under climate change remains to be seen. Colour or nutrition levels of produce could be impacted by climate change, with some of the possible aesthetic impacts identified being re-greening of oranges, and yellowing of tomatoes and capsicums.

Risks of maladaptation

There is a risk of maladaptation if changes to production are implemented without considering the effect on marketing. Fruit and vegetable production and marketing are very closely linked. A mismatch resulting in oversupply or undersupply can have a dramatic effect on prices, with consequent effects on both producer and consumer. Therefore any adaptation strategies that result in changes to production timing and location will have consequences for marketing of produce (e.g. Howden *et al.* 2005).

Selection of peach varieties with medium chilling requirements (e.g. T204) has occurred in the Goulburn Valley where growers have been concerned that high chill varieties will not set fruit in seasons following a mild winter. Medium chill varieties can, however, flower earlier and hence have a higher risk of frost damage (Carbone and Schwartz 1993).

Drought-relief packages that act to subsidise the growing of crops in climates not suited to them could be considered a maladaptation cost. As an example, the 1991–95 drought was estimated to cost the Australian economy $5 billion, and the Commonwealth Government provided $590 million in drought relief between September 1992 and December 1995 (Australian Bureau of Meteorology 2007).

Costs and benefits

Fruit orchards have a life of 20 to over 60 years, so current plantings will be directly affected by the warming climate. It is prudent to consider the impact of warmer temperatures on phenology when replacing older trees in established orchards, or as new regions and sites are identified, so that a better matching of varieties with future climates is made. This strategy is also pertinent with respect to future matching of chilling requirements of deciduous species.

There is the potential that perennial fruit and vegetable growers will not be able to change their land use easily in response to increasing temperatures and other associated effects. This may lead to lower returns and higher costs as growers 'hang on' and try to produce crops in environments that have become suboptimal.

Validation of climate indices that determine locations of current horticultural enterprises can be modelled relatively quickly so that prediction of potentially suitable crops for each region, or potentially suitable sites for particular crops, can be assessed. Risks associated with climate extremes can be assessed at the same time as suitable macroclimates and mesoclimates are being evaluated. Substantial benefits can be achieved if damaging temperature threshold events can be avoided in future.

Similarly, calculating the risks associated with possible reduced water availability may alleviate any adverse impacts with regard to allocating water to various agricultural sectors in the future. It will be possible to assess the water requirements for horticultural crops more accurately if the effect of rising CO_2 on crop water use is better understood. It will remain necessary to continue improvements in irrigation technology.

Discussion

The Australian horticulture industry is currently engaged in assessing the regional impacts of projected climate change. Adaptation options are listed in Table 8.3, with the most urgent options being given the highest priority ranking.

After identifying relevant, crop-specific, climatic thresholds, it will be possible to estimate whether climate change will increase or decrease the risks of growing crops in particular geographic locations. Some new regions may be identified as becoming more suitable for horticultural production, while others may become less suitable. An increase in production from subtropical and tropical regions could see new products being introduced, or a changed accessibility of produce. This change will most likely be gradual and marketing strategies will adjust accordingly.

Effects on crop quality will vary. In some cases positive impacts will be experienced but negative impacts will also occur. The magnitude and

Table 8.3: Adaptation options for the Australian horticulture industry with the most urgent options being given the highest priority ranking (1).

Adaptation options	Priority
Temperature increase	
Determine climatic thresholds to plant growth and product quality	1
Re-assess location in regional terms to optimise reduction of climatic risk	1
Invest in conventional breeding and biotechnology to address future adaptation capacity	1
Tailor seasonal climate forecasts to horticultural requirements	2
Develop and modify markets for new crops and crop schedules	3
Change crop production schedules to align with new climate projections	3
CO_2	
Ascertain the crop-specific interactive effect of increased CO_2, temperature and water use	1
Determine the effect of CO_2 on pests, diseases and weed species	1
Rainfall	
Integrate catchment management and climate change projections to assess future water availability	1
Constantly benchmark irrigation management to increase efficiency	1
Pests and diseases	
Geographically sensitive pest and disease risk assessments using projected climate data	2

extent of these impacts will be governed by the present day climate of the region, the rate of projected change of climate and crop type. Management inputs available to ameliorate any adverse climatic impacts will be determined by the value of the crop to some extent. Since overall vulnerability will be crop and site specific, it may be difficult to determine this for the horticulture industry as a whole.

Once some of the potential vulnerabilities of horticultural crops have been identified, it will be possible to focus on conventional breeding programs and biotechnology to build resilient cultivars that are more suited to future climates, e.g. drought-tolerant cultivars. Considering the impacts of rising CO_2 on crops (and the interaction of CO_2 with increased temperature) would assist this endeavour.

Adapting to a changing climate may occur autonomously. For instance, farm managers currently adjust to gradually increasing temperatures by modifying timing of planting on a year-by-year basis. In some instances management may become easier, e.g. frost avoidance mechanisms may not need to be implemented. Glasshouses may not require as much heating, though cooling requirements may be increased.

Seasonal forecasts can be utilised to address inter-annual climate variability to enable effective use of resources (e.g. labour, planting material). As knowledge of climate systems improve these forecasts should become more accurate and sophisticated, and could include considerations of climate change trends.

Quantifying potential future water availability is critical in informing possible altered access to irrigation supplies. Integrated catchment management has become increasingly important with a range of water monitoring, forecasting and management initiatives being implemented. Temporary water trading will probably favour the high value horticultural crop producers over broadacre farming or pasture production as users of water allocations in the short term, provided existing and future markets allow this. With permanent water trading, water purchasing may not remain as affordable into the future as lower value ($ per ML) users gradually leave the market. Regardless of the type of water right accessed, increased water costs will need to be borne by both farmers and consumers. At the farm level, methods for conserving water once it reaches the farm will become critical in an environment of increasing demand and decreasing supply.

Acknowledgements

The authors wish to acknowledge the input and comments from Peter Deuter (Senior Principal

Horticulturist, Queensland Department of Primary Industries and Fisheries), Phillip Wilk (District Horticulturist, NSW Department of Primary Industries), Dr Gordon Rogers (Applied Horticultural Research, University of Sydney), Dr Ian Goodwin (Senior Irrigation Scientist – Fruit Trees and Winegrapes, Department of Primary Industries Tatura), Dr Craig Hardner (School of Land, Crop and Food Sciences, University of Queensland), Professor John A Considine (Snr Honorary Research Fellow: Horticulture & Viticulture, The University of Western Australia), Peter Clingeleffer (CSIRO Plant industry, Merbein), Associate Professor Gregory Dunn and Professor Snow Barlow (The University of Melbourne), Dr Ian Smith and Ian Macadam (CSIRO Marine and Atmospheric Research).

References

ABGC (2006). Cyclone Larry. The Australian Banana Growers Council Inc. http://www.abgc.org.au/pages/media/cyclonelarry.asp

Ainsworth EA, Long SP (2005). What have we learned from 15 years of free-air CO_2 enrichment (FACE)? A meta-analytic review of the responses of photosynthesis, canopy properties and plant production to rising CO_2. *New Phytologist* **165**, 351–371.

Alique R, Martínez MA, Alonso J (2006). Metabolic response to two hydrocooling temperatures in sweet cherries cv Lapins and cv Sunburst. *Journal of the Science of Food and Agriculture* **86**, 1847–1854.

Al-Jamal MS, Ball S, Sammis TW (2001). Comparison of sprinkler, trickle and furrow irrigation efficiencies for onion production. *Agricultural Water Management* **46**, 253–266.

ANRDL (2007). Land Use of Australia, Version 3 – 2001/2002. Australian natural resources data library. http://adl.brs.gov.au/anrdl/php/full.php?fileidentifier=http://adl.brs.gov.au/findit/metadata_files/a_luav3r9eg__00112a06.xml

Aurambout J-P, Constable F, Luck J, Sposito V (2006). *The Impacts of Climate Change on Plant Biosecurity – Literature Review*. Department of Primary Industries and Cooperative Research Centre for Plant Biosecurity, Melbourne.

Australian Bureau of Meteorology (2006). Severe Tropical Cyclone Larry. Queensland Regional Office, Australian Bureau of Meteorology. http://www.bom.gov.au/weather/qld/cyclone/tc_larry/

Australian Bureau of Meteorology (2007). Living with drought. Australian Government. http://www.bom.gov.au/climate/drought/livedrought.shtml

Australian Bureau of Meteorology (2008). http://www.bom.gov.au/climate/

Australian Bureau of Statistics (2005). 'Value of agricultural commodities produced (VACP) – year ended 30 June 2005.' http://www.ausstats.abs.gov.au/ausstats/subscriber.nsf/0/758410C5AB68DCA9CA2571E6001C8F0F/$File/75030_2004-05.pdf

Australian Bureau of Statistics (2007). '7111.0 – Principal agricultural commodities, Australia, preliminary, 2005–06.' http://www.abs.gov.au/AUSSTATS/abs@.nsf/DetailsPage/7111.02005-06?OpenDocument

Austin PT, Hall AJ, Snelgar WP, Currie MJ (2000). Earlier maturity limits increasing apple fruit size under climate change scenarios. In *New Zealand Institute of Agricultural Science and the New Zealand Society for Horticultural Science Annual Convention 2000*.

Baxter PM (1997). *Growing Fruit in Australia: For Profit or Pleasure*. Pan Macmillan, Sydney, NSW.

Beal PR (1976). Local mango varieties. *Queensland Agriculture Journal* **102**, 583–588.

Berry J, Bjorkman O (1980). Photosynthetic response and adaptation to temperature in higher plants. *Annual Review of Plant Physiology* **31**, 491–543.

Bjornlund H, Rossini P (2005). Fundamentals determining prices and activities in the market for water allocations. *International Journal of Water Resources Development* **21**, 355–369.

Carbone GJ, Schwartz MD (1993). Potential impact of winter temperature increases on South Carolina peach production. *Climate Research* **2**, 225–233.

Chakroborty S (2004). Potential impact of climate change on plant pathogen interactions. *Australasian Plant Pathology* **34,** 443–448.

Coakley SM, Scherm H, Chakraborty S (1999). Climate change and plant disease management. *Annual Review of Phytopathology* **37**, 399–426.

Coombs B (1995). *Horticulture Australia*. Morescope Publishing, Hawthorn, Vic.

CSIRO and Australian Bureau of Meteorology (2007). 'Climate change in Australia'. CSIRO and Australian Bureau of Meteorology. http://www.climatechangeinaustralia.gov.au

Damasco OP, Smith MK, Godwin ID, Adkins SW, Smillie RM, Hetherington SE (1997). Micropropagated dwarf off-type Cavendish bananas (*Musa* spp., AAA) show improved tolerance to suboptimal temperatures. *Australian Journal of Agricultural Research* **48**, 377–384.

Daymond AJ, Wheeler TR, Hadley P, Ellis RH, Morison JIL (1997). The growth, development and yield of onion (*Allium cepa* L.) in response to temperature and CO_2. *Journal of Horticultural Science* **72**, 135–145.

DeJong TM, Grossman YL, Vosburg SF, Pace LS (1996). PEACH: a user friendly peach tree growth and yield simulation model for research and education. *Acta Horticulturae* **416**, 199–206.

Deuter P (2006). 'Scoping study into climate change and climate variability.' Horticulture Australia Ltd, Project Number: VG05051.

Deuter PL (1989). The development of an insecticide resistance strategy for the Lockyer Valley, Queensland. *Acta Horticulturae* **247,** 267–272. http://www.actahort.org/books/247/247_51.htm

Deuter PL (1995). Broccoli. In *Horticulture Australia*. (Ed. B Coombes). Morescope Publishing, Hawthorn, Vic.

Diczbalis Y, Menzel CM (1998). Low temperatures decrease CO_2 assimilation and growth in the tropical rambutan. *Journal of Horticultural Science and Biotechnology* **73**, 65–71.

Dioguardi D (1995). Lettuces. In *Horticulture Australia*. (Ed. B Coombes). Morescope Publishing, Hawthorn, Vic.

Doi H, Gordo O, Katano I (2008). Heterogeneous intra-annual climatic changes drive different phenological responses at two trophic levels. *Climate Research* **36**, 181–190.

Drake BG, Gonzalez-Meler MA, Long SP (1997). More efficient plants: A consequence of rising atmospheric CO_2. *Annual Review of Plant Physiology and Molecular Biology* **48**, 609–639.

Easterling WE, Aggarwal PK, Batima P, Brander KM, Erda L, Howden SM, Kirilenko A, Morton J, Soussana J-F, Schmidhuber J, Tubiello FN (2007). Climate change 2007: impacts, adaptation and vulnerability. Food, fibre and forest products. In *Contribution of Working Group II to the Fourth Assessment Report of the Intergovernmental Panel on Climate Change*. Cambridge University Press, Cambridge.

Estrella N, Sparks TH, Menzel A (2007). Trends and temperature response in the phenology of crops in Germany. *Global Change Biology* **13**, 1737–1747.

Ewa Ubia B, Hondaa C, Besshoc H, Kondod S, Wadac M, Kobayashia S, Moriguchia T (2006). Expression analysis of anthocyanin biosynthetic genes in apple skin: Effect of UV–B and temperature. *Plant Science* **170**, 571–578.

Fereres E, Goldhamer DA, Parsons LR (2003). Irrigation water management of horticultural crops. *HortScience* **38**, 1036–1042.

George AP, Nissen RJ (1990). Effects of hydrogen cyanamide on yield, growth and dormancy release of table grapes in subtropical Australia. *Acta Horticulturae* **279**, 427–436.

Giddings J, Kelly S, Chalmers Y, Cook H (2002). Winegrape irrigation benchmarking Murray–Darling and Swan Hill 1998–2002. In *Managing Water*. (Eds C Dundon, R Hamilton, R Johnstone, S Partridge) pp. 15–18. Australian Society of Viticulture and Oenology, Mildura.

Glenn DM, Puterka GJ, Drake SR, Unruh TR, Knight AL, Baherle P, Prado E, Baugher TA (2001). Particle film application influences apple leaf physiology, fruit yield, and fruit quality. *Journal of the American Society for Horticultural Science* **126**, 175–181.

Grant RF, Kimball BA, Wall GW, Triggs JM, Brooks TJ, Pinter PJ, Conley MM, Ottman MJ, Lamorte RL, Leavitt SW, Thompson TL, Matthias AD (2004). Modeling elevated carbon dioxide effects on water relations, water use, and growth of irrigated sorghum. *Agronomy Journal* **96,** 1693–1705.

Hamilton AJ, Boland A, Stevens D, Kelly J, Radcliffe J, Ziehrl A, Dillonc P, Pauline B (2005). Position of the Australian horticultural industry with respect to the use of reclaimed water. *Agricultural Water Management* **71**, 181–209.

Hanninen H (1991). Does climatic warming increase the risk of frost damage in northern trees? *Plant Cell and Environment* **14**, 449–454.

Hatch RL, Ruiz M (1987). Influence of pruning date on budbreak of desert table grapes. *American Journal of Enology and Viticulture* **38**, 326–328.

Hennessy KJ, Clayton-Greene K (1995). Greenhouse warming and vernalisation of high chill fruit in Southern Australia. *Climatic Change* **30**, 327–348.

Hetherington SE (2005). Integrated pest and disease management for Australian summerfruit, NSW Department of Primary Industries and Summerfruit Australia Inc. http://www.dpi.nsw.gov.au/__data/assets/pdf_file/0008/184526/summerfruit-fulla.pdf

Hobbs RJ, McIntyre S (2005). Categorizing Australian landscapes as an aid to assessing the generality of landscape management guidelines. *Global Ecology and Biogeography* **14**, 1–15.

Hood A, Hossain H, Sposito V, Tiller L, Cook S, Jayawardana C, Ryan S, Skelton A, Whetton P, Cechet B, Hennessy K, Page C (2002). Options for Victorian agriculture in a 'new' climate – a pilot study linking climate change scenario modelling and land suitability modelling. Volume One: Concepts and Analysis; Volume Two: Modelling Outputs. Department of Natural Resources and Environment, Victoria.

Houerou HN, Jeftic L, Milliman JD, Sestini G (1992). Vegetation and land use in the Mediterranean basin by the year 2050: a prospective study. In *Climate Changes and the Mediterranean: Environmental and Societal Impacts of Climatic Change and Sea-level Rise in the Mediterranean Region*. (Eds L Jeftic, JD Milliman and G Sestini) pp. 175–232. Edward Arnold, London.

Howden M, Newett S, Deuter P (2005). Climate change – risks and opportunities for the avocado industry. New Zealand and Australia Avocado Growers' Conference '05, Tauranga, New

Zealand. http://www.avocadosource.com/Journals/AUSNZ/AUSNZ_2005/HowdenMark2005.pdf

Jones PA (1991). Historical records of cloud cover and climate for Australia. *Australian Meteorological Magazine* **39**, 181–189.

Krug H (1997). Environmental Influences on development growth and yield. In *The Physiology of vegetable crops*. (Ed HC Wien) pp. 101–180. CAB International, Wallingford, UK.

Lavee S, May P (1997). Dormancy of Grapevine buds – facts and speculation. *Australian Journal of Grape and Wine Research* **3**, 31–46.

Lobell DB, Field CB, Cahill KN, Bonfils C (2006). Impacts of future climate change on California perennial crop yields: model projections with climate and crop uncertainties. Agricultural and Forest Meteorology **141**, 208–218.

Lopez G, Johnson RS, DeJong TM (2007). High spring temperatures decrease peach fruit size. http://CaliforniaAgriculture.ucop.edu, pp. 31–34.

Maltby JE (1995). Tomatoes. In *Horticulture Australia*. (Ed. B Coombes). Morescope Publishing, Hawthorn, Vic.

Manrique LA, Bartholomew DP (1991). Growth and yield performance of potato grown at three elevations in Hawaii: II Dry matter production and efficiency in partitioning. *Crop Science* **31**, 367–372.

McFadyen R (2008). 'Briefing notes.' CRC for Australian Weed Management. http://www.weedscrc.org.au/documents/bn_climate_change_2007.pdf

MDBC (2007). 'River Murray System drought update no. 10 October 2007'. Murray–Darling Basin Commission. http://www.mdbc.gov.au/__data/page/1366/RMSystem_Drought_Update10_October07.pdf

Meinke H, Hammer GL (1997). Forecasting regional crop production using SOI phases: an example for the Australian peanut industry. *Australian Journal of Agricultural Research* **48**, 789–793.

Middleton S, McWaters A (2002). Hail netting of apple orchards – Australian experience. *Compact Fruit Tree* **35,** 51–55.

Miglietta F, Bindi M, Vaccari FP, Schapendonk AHCM, Wolf J, Butterfield R (2000). Crop ecosystem responses to climatic change: root and tuberous crops. In *Climate Change and Global Crop Productivity*. (Eds KR Reddy, HF Hodges) pp. 189–212. CABI Publishing, Mississippi State University, USA.

Monk PR, Hook D, Freeman BM (1986). Amino Acid metabolism by yeasts. In *Proceedings of the Sixth Australian Wine Industry Technical Conference*. (Ed. T Lee).14–17 July, Adelaide, South Australia.

Morison JL, Lawlor DW (1999). Interactions between increasing CO_2 concentration and temperature on plant growth. *Plant, Cell and Environment* **22**, 659–682.

Morrison BJ (1995). Strawberries. In *Horticulture Australia*. (Ed. B Coombes). Morescope Publishing, Vic, Australia.

Murison J (1995). Capsicums and chillies. In *Horticulture Australia*. (Ed. B Coombes). Morescope Publishing, Vic, Australia.

Nir G, Klien I, Lavee S, Spieler G, Barak U (1988). Improving grapevine budbreak and yields by evaporative cooling. *Journal of the American Society for Horticultural Science* **113**, 512–517.

NLWRA (2001). 'Australian water resources assessment 2000'. National Land and Water Resources Audit, Land and Water Australia. http://www.anra.gov.au/topics/irrigation/production/index.html

Olesen JE, Friis E, Grevsen K (1993). Simulated effects of climate change on vegetable crop production in Europe. In *The Effect of Climate Change on Agricultural and Horticultural Potential in Europe*. (Ed. GJ Kenny, PA Harrison, ML Parry). Environmental Change Unit, University of Oxford, Oxford.

Olesen JE, Grevsen K (1993). Simulated effects of climate change on summer cauliflower production in Europe. *European Journal of Agronomy* **2**, 313–323.

Pearson S, Wheeler TR, Hadley P, Wheldon AE (1997). A validated model to predict the effects of environment on the growth of lettuce (Lactuca sativa L): Implications for climate change. *Journal of Horticultural Science* **72**, 503–517.

Peet MM, Wolfe DW (2000). Crop ecosystem responses to climate change: vegetable crops. In *Climate Change and Global Crop Productivity*. (Eds KR Reddy, HF Hodges) pp. 213–243. CABI Publishing, Mississippi State University, USA.

Piskolczi M, Varga C, Racskó J (2004). A review of the meteorological causes of sunburn injury on the surface of apple fruit (*Malus domestica* Borkh). *Journal of Fruit and Ornamental Plant Research* **12**, 245–251.

Possingham JV (2008). Developments in the production of table grapes, wine, and raisins in tropical regions of the world. *Acta Horticulturae* **785**, 45–50.

Potter NJ, Chiew FHS, Frost AJ, Srikanthan R, McMahon TA, Peel MC, Austin JM (2008). 'Characterisation of recent rainfall and runoff in the Murray–Darling Basin, A report to the Australian Government from the CSIRO Murray–Darling Basin Sustainable Yields Project'. Water for a Healthy Country Flagship. http://www.csiro.au/files/files/pmax.pdf

Richards RA (2002). Current and emerging environmental challenges in Australian agriculture – the role of plant breeding. *Australian Journal of Agricultural Research* **53**, 881–892.

Richardson AC, Marsh KB, Boldingh HL, Pickering AH, Bulley SM, Frearson NJ, Ferguson AR, Thornber SE, Bolitho KM, Macrae EA (2004). High growing temperatures reduce fruit carbohydrate and Vitamin C in fruit. *Plant, Cell and Environment* **27**, 423–435.

Rigby JR, Porporato A (2008). Spring frost risk in a changing climate. *Geophysical Research Letters* **35**, L12703.

Riveroa RM, Sánchez E, Ruiza JM, Romeroa L (2003). Influence of temperature on biomass, iron metabolism and some related bioindicators in tomato and watermelon plants. *Journal of Plant Physiology* **160**, 1065–1071.

Rogers GS (2007). 'Post harvest improvement in iceberg and cos lettuce to extend shelf life for fresh cut salads.' Horticulture Australia Final Report, Project Number VG03092.

Rosenzweig C, Phillips J, Goldberg R, Carroll J, Hodges T (1996). Potential impacts of climate change on citrus and potato production in the US. *Agricultural Systems* **52**, 455–479.

Salinger MJ, Kenny GJ (1995). Climate and kiwifruit cv. 'Hayward' 2. Regions in New Zealand suited for production. *New Zealand Journal of Crop and Horticultural Science* **23**, 173–184.

Salvestrin J (1995). Onions. In *Horticulture Australia*. (Ed. B Coombes). Morescope Publishing, Hawthorn, Vic.

Sato S, Peet MM, Thomas JF (2000). Physiological factors limit fruit set of tomato (*Lycopersicon esculentum Mill.*) under chronic, mild heat stress. *Plant, Cell and Environment* **23**, 719–726.

Saure MC (1985). Dormancy release in deciduous fruit trees. *Horticultural Review* **7**, 239–300.

Schaffer B, Searle C, Whiley AW, Nissen RJ (1996). Effects of atmosphertic CO_2 enrichment and root restriction on leaf gas exchange and growth of banana (*Musa*). *Physiologia Plantarum* **97**, 685–693.

Schaffer B, Whiley AW, Wolstenholme BN (2002). *Avocado: Botany, Production and Uses*. CAB International, Wallingford, UK.

Schulze RE, Kunz RP (1995). Potential shifts in optimal growth areas of selected commercial tree species and subtropical crops in southern Africa due to global warming. *Journal of Biogeography* **22**, 679–688.

Scott C (1995). Pineapples. In *Horticulture Australia*. (Ed. B Coombes). Morescope Publishing, Vic, Australia.

Sport and Recreation Victoria (2007). Sport and Recreation Victoria Country Football Grounds Assistance Program Stage 2, Outcomes Report. Sport and Recreation Victoria Department for Victorian Communities, Melbourne. http://www.sport.vic.gov.au/web9/rwpgslib.nsf/GraphicFiles/CountryFootyOutcomeFinal/$file/CountryFootyOutcomeFinal.pdf (Accessed 22 August 2008).

Sukhvibul N, Whiley AW, Smith MK, Hetherington SE (2000). Susceptibility of mango (*Mangifera indica* L.) to cold induced photoinhibition and recovery at different temperatures. *Australian Journal of Agricultural Research* **51**, 503–13.

Sutherst RW, Collyer BS, Yonow T (2000). The vulnerability of Australian horticulture to the Queensland fruit fly, *Bactrocera* (*Dacus*) *tryoni*, under climate change. *Australian Journal of Agricultural Research* **51**, 467–480.

Thomaia T, Sfakiotakis E, Diamantidis G, Vasilakakisa M (1998). Effects of low preharvest temperature on scald susceptibility and biochemical changes in 'Granny Smith' apple peel. *Scientia Horticulturae* **76**, 1–15.

Thomas AL, Muller ME, Dodson BR, Ellersieck MR, Kapš M (2004). A kaolin-ased particle film suppresses certain insect and fungal pests while reducing heat stress in apples. *Journal of American Pomological Society* **58**, 42–51.

Tilbrook J, Tyerman SD (2008). Cell death in grape berries: varietal differences linked to xylem pressure and berry weight loss. *Functional Plant Biology* **35**, 173–184.

Tindall JA, Beverly RB, Radcliffe DE (1991). Mulch effect on soil properties and tomato growth using micro-irrigation. *Agronomy Journal* **83**, 1028–1034.

Titley M (2000). Australian lettuce production and processing – an overview. *Australian Lettuce Industry Conference.* 6-8 June, Hay, New South Wales, Australia, pp.12–31.

Topp B, Wilk P, Bignell G, Russell D (2008). Summary of the DPI&F subtropical peach breeding from 2002–2007. *Low Chill Stonefruit Grower* **1** (08), 4–5.

Topp BL, Sherman WB (2000). Breeding strategies for developing temperate fruits for the sub-tropics, with particular reference to Prunus. *Acta Horticulturae* **522**, 235–240.

Trought MCT, Howell GS, Cherry N (1999). 'Practical considerations for reducing frost damage in vineyard'. Lincoln University, New Zealand. http://www.nzwine.com/assets/frost_review. pdf

Uhlig BA, Clingeleffer PR (1998). Ripening characteristics of the fruit from *Vitis vinifera* L. drying cultivars Sultana and Merbein Seedless under furrow irrigation. *American Journal for Enology and Viticulture* **49**, 375–382.

Victorian Department of Sustainability and Environment (2008). 'Northern region sustainable water strategy discussion paper: documents & fact sheets.' http://www.ourwater.vic.gov.au/ programs/sws/northern/submissions/northern_region_sws_discussion_paper_submissions

Wand G (2007). Climate-ameliorating measures influence photosynthetic gas exchange of apple leaves. *Annals of Applied Biology* **150**, 75–80.

Waterwise Ways for WA (2007). What are my watering days? Water Corporation http://www. waterwisewaysforwa.com.au/go/secondary-navigation/what-are-my-watering-days

Wien HC (1997). *The Physiology of Vegetable Crops*. CAB International, Wallingford, UK.

Wilk P (2005). 'Low chill stone fruit varieties 2005.' Report No. Agnote DPI–479, 3rd edition. NSW Department of Primary Industries. http://www.dpi.nsw.gov.au/__data/assets/pdf_ file/0020/138125/Low-chill-stone-fruit-varieties-2005.pdf

Wurr DCE, Edmonsen RN, Fellows JR (2000). Climate change: a response surface study of the effects of CO_2 and temperature on the growth of French beans. *Journal of Agricultural Science* **135**, 379–387.

Wurr DCE, Hand DW, Edmondson RN, Fellows JR, Hannah MA, Cribb DM (1998). Climate change: a response surface study of the effects of CO_2 and temperature on the growth of beetroot, carrots and onions. *Journal of Agricultural Science* **131**, 125–133.

Zegbe-Domínguez JA, Behboudian MH, Lang A, Clothier BE (2003). Deficit irrigation and partial rootzone drying maintain fruit dry mass and enhance fruit quality in 'Petopride' processing tomato (*Lycopersicon esculentum*, Mill.). *Scientia Horticulturae* **98**, 505–510.

Ziska LH, Teasdale JR (2000). Sustained growth and increased tolerance to glyphosate observed in a C3 perennial weed, quackgrass (*Elytrigia repens*), grown at elevated carbon dioxide. *Australian Journal of Plant Physiology* **27**, 159–166.

9
FORESTRY

TH Booth, MUF Kirschbaum and M Battaglia

KEY MESSAGES:

- Australia's native forests, which cover about 149 million hectares, include more than two thousand tree species many of which are highly vulnerable to climate change because their current distributions cover only narrow climatic ranges.

- Australia's plantation forests are dominated by *Pinus radiata* (radiata pine) and *Eucalyptus globulus* subsp. *globulus* (blue gum), which together account for about 69% of the total plantation area.

- Any reduction in rainfall, as seen particularly for southern Australia in several climate change scenarios, coupled with increased water requirements in a warmer climate, is likely to lead to increased risk of tree mortality.

- Bioclimatic analyses can identify plantations that currently experience particularly hot and/or dry conditions. These sites could be monitored to provide an early warning if conditions become unsuitable for particular species in particular regions.

- Plantation productivity may be increased by rising levels of atmospheric carbon dioxide, but may be reduced by increased temperature and increased water loss, that will be compounded if rainfall is reduced. There are also potential problems from increased pest and disease risks, as well as more frequent and severe bushfires.

- There is an urgent need to improve the understanding of the effects of increased levels of atmospheric carbon dioxide and changes in temperature and rainfall on tree growth. It is important to assess whether growth rates of particular species are likely to be increased or decreased at particular sites, and how trees respond to stress from a combination of multiple climatic changes.

- As both *P. radiata* and *E. globulus* subsp. *globulus* are grown over relatively wide climatic ranges, the core areas should not be highly vulnerable to climate change in the short to medium term, unless serious pest and/or disease problems arise. If climatic conditions do become limiting at particularly hot and/or dry sites, alternative species better suited to warmer and/or drier conditions could be considered as well as changed management (e.g. planting at wider spacings at drier sites).

Introduction

Forestry involves the study and management of forests and plantations for a wide range of commercial and environmental values, including wood production, watershed protection, biodiversity conservation and carbon sequestration.

Forestry may be best considered as a continuum with native forests at one extreme and large-scale industrial plantations at the other (Donaldson and Pritchard 2000). Farm forestry refers to the management of trees in stands or woodlots for the production of wood products, or to provide other benefits such as shade, nature conservation and

Table 9.1: Total plantation areas in major regions (ha).

Region	Hardwoods	Softwoods	Total
a) Western Australia (south-west)	270 813	104 480	375 293
b) Green Triangle (Mt Gambier region)	130 145	166 650	296 795
c) South-east Queensland	31 675	161 052	192 727
d) Murray Valley	6380	178 100	184 480
e) Central Gippsland	33 298	58 803	92 101
f) Tasmania	155 500	71 600	227 100
Other regions			370 954
Total (2005)			1 739 450

Data from National Forest Inventory (2006).

dryland salinity amelioration (Abel *et al.* 1997). Both forestry and farm forestry can involve management of native stands, but apart from some brief background information on native forests, this review concentrates mainly on plantations.

The formal definition of forests refers to vegetation of a mature or potentially mature height exceeding two metres with a potential crown cover of 20% or more. There are 149 million hectares (Mha) of forest in Australia (MPIGA 2008), but only about 7% of that area, including 1.8 Mha of plantations, is managed for timber production. The plantations produce 62% of logs harvested from all forests. Farm forestry involves plantations of less than 1000 hectares under single ownership and is growing in importance. About a third of farm forest plantations have been planted since 1995 and now account for about 20% of plantations (National Forest Inventory 2007a). About 92 Mha of forests are classified as woodlands, and are primarily used for extensive grazing rather than wood production.

Climate change will pose particular risks for many unmanaged native forests. Australian forests include many hundreds of different tree species. Many of these species have comparatively limited distributions and hence occupy narrow climatic ranges. For example, a 2.5°C rise in mean annual temperature would result in about half of Australia's approximately 800 eucalypt species having their entire distributions shifted outside their current climatic range (Hughes *et al.* 1996; Kirschbaum 2000). This does not necessarily mean that these species would all die outside their current climatic ranges, but given that they do not currently persist under those conditions, it is likely that they will be at greater risk of extinction (Booth 2007).

Hennessy *et al.* (2007) have reviewed the likely effects of climate change on Australia and New Zealand for the Intergovernmental Panel on Climate Change (IPCC). They suggested that 'productivity of exotic softwoods and native hardwood plantations are likely to be increased by CO_2 fertilisation effects, although the amount of increase will be limited by projected increases in temperature, changes in rainfall and by feedbacks such as nutrient cycling'. Where tree growth is not water-limited, warming could additionally expand the length of the growing season in southern Australia, but they also cautioned that increased pest, disease and fire damage may negate some gains so that productivity declines are also possible. It is important to recognise that climatic and atmospheric changes are likely to have both positive and negative impacts on forestry, which we describe in further detail below.

Plantation areas are located mainly in southern and eastern regions of Australia. About 44% of plantations are hardwoods and 56% softwoods (National Forest Inventory 2007b). The area of hardwood plantations has increased rapidly since 1994, while the area of softwood plantations has remained stable. Of the hardwood plantations, about 61% are *Eucalyptus globulus* subsp. *globulus* (blue gum) while 75% of the softwood plantations are planted with *P. radiata* (radiata pine), both grown mainly in temperate regions. *E. globulus* is grown mainly for paper products, while *P. radiata* is grown mainly for sawn timber, posts and poles, with residues being used for paper, particle board and other panels. In subtropical parts of Australia, such as south-east Queensland, hardwood plantations mainly use *Eucalyptus grandis* (flooded gum) and *Eucalyptus dunnii* (Dunn's white gum) and softwoods used are *Pinus elliottii* (slash pine),

Pinus caribaea (caribbean pine) and their hybrids. Table 9.1 shows total plantation areas for the six main forestry regions in Australia, which account for almost 80% of the total (e.g. National Forest Inventory 2006, 2007a).

Impacts of projected climate change

Several studies have examined possible impacts of climate change on Australia's forests (e.g. Booth and McMurtrie 1988; Howden and Gorman 1999; Kirschbaum 1999a). It is generally recognised that in Australia's dry environment, water limitations and the possibility of increasing water shortages are the greatest concern for the future (Pittock *et al.* 2001). Any reduction in rainfall, as seen in various climate change scenarios, coupled with increased water requirements in a warmer climate (Kirschbaum 2000) can potentially lead to increased tree mortality. Any series of years with well below-average rainfall may lead to tree mortality unless trees have access to sufficient water reserves deep within the soil profile (Smettem *et al.* 1999). Tree deaths, especially of *E. globulus*, have already occurred in some plantations.

Experimental evidence

Most studies have recognised the importance of assessing the effects of atmospheric as well as climatic change. Atmospheric CO_2 is the basic substrate for photosynthesis, which underlies plant growth. Increasing CO_2 concentration can affect tree growth directly through increased photosynthetic rates and indirectly through improved water use efficiency (Steffen and Canadell 2005). It is well established that short-term photosynthetic rates in C_3 plants increase by 25–75% for a doubling of CO_2 concentration (Kimball 1983; Eamus and Jarvis 1989; Luxmoore *et al.* 1993; Drake *et al.* 1997). It is also recognised that the sensitivity of C_3 photosynthesis to CO_2 concentration increases with rising temperature (Kirschbaum 1994), and hence the stimulation of plant growth by increasing CO_2 concentration is likely to be larger at higher temperatures (Rawson 1992), with little stimulation and sometimes even inhibition at low temperatures (Kimball 1983). Limited observational evidence from international research suggests that the effects of elevated CO_2 may decrease as trees age (Steffen and Canadell 2005).

One of the key questions for future water use from plant canopies is the response of stomata to increasing CO_2 concentration. Morison (1985) and Allen (1990) compiled a range of observations from the literature, and showed that stomatal conductance was reduced by about 40% when CO_2 concentration was doubled. Drake *et al.* (1997), however, found stomata closed by only 20%, and Medlyn *et al.* (2001) found a 21% decrease in stomatal conductance in studies on European trees. Curtis and Wang (1998) found even less stomatal closure in their review of studies on woody plant responses to CO_2. Reduced stomatal conductance has also been deduced from analysis of herbarium specimens that has shown that the number of stomata on leaves has decreased with historical increases in global CO_2 concentration (Woodward 1987; Rundgren and Björck 2003; Kouwenberg *et al.* 2003).

The carbon isotope discrimination between $^{13}CO_2$ and $^{12}CO_2$ can also be used to infer changes in the intercellular CO_2 concentration (Farquhar *et al.* 1982; Korol *et al.* 1999) during historical changes in atmospheric CO_2 (Dawson *et al.* 2002). Using this approach, Arneth *et al.* (2002) and Duquesnay *et al.* (1998) reported some stomatal closure in response to increasing atmospheric CO_2, but Marshall and Monserud (1996) and Monserud and Marshall (2001) found no evidence of stomatal closure in their data sets.

These existing observations and reviews thus still lead to conflicting conclusions, yet this is an area of key importance for understanding the future response of ecosystems to climatic changes. Hence, there is an important need to further review the existing observations, find the commonalities and differences in plant responses and generate a generalised global understanding.

In addition, these studies all report on the relative response of plants to climate change when they are either unstressed or only mildly stressed. Yet it is of particular importance to understand how plants will respond to episodes of severe stress. Episodic periods of drought could become more severe as temperatures rise. Under such conditions, plants are likely to close their stomata to conserve water. Nonetheless, water shortages may intensify further under intense atmospheric evaporative demand. It is not known whether elevated CO_2 can confer some kind of protection under these extreme conditions, or whether trees instead become more vulnerable owing to possibly greater leaf area development. This tree response is critical but extremely difficult to investigate, and there is very little available research information.

Increased photosynthetic rate and decreased water requirement translate into increased tree seedling growth (Luxmoore *et al.* 1993) which has also been observed for mature trees in free air CO_2 enrichment (FACE) experiments in largely undisturbed forests (Herrick and Thomas 2001; Gunderson *et al.* 2002). Based on summarising the available evidence, Gielen and Ceulemans (2001) concluded that the growth of poplar trees may be stimulated by about 30% by doubling CO_2 concentration. As water use efficiency can be greatly enhanced by increased CO_2 concentration (Eamus and Jarvis 1989), relative plant responses to increases in CO_2 are generally found to be more pronounced under water-limited conditions (e.g. Gifford 1979; Allen 1990). CO_2 responses are also evident under nutrient-limited conditions (e.g. Idso and Idso 1994) but these tend to be less than they are when trees have adequate nutrition (Drake *et al.* 1997).

Of the plantation species widely planted in Australia, *P. radiata* is the only one that has been the subject of experimental exposure to prolonged periods of elevated CO_2 (Turnbull *et al.* 1998; Griffin *et al.* 2000; Tissue *et al.* 2001; Greenep *et al.* 2003). These studies showed gains of 30–50% photosynthetic rates in response to doubling atmospheric CO_2. Experiments on the effects of elevated CO_2 on eucalypts to date have only involved short-term exposure of small plants (e.g. Roden *et al.* 1999; Atwell *et al.* 2007), but new experiments investigating the effect of sustained levels of CO_2 on tree functioning are currently under way. Up-regulation of photosynthesis by temperate Australia plantation eucalypt species, *E. globulus* and *E. nitens* following partial defoliation (Pinkard *et al.* 1998, 2007), suggests that these trees are often sink- rather than source-limited. As has been shown for *P. radiata* (Greenep *et al.* 2003), the photosynthetic rates of trees are only likely to respond to elevated CO_2 if there are sinks for available substrate. At present, modelling and simulation are the only tools available to anticipate the likely response of many of Australia's forest species.

Future work

A range of major experiments are either under way or have been proposed to address a number of specific questions about plant response to atmospheric change. The Hawkesbury eucalypt experiment is examining the effects of increased CO_2 on eucalypts using whole-tree chambers. While the enclosures allow CO_2 levels to be increased, the facility cannot reproduce the conditions that are experienced by trees growing in the open. Hence, there has been growing interest to conduct a FACE experiment in Australia where the atmospheric CO_2 concentration is raised around a group of plants under unenclosed, open air conditions. Raison *et al.* (2007) reviewed the feasibility of conducting FACE experiments in Australia. Indicative annual running costs are approximately \$2–\$4.5 million per year for each experiment. They recommend that an initial FACE study should be established either in open dry sclerophyll forest or woodland that typically has a nitrogen fixing understorey. A FACE study could be established in a plantation at a later time with the support of appropriate stakeholders.

Modelling analyses

Kirschbaum (1999b) used CenW to investigate the likely response of forest growth to climate change in Australia. The model was initialised under current climatic and fertility conditions. Temperature, CO_2 concentration and/or precipitation were then changed, and the subsequent growth response was recorded within the constraints set by the limitations that had been established under steady-state conditions in the current climate. The simulated responses to climate change differed greatly across the continent. For example, positive growth responses to increasing temperature were found in wet and nutrient-limited regions because the higher temperature led to increased nitrogen mineralisation, which enhanced growth. These same regions showed only slight responses to increased CO_2 concentration because increased carbon fixation led to immobilisation of associated nitrogen in soil organic matter. It highlighted an important problem in trying to adapt to climatic changes; not only is the nature of climate change not yet adequately known, but plant responses are likely to differ in interaction with current conditions even where the nature of external changes is known.

While Kirschbaum (1999b) simulated the effects on generic plant growth, Battaglia and Bruce (2008) are carrying out a major project assessing the impacts of climate change on a selection of hardwood and softwood plantations in Australia. To anticipate the interactive effect of climate change and elevated CO_2, Battaglia and Bruce (2008) have used the process-based model CABALA (Battaglia *et al.* 2004). There is considerable uncertainty about whether stimulated levels of photosynthesis under elevated CO_2 are sustained in the long term, and some evidence suggests that enhanced levels of photosynthesis may diminish over time as a result of nitrogen dilution in leaves and/or down-regulation of photosynthesis (after Medlyn *et al.* 2001; Ainsworth and Long 2005; Buckley 2008; Rogers and Ellsworth 2002). While the case of nitrogen limitation is accommodated in models with linked carbon-nitrogen modules, the later case led Battaglia and Bruce (2008) to consider both the possibilities that photosynthesis may not be stimulated by increased CO_2 and the possibility that the short-term responses to elevated CO_2 are sustained. This covered both a best and a worst case for plantation responses.

In evaluating the effects of climate change on wood production, the use of scenarios in the way developed here must be approached with caution. The outcomes can be influenced markedly by the climate model selected, the sites chosen and the modelling assumptions made. To overcome these limitations, Battaglia and Bruce (2008) have used a range of climate models and down-scaling techniques and a sensitivity analysis of assumptions of tree physiological responses, such as photosynthetic down-regulation under elevated CO_2.

Preliminary results suggest that in cold, wet parts of the country, such as Tasmania, production gains for *E. globulus* and *E. nitens* may be positive and markedly so (by 30–80%) if a photosynthetic response is maintained. In places such as the Green Triangle region (i.e. south-east South Australia and south-west Victoria), central Victoria and south-western Western Australia where existing rainfall already imposes strong production limitations and where species are growing closer to their optimum growing temperature, production gains may not eventuate or may be reduced particularly under more severe climate change scenarios (–5 to –10%). Without photosynthetic responses to elevated CO_2, production gains across the *P. radiata* plantation estate are likely to be realised only in the colder and wetter parts of the estate. Across other parts of Australia's plantation estate, however, the risk of drought death is likely to increase. If the coincidence of high temperature and dry soil is taken as an indicator of drought death risk, then across the Victorian hardwood plantation estate, the number of high risk days may increase by 50–100% by 2030.

These preliminary results suggest, in accord with the work of Kirschbaum (1999b), that the impacts of climate change on forest production may be highly site-specific and will depend, in part, on the balance between the detrimental effect of temperature and rainfall changes versus the potential productivity gains through elevated atmospheric CO_2.

Interactions between CO_2 concentration and plant physiological functioning not only affect tree productivity, but sometimes also produce some surprising physiological effects. For example, work with seedlings of *Eucalyptus pauciflora* (snow gum) grown in open-topped chambers showed that seedlings grown under elevated CO_2 levels suffered 10 times as much frost damage as seedlings grown under ambient conditions (Barker *et al.* 2005).

In addition to the importance of atmospheric change, climatic factors are important when selecting appropriate species for planting in different regions. Booth and Jovanovic (2005) analysed the climatic requirements of 31 tree species including the currently grown species, such as *E. globulus* and *P. radiata*, as well as lesser-known species, such as *Eucalyptus argophloia* (Chinchilla white gum) and *Eucalyptus kartoffiana* (Araleun gum), which may have potential for farm forestry in the future. Their climatic requirements were assessed by bioclimatic analysis of their natural distributions and from their growth performance at trial sites outside their natural distributions both in Australia and overseas. These descriptions were used to generate maps indicating climatically suitable areas under current conditions and under future scenarios developed for 2030 and 2070 (McGregor 2003). The main features of the scenario used here are a mean temperature increase of 0.85°C by 2030 and 2.3°C by 2070, combined with a reduction in precipitation of 3.9% by 2030 and 10.7% by 2070.

Figure 9.1 shows a simplified version of possible changes in suitable growing regions for *P. radiata* (Booth and Jovanovic 2005). Booth and McMurtrie (1988) showed that most *P. radiata* plantations in Australia are located in temperate medium to high rainfall areas defined by a mean annual temperature of 10–18°C and mean annual rainfall

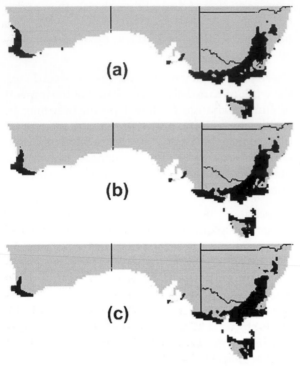

Figure 9.1: Black shaded areas are temperate medium to high rainfall areas suitable for growing *P. radiata* under (a) current climatic conditions, (b) 2030 and (c) 2070. Prepared using a climate-matching program developed by Booth and Jovanovic (2005) using a CSIRO climate change scenario described by McGregor (2003).

of 600–1800 mm received mainly in winter or uniformly throughout the year (see Figure 9.1.a). This includes the Western Australia, Green Triangle, Central Tablelands, Southern Tablelands, Murray Valley, central Victoria, central Gippsland, East Gippsland-Bombala and Tasmania regions. Figures 9.1b (2030) and 9.1c (2070) show the location of the climatically suitable regions for *P. radiata* in future years using a particular climate change scenario. Differences between current conditions and 2030 are relatively slight. There is some reduction in the climatically suitable area in Western Australia and the unsuitable area in Central Victoria becomes larger. The changes become more pronounced by 2070, with the most obvious being further reductions in the suitable areas in Western Australia and western Victoria. However, most of the major *P. radiata* plantation areas remain climatically broadly suitable.

Pests and diseases

Pests and diseases provide other potentially complex interactions between climate change and plant responses. Podger *et al.* (1990) investigated climatic factors affecting the distribution of the soil-borne fungus *Phytophthora cinnamomi* in Tasmania, and Chakraborty *et al.* (1998) published a broader review of potential climate change impacts on plant diseases including those that affect tree species, such as *P. radiata* and *E. globulus*. Booth *et al.* (2000a, b) described the climatic requirements of two pathogens that either are already present in Australia or might pose a potential threat to trees in Australia. The CLIMEX model (Sutherst and Maywald 1985) is currently being used to study the effects of climate change on pests and diseases important for forestry. A study for the Department of Climate Change (DCC) is using CLIMEX to assess the potential effects of climate change on the distribution of a eucalypt pest and disease (*Mnesampela privita*, Autumn gum moth and *Mycosphaerella* spp., crinkle leaf disease) as well as a pine pest and disease (*Essigella californica*, pine aphid and *Dothistroma septospora*, Dothistroma needle blight). For example, the CLIMEX model suggests that by 2070 the distribution of *Mycosphaerella* spp. leaf disease may shift south and increase in altitude. The projected distribution of the disease is being compared with the current and likely future distribution of *E. globulus* plantations in order to examine any potential risks.

Options for dealing with climate variability

Managing for climatic variability is particularly important for forestry as it can take many years for trees to produce a commercial product. For example, *E. globulus* requires about 12 years to produce a pulpwood crop, while *P. radiata* typically takes about 30 years to produce a final sawlog crop and *Acacia melanoxylon* (blackwood) may take 50 or more years to produce high-value timber. Trees attain their full value only if stands can grow to full maturity. In the past, forest managers have tended to minimise risks associated with climatic variability as far as possible by planting trees in medium to high rainfall regions. They have also tended to plant trees at densities which under normal circumstances would avoid significant losses during droughts.

Managing for rainfall variability is likely to become more important for forestry as many existing plantations experience unusually prolonged drought conditions and new plantations are increasingly being located in lower rainfall zones to meet natural resource management aims such as salinity control and carbon sequestration (Consortium 2001). For example, Neumann *et al.* (2006) describe how the FloraSearch project is evaluating forestry and agroforestry systems suitable for the 250–650 mm rainfall zone of southern Australia.

Recent severe droughts have already caused tree deaths in some areas of *P. radiata* and *E. globulus* plantations. Increasingly, growers can anticipate drought problems and adapt their actions accordingly. Tools such as the Queensland Department of Primary Industries RainMan (www.dpi.qld.gov.au/rainman/), Bureau of Meteorology products on the SILO website (www.nrw.qld.gov.au/silo/), Queensland Department of Natural Resources products on the LongPaddock website (www.longpaddock.qld.gov.au/) and the Bureau of Rural Sciences Rainfall Reliability Wizard (www.brs.gov.au/rainfall/) can assist these analyses. Information on likely El Niño conditions, particularly the Southern Oscillation Index (SOI), is also available to assist decision-making. Another adaptation to reduced rainfall is the selection of species with superior drought tolerance. The Australian Low Rainfall Tree Improvement Group (ALRTIG) is a cooperative involving CSIRO, the Australian National University and organisations from New South Wales, South Australia, Tasmania and Victoria. It is developing improved germplasm for low rainfall (400–600 mm) environments.

Options for dealing with projected climate change

Forest managers have a relatively wide range of options for adapting to climate change before plantation establishment, but fewer options once plantations are established. The following sections explore some of the adaptation strategies available before and after tree establishment.

Genotypes

Choice of species is one of the most fundamental decisions in plantation establishment. Though *P. radiata* has been grown in small areas in southern Queensland, species more suitable to warmer conditions such as *P. caribaea*, *P. elliottii* and the hybrid between these two species are the important plantation pines grown in Queensland. *P. elliottii* and the hybrid with *P. caribaea* would be obvious options for consideration at sites that became too hot for *P. radiata*.

Even though species selection is still important, the choice of trees increasingly involves the selection of appropriate provenances (i.e. seed of a species from a particular location), hybrids or clones. In future, the use of genetically modified (GM) material may also be considered, although GM material based on native species would need to be sterile to avoid transfer of modified genes to native stands. The quality of information available on the uses and environmental requirements of particular trees is rapidly improving. For example, CAB International (2005) has developed a CD-based Forestry Compendium that allows users to select species on the basis of information on natural distribution, environmental requirements, end uses and silvicultural requirements. The Compendium uses the same descriptions of climatic requirements used in climatic mapping programs (e.g. Booth and Jovanovic 2005). Generally, the cost of material for planting is only a small fraction of establishment costs, but selection of appropriate genetic material has a strong and long-term effect on the stand's potential biological performance.

In most cases, pulp plantations will probably be able to complete their current rotation before they are significantly impacted by climate change, and different genetic material could be chosen for the next rotation if that seems warranted. Potential climate change poses a more serious problem for longer-term sawlog plantations, as they may well grow for long enough to experience a considerable extent of climate change over a single rotation. In this case, it becomes important to choose appropriate genetic material that can grow not only under current conditions, but also under the possibly changed conditions in 30 or 40 years time.

Spacing and thinning

Simple adaptations to reduced rainfall include planting trees at wider spacings (Smettem *et al.* 1999) or thinning existing stands. Wider spacing of trees reduces the competition between trees and allows each individual tree to exploit a larger volume of soil. This is particularly effective as a safeguard against drought-induced mortality

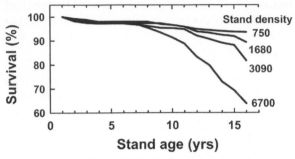

Figure 9.2: The effect of initial planting density (stems per ha) on survival of trees in a loblolly pine plantation.
Redrawn from Sharma *et al.* 2002

(see Figure 9.2). However, spacing at wider than optimal density may reduce stand growth, as trees only partly utilise available site resources. Battaglia *et al.* (2004) have described the use of computer modelling of tree growth and drought risk to recommend appropriate spacings for different environments.

Watering

Trees are sometimes watered just after planting to assist in the early establishment phase. There may be some potential for the wider use of effluent irrigation (Myers *et al.* 1999), which is seen as a promising means of disposing of effluent while creating a useful product. In contrast, extensive use of freshwater irrigation for trees seems unlikely as water will probably become even more expensive in the future and there are likely to be more profitable uses for any water available.

Nutrients

The rate of water loss from plantations is strongly related to the total leaf area of trees. Total stand leaf area can be modified through tree spacing as discussed above or, more subtly, through adjusting fertility levels. Nutrient-limited stands produce fewer leaves, and that makes them less vulnerable to developing water stress. Choosing to withhold fertiliser in order not to increase leaf area, and hence water loss, is a possible adaptation option to drier conditions though in practice, it is very rarely used in this way. Better nutrition also ensures good use of site resources when they are available, and improved plant vigour can enhance tree resilience to adverse conditions. These opposing effects of nutrition make fertiliser management a difficult adaptive strategy to use.

Site selection

In regions where climate change is likely to bring reduced rainfall, an obvious adaptation would be to restrict the planting of trees to wetter environments. However, as mentioned above, forestry is currently tending to move into lower rainfall regions because of the availability of cheap land and because of perceived associated environmental benefits of planting trees in drier, and more marginal, agricultural areas (Consortium 2001). Plantings in high rainfall (>800 mm mean annual rainfall) environments may also become restricted in future because of concerns about their impact in reducing catchment runoff (Nambiar and Brown 2001; Keenan *et al.* 2006). Forestry is thus increasingly moving into environments where trees are becoming more vulnerable to climatic changes.

Fire management

Any form of forestry is a long-term investment, which makes it particularly vulnerable to loss by bushfire. Changed climatic conditions, particularly warmer, drier and potentially windier periods are of concern. The small-scale and relatively isolated nature of most farm forestry plantings provides some natural fire protection as it makes it more difficult for fires to spread from one isolated stand to another, while larger-scale commercial plantings can maintain good access networks and staff trained in fire fighting. Adaptation options to increased fire risks include widening firebreaks and increasing the frequency of hazard-reduction burns where appropriate.

Pest, disease and weed management

Climate change may change the range of existing pests, diseases and weeds so that new challenges may be faced by forest managers. It will be important to consider the potential for these new problems in the selection of appropriate trees for plantings, and to plan suitable control strategies for emerging pests, diseases and weeds in existing plantations.

Good weed management is particularly important to ensure good early tree growth. If weeds are present in plantations, they can compete with trees for access to soil resources, especially water. If wider tree spacing is used as an adaptation to lower rainfall, it will allow more light to penetrate stands and it will take longer for trees to outcompete and suppress weeds. This could lead to a

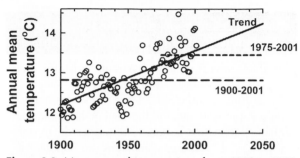

Figure 9.3: Mean annual temperature from 1900 to 2001 for Low Head, Tasmania. This data set is chosen as an example of a site for which a long climate record is available. The solid line gives a linear fit to the data. The long-dashed lines gives the mean of all data from 1900 to 2001, and the shorter dashes give the mean temperature from 1975 to 2001.

greater proliferation of weeds and partly negate the benefits of wider spacings.

Establishment strategies

Mortality during establishment is a significant risk, particularly for plantings in lower rainfall regions. There is a wide range of establishment strategies available, including the use of tubestock, direct seeding, assisted and natural regeneration, in combination with different site preparation and management regimes. In principle, matching combinations of these to seasonal climate forecasting may provide useful ways of managing climatic risks while minimising financial risks. However, more research on the optimal combination of these options is needed.

Assessment of climatic conditions

Many of the issues described in the previous sections require some consideration of current and future climatic conditions. When carrying out these analyses it would be prudent to place increasing reliance on more recent climatic records. For example, it may be advisable to consider frost risks on the basis of data from the last 25 years rather than the last 100 years, to fit a linear trend of temperature over the past record to provide some guidance to possible conditions in the future (see Figure 9.3) or consult trend maps prepared by the Bureau of Meteorology (see Chapter 2).

Risks of maladaptation

Some forest managers would like to be more conservative in their management to be prepared for climatic change, but the imperative to achieve

high growth rates and rapid economic returns makes such a cautious strategy difficult to implement. An example is seen in the rapid expansion of *E. globulus* plantations. This has led plantations to be pushed to, and sometimes beyond, reasonable rainfall limits as cheaper and more readily available land was sought for further expansion of the plantation estate. In some extreme locations, these economic imperatives make it difficult to cope even with current climatic variability, let alone provide the extra flexibility of dealing with future climate change.

The worst-case scenario for a tree species under climate change would be its extinction if its present environment became totally climatically unsuitable. This may constitute an important risk for many native species that currently occur over very limited climatic ranges (Hughes *et al.* 1996), although many species may be more adaptable than their natural distribution suggests (Kirschbaum 2000; Booth 2007).

Figure 9.1 and the analyses of Booth and Jovanovic (2005), suggest that commercial species are not generally in immediate danger of their most commercially important centres becoming climatically unsuitable. This may be because commercial species tend to be planted in relatively good climatic locations and are not generally planted at the margins of their climatically suitable ranges. For example, most *P. radiata* plantations are in locations with mean annual temperature below 14°C, although some plantations are situated in locations above 16°C (Booth and McMurtrie 1988). Mean annual temperatures would probably need to rise by well over 2°C before the major commercially important regions would become unsuitable for species such as *P. radiata* and *E. globulus*. However, changing climatic conditions may affect pest, disease and fire risk conditions at particular sites and reduce growth rates long before the sites become completely climatically unsuitable for a particular species.

Costs and benefits

The Allen Consulting Group prepared a report for the Australian Greenhouse Office (AGO 2005) that discussed possible costs and benefits of climate change impacts and adaptation for several agricultural sectors and forestry. Though the report did not provide a quantitative analysis itself, it suggested a framework for this type of analysis and identified some of the key issues that would need to be considered.

The current annual value of the wood and wood products industries is about $18 billion, and it employs about 83 000 people (BRS 2007). Vulnerability is a function of the exposure to climate change, of the sensitivity to change and of the capacity to respond and adapt to that change. Forestry has significant exposure to climate change as described above, with immature forests being particularly susceptible to drought. The sensitivity of the plantation industry is a measure of its response to climatic influences. Sensitive systems respond strongly to climatic drivers and can be significantly affected by small climatic changes.

The AGO (2005) report assessed forestry to be moderately sensitive to the effects of climate change. Even though the two major plantation species have broad climatic ranges, many native species have narrow climatic ranges. Forestry was, however, considered to have a high adaptive capacity as alternative species could be planted if required, which reduces the industry's assessed vulnerability, although long planning horizons, including long-term supply contracts, could make some adjustments difficult. The report concluded that forestry could minimise any problems from early attention to adaptation planning through better selection of species that takes climate change into account.

While climate change creates some potential problems for forest managers, it also provides new opportunities. There is increasing interest in biosequestration, i.e. the planting of trees to remove CO_2 from the atmosphere and store it in growing trees or wood products (see Chapter 14 for more details). The Prime Ministerial Task Group on Emissions Trading (2007) advised that 'undertaking low-cost measures to reduce deforestation and promote carbon sinks, both within Australia and internationally, should be an immediate priority'. The Australian Bureau of Agricultural Research Economics (ABARE) has examined climate change issues for agriculture and forestry (Gunasekara et al. 2007), including a consideration of costs and benefits of forest-based carbon offsets.

There is also potential for forest products and residues to be used for the production of bioenergy. Raison (2006) estimated that there are about 14–16 million cubic metres a year of forest waste

Box 9.1: Example of adaptation in Blue gum plantations in south-west Western Australia

The south-west region of Western Australia has more than 250 000 ha of blue gum plantations. Since 1980, annual rainfall has been consistently below the average for the region and climate change scenarios suggest that further decreases are likely. *E. globulus* plantations are capable of rapid growth, but are susceptible to drought death when demand for water exceeds the supply from the soil. This has occurred repeatedly during severe drought periods over the last decade resulting in areas of tree death. Reducing stocking at planting or by thinning increases the amount of soil available to each tree and reduces the risk of drought deaths. Applying nutrients to match supply with demand will maximise growth and the water use efficiency of wood production for a given stocking density. The industry is now adapting standard silviculture to balance plantation productivity with drought risk, producing the same wood volumes at lower risk and with reduced harvest costs. This presents a clear example of how the forestry industry has adapted to changing climatic conditions. Successful adaptation has required the development of an understanding of risk stages in the crop life cycle of trees and an ability to anticipate the supply and demand for water. For example, the industry no longer plants at the same planting density at all sites, but instead applies a simple rule for determining planting densities for each hectare; one seedling per mm of rainfall down to 600 mm. However, there is still a need for wider application of more sophisticated management tools involving a better characterisation of sites and a tailoring of site management to site attributes.

products that could be used for energy generation. There is also increasing interest in the production of cellulosic ethanol as a biofuel from forest products or residues, as these can often be produced on lower-value land than agricultural alternatives.

Discussion

Forestry has successfully implemented many significant changes in recent years in response to industry needs. For example, the area of hardwood

Table 9.2: Summary of climate change adaptation options for the forestry industry. Priority 1 (high), 2 (medium) and 3 (low).

Adaptation options	Priority
Policy level	
Use bioclimatic analysis to identify plantations of key species already going under extreme climatic conditions. Monitor these sites to provide early warning of potential problems	1
Support studies of impacts of increased levels of CO_2 on key exotic and native species and incorporation of results into simulation models to assist adaptation planning	1
Undertake further adaptation studies which include costs/benefit analyses	2
Support continuing commitment from all levels of government for pest, disease and weed control	2
Promote introduction of climate change adaptation into routine forest management	2
Forest management	
Take climate change into account when planning establishment of new plantations	1
Use updated fire behaviour prediction systems to plan for a changing fire regime	1
Use tree growth models and permanent sample plot (PSP) records from existing plantations to plan alternative management strategies	1
Use improved tools to assess risk of weed, pest and disease problems under climate change	2
Provide advice on opportunities to use nutritional adjustments as an adaptation to climate change	3
Climate information and use	
Improve regional level climate change modelling to provide more reliable scenarios to assist decision-making	1
Give greater weight to more recent climatic data when planning	2

plantations has increased from about 175 000 ha in 1994 to 740 000 ha in 2005. The high cost of land in relatively high rainfall areas has led to the expansion of plantations in low to medium rainfall regions. Genotypes of species such as *Pinus pinaster* have been bred to suit low to medium rainfall conditions (approximately 400–600 mm annual rainfall). New silvicultural systems have been developed for planting oil mallee eucalypts in 200–400 mm rainfall wheatbelt regions. These have involved planting trees in alleys between cropped areas. At the same time, forest managers have had to cope with unusually dry conditions in many of the well-established plantation regions. The experience of introducing new genotypes and silvicultural systems, as well as the move to drier areas and the need to adjust to drought conditions in many usually medium–high rainfall areas, has shown that the forestry sector can successfully adapt to changing climatic conditions. Table 9.2 outlines some possible adaptations to future climate change.

Effective adaptation to future climatic changes will rely on good knowledge of anticipated changes. Therefore, a high priority for research must be the further refinement of projections of likely climatic changes at the regional level. However, it must be acknowledged that even with perfect understanding of the global climate

systems, there will always remain fundamental uncertainty in climate change scenarios due to uncertainty in future emissions. Future emissions will largely depend on technical, political, social and economic factors and are thus inherently difficult to predict. Furthermore, actual future emissions and consequential climatic changes will depend in part on the success or failure of international attempts at emission control. Despite these limitations, forest managers would be well advised to consult the latest climate change projections as a routine part of planning plantation establishment. Climatic conditions at the potential planting site should be compared with other sites where particular trees are being grown. Occasional monitoring of plantations of particular species or genotypes that are already growing under hotter or drier conditions could provide early warning of potential problems.

The actual growing conditions experienced by trees are a result of interactions between climatic and site conditions, such as soil depth, texture and fertility. Commercial operators already carefully assess site conditions before plantation establishment. Such assessments will become even more important under climate change. Simulation models may be used to recommend spacing and thinning treatments that are well suited to future as well as current conditions.

In future, these models will be able to make use of improved knowledge of the impacts of high CO_2 and drought risk on tree mortality, including identification of optimal strategies between high growth (e.g. dense stands with high leaf area) and risk aversion (e.g. sparse stands with low leaf area) for particular sites and particular trees/products.

Having selected the appropriate genotype and silvicultural system, forest managers will be able to make use of soil water tracking and rainfall outlook systems to help select an optimal time for planting. However, successful establishment under variable and changing climatic conditions will continue to rely on good site preparation and effective weed control.

As the plantation develops, changing climatic conditions may affect pest, disease and weed risks. Computer simulation models will provide increasingly better indications of where and when particular organisms may create problems and what measures need to be taken to deal with them. Improved assessments of bushfire risk will also indicate particular times and places at high risk. A changing climate may require adaptations, such as preparation of wider firebreaks or more frequent hazard reduction burns.

Changing atmospheric as well as climatic conditions may well bring increased productivity to some forests. Growing interest in carbon sequestration is already bringing increased opportunities for establishment of both commercial and environmental forests. The forest industry has shown itself very willing to adapt to changing conditions in the past and this experience equips it well to deal with future climate change.

Acknowledgements

Thanks to Don White (CSIRO Sustainable Ecosystems) for his comments on blue gum plantations in south-west Western Australia.

References

Abel N, Baxter J, Campbell A, Cleugh H, Fargher J, Lambeck R, Prinsley R, Prosser M, Reid R, Revell G, Schmidt C, Stirzaker R, Thorburn P (1997). Design principles for farm forestry. RIRDC/LWRRDC/FWPRDC Joint Venture Agroforestry Program, Canberra.

AGO (2005). 'Climate change risk and vulnerability'. Report to the AGO by the Allen Consulting Group. Australian Greenhouse Office, Canberra.

Ainsworth EA, Long SP (2005). What have we learned from 15 years of free-air CO_2 enrichment (FACE)? A meta-analytic review of the responses of photosynthesis, canopy properties and plant production to rising CO_2. *New Phytologist* **165**, 351–372.

Allen LH (1990). Plant responses to rising carbon dioxide and potential interactions with air pollutants. *Journal of Environmental Quality* **19**, 15–34.

Arneth A, Lloyd J, Santruckova H, Bird M, Grigoryev S, Kalaschnikov YN, Gleixner G, Schulze ED (2002). Response of central Siberian Scots pine to soil water deficit and long-term trends in atmospheric CO_2 concentration. *Global Biogeochemical Cycles* **16**, 1–13.

Atwell BJ, Henery ML, Rogers GS, Seneweera SP, Treadwell M, Conroy JP (2007). Canopy development and hydraulic function in *Eucalyptus tereticornis* grown in drought in CO_2-enriched atmospheres. *Functional Plant Biology* **34**, 1137–1149.

Barker DH, Loveys BR, Egerton JGJ, Gorton H, Williams WE, Ball MC (2005). CO_2 enrichment predisposes foliage of a eucalypt to freezing injury and reduces spring growth. *Plant, Cell and Environment* **28**, 1506–1515.

Battaglia M, Bruce J (2008). 'Climate change and Australia's Plantation Estate. Milestone report four for project Pn 07.4021'. Forest and Wood Products Australia, Ensis, Hobart.

Battaglia M, Sands PJ, White D, Mummery D (2004). CABALA: a linked carbon, water and nitrogen model of forest growth for silvicultural decision support. *Forest Ecology and Mangement* **193**, 251–282.

Booth TH (2007). Are all Acacias in the south west headed for extinction by 2030? *Australian Forest Grower* **Autumn 2007**, 44–48.

Booth TH, Jovanovic T (2005). 'Tree species selection and climate change in Australia'. Report for the Australian Greenhouse Office, Ensis, Canberra.

Booth TH, McMurtrie RE (1988). Climatic change and *Pinus radiata* plantations in Australia. In *Greenhouse: Planning for Climate Change*. (Ed. GI Pearman) pp. 534–545. E.J. Brill, Lieden.

Booth TH, Jovanovic T, Old KM, Dudzinski MJ (2000a). Using climatic mapping programs to identify high risk areas for *Cylindrocladium quinqueseptatum* leaf blight on eucalypts in mainland South East Asia and around the world. *Environmental Pollution* **108**, 365–372.

Booth TH, Old KM, Jovanovic T (2000b). A preliminary assessment of high risk areas for *Puccinia psidii* (Eucalyptus rust) in the Neotropics and Australia. *Agriculture, Ecosystems and Environment* **82**, 295–301.

BRS (2007). 'Australia's forests at a glance 2007.' Bureau of Rural Sciences, Canberra.

Buckley TN (2008). The role of stomatal acclimation in modelling tree adaptation to high CO_2. *Journal of Experimental Botany* **59**, 1951–1961.

CAB International (2005). Forestry Compendium CD-ROM. CAB International, Wallingford, UK.

Chakraborty S, Murray GM, Magarey PA, Yonow T, O'Brien RG, Croft BJ, Sivasithamparam K, Old KM, Dudzinski MJ, Sutherst RW, Penrose LJ, Archer C, Emmett RW (1998). Potential impact of climate change on plant diseases of economic significance to Australia. *Australasian Plant Pathology* **27**, 15–35.

Consortium (2001). The contribution of mid to low rainfall forestry and agroforestry to greenhouse and natural resource management outcomes. Australian Greenhouse Office and Murray–Darling Basin Commission.

Curtis PS, Wang X (1998). A meta-analysis of elevated CO_2 effects on woody plant mass, form and physiology. *Oecologia* **113**, 299–313.

Dawson TE, Mambelli S, Plamboeck AH, Templer PH, Tu KP (2002). Stable isotopes in plant ecology. *Annual Review of Ecology and Systematics* **33**, 507–559.

Donaldson JD, Pritchard P (2000). Farm forest policy in Australia – a working paper. *Proceedings of the IUFRO 2000 Conference, Cairns.*

Drake BG, Gonzalez-Meler MA, Long SP (1997). More efficient plants: a consequence of rising atmospheric CO_2? *Annual Review of Plant Physiology and Plant Molecular Biology* **48**, 609–639.

Duquesnay A, Breda N, Stievenard M, Dupouey JL (1998). Changes of tree-ring $\delta^{13}C$ and water-use efficiency of beech (Fagus sylvatica L.) in north-eastern France during the past century. *Plant, Cell and Environment* **21**, 565–572.

Eamus D, Jarvis PG (1989). The direct effects of increase in the global atmospheric CO_2 concentration on natural and commercial temperate trees and forests. *Advances in Ecological Research* **19**, 1–55.

Farquhar GD, O'Leary MH, Berry JA (1982). On the relationship between carbon isotope discrimination and the inter-cellular carbon-dioxide concentration in leaves. *Australian Journal of Plant Physiology* **9**, 121–137.

Gielen B, Ceulemans R (2001). The likely impact of rising atmospheric CO_2 on natural and managed Populus: a literature review. *Environmental Pollution* **115**, 335–358.

Gifford RM (1979). Carbon dioxide and plant growth under water and light stress: Implications for balancing the global carbon budget. *Search* **10**, 316–318.

Greenep H, Turnbull MH, Whitehead D (2003). Response of photosynthesis in second-generation *Pinus radiata* trees to long-term exposure to elevated carbon dioxide partial pressure. *Tree Physiology* **23**, 569–576.

Griffin KL, Tissue DT, Turnbull MH, Whitehead D (2000). The onset of photosynthetic acclimation to elevated CO_2 partial pressure in field-grown *Pinus radiata* D. Don. after 4 years. *Plant, Cell and Environment* **23**, 1089–1098.

Gunasekera D, Ford M, Tulloh C (2007). Climate change – issues and challenges for Australian agriculture and forestry. *Australian Commodities* **07.3**, 493–515.

Gunderson CA, Sholtis JD, Wullschleger SD, Tissue DT, Hanson PJ, Norby RJ (2002). Environmental and stomatal control of photosynthetic enhancement in the canopy of a sweetgum (*Liquidambar styraciflua* L.) plantation during 3 years of CO_2 enrichment. *Plant, Cell and Environment* **25**, 379–393.

Hennessy K, Fitzharris B, Bates BC, Harvey N, Howden SM, Hughes L, Salinger J, Warrick R (2007). Australia and New Zealand. Climate change 2007: impacts, adaptation and vulnerability. In *Contribution of Working Group II to the Fourth Assessment Report of the Intergovernmental Panel on Climate Change*. (Eds ML Parry, OF Canziani, JP Palutikof, PJ van der Linden, CE Hanson) pp. 507–540. Cambridge University Press, Cambridge.

Herrick JD, Thomas RB (2001). No photosynthetic down-regulation in sweetgum trees (*Liquidambar styraciflua* L.) after three years of CO_2 enrichment at the Duke FACE experiment. *Plant, Cell and Environment* **24**, 53–64.

Howden SM, Gorman JT (Eds) (1999). 'Impacts of global change on Australian temperate forests'. Working Paper Series, 99/08. CSIRO Wildlife and Ecology, Canberra.

Hughes L, Cawsey EM, Westoby M (1996). Climatic range sizes of Eucalyptus species in relation to future climate change. *Global Ecology and Biogeography Letters* **5**, 23–29.

Idso KE, Idso SB (1994). Plant responses to atmospheric CO_2 enrichment in the face of environmental constraints: a review of the past 10 years' research. *Agricultural and Forest Meteorology* **69**, 153–203.

Keenan RJ, Gerrand A, Nambiar EKS, Parsons M (2006). Plantations and water – plantation impacts on streamflow. Bureau of Rural Sciences, Canberra.

Kimball BA (1983). Carbon-dioxide and agricultural yield: an assemblage and analysis of 430 prior observations. *Agronomy Journal* **75**, 779–788.

Kirschbaum MUF (1994). The sensitivity of C_3 photosynthesis to increasing CO_2 concentration. A theoretical analysis of its dependence on temperature and background CO_2 concentration. *Plant, Cell and Environment* **17**, 747–754.

Kirschbaum MUF (1999a). The effect of climate change on forest growth in Australia. In: *Impacts of Global Change on Australian Temperate Forests*. (Eds SM Howden, JT Gorman). Working Paper Series, 99/08, pp. 62–68.

Kirschbaum MUF (1999b). Modelling forest growth and carbon storage with increasing CO_2 and temperature. *Tellus B* **51**, 871–888.

Kirschbaum MUF (2000). Forest growth and species distributions in a changing climate. *Tree Physiology* **20**, 309–322.

Korol RL, Kirschbaum MUF, Farquhar GD, Jeffreys M (1999). The effect of water status and soil fertility on the C-isotope signatures in *Pinus radiata*. *Tree Physiology* **19**, 551–562.

Kouwenberg LLR, McElwain JC, Kurschner WM, Wagner F, Beerling DJ, Mayle FE, Visscher H (2003). Stomatal frequency adjustment of four conifer species to historical changes in atmospheric CO_2. *American Journal of Botany* **90**, 610–619.

Luxmoore RJ, Wullschleger SD, Hanson PJ (1993). Forest responses to CO_2 enrichment and climate warming. *Water, Air, and Soil Pollution* **70**, 309–323.

Marshall JD, Monserud RA (1996). Homeostatic gas-exchange parameters inferred from $^{13}C/^{12}C$ in tree rings of conifers. *Oecologia* **105**, 13–21.

McGregor JL (2003). 'A new convection scheme using simple closure. Current issues in the parameterization of convection: extended abstracts of presentations at the fifteenth annual BMRC Modelling Workshop, Bureau of Meteorology Research Centre.' (Eds PJ Meighen, AJ Hollis) pp. 33–36. BMRC Research Report No. 93. BMRC, Melbourne.

Medlyn BE, Barton CVM, Broadmeadow MSJ, Ceulemans R, De Angelis P, Forstreuter M, Freeman M, Jackson SB, Kellomäki S, Laitat E, Rey A, Sigurdsson BD, Strassemeyer J, Wang K, Curtis PS, Jarvis PG (2001). Stomatal conductance of forest species after long-term exposure to elevated CO_2 concentration: a synthesis. *New Phytologist* **149**, 247–264.

Monserud RA, Marshall JD (2001). Time-series analysis of $\delta^{13}C$ from tree rings. I. Time trends and autocorrelation. *Tree Physiology* **21**, 1087–1102.

Morison JIL (1985). Sensitivity of stomata and water use efficiency to high CO_2. *Plant, Cell and Environment* **8**, 467–474.

MPIGA (2008). 'Australia's State of the Forests Report.' Montreal Process Implementation Group for Australia, Bureau of Rural Sciences, Canberra.

Myers BJ, Bond WJ, Benyon RG, Falkiner RA, Polglase PJ, Smith CJ, Snow VO, Theiveyanathan S (1999). *Sustainable Effluent-Irrigated Plantations: An Australian Guideline.* CSIRO Forestry and Forest Products, Canberra.

Nambiar EKSN, Brown AG (Eds) (2001). Plantations, farm forestry and water. *Proceedings of a National Workshop*, 20–21 July 2001, Melbourne. Water and Salinity Issues in Forestry. No. 7. RIRDC, Canberra.

National Forest Inventory (2006). National Forest Inventory 2006. Bureau of Rural Sciences, Canberra.

National Forest Inventory (2007a). Australia's forests at a glance. Bureau of Rural Sciences, Canberra.

National Forest Inventory (2007b). National Forest Inventory 2007 Update. Bureau of Rural Sciences, Canberra.

Neumann CR, Hobbs TJ, Bennell M, Huxtable D, Bartle J, George B, Grundy I (2006). FloraSearch – developing broadscale commercial revegetation industries in low rainfall regions of southern Australia. *VegFutures Conference*. 22 March 2006, Albury.

Pinkard EA, Beadle CL (1998). Regulation of photosynthesis in *Eucalyptus nitens* (Deane and Maiden) Maiden following green pruning. *Trees* **12**, 366–376.

Pinkard EA, Battaglia M, Mohammed CL (2007). Defoliation and nitrogen effects on photosynthesis and growth of *Eucalyptus globulus*. *Tree Physiology* **27**, 1053–1063.

Pittock B, Wratt D, Basher R, Bates B, Finlayson M, Gitay H, Woodward A (2001). Australia and New Zealand. In *Climate Change 2001: Impacts, Adaptation and Vulnerability*. (Eds JJ McCarty, OF

Canziani, NA Leary, DJ Dokken, KS White), IPCC Working Group II, Third Assessment Report, pp. 591–639. Cambridge University Press, Cambridge.

Podger FD, Mummery DC, Palzer CR, Brown MJ (1990). Bioclimatic analysis of the distribution of damage to native plants in Tasmania by *Phytophthora cinnamomi*. *Australian Journal of Ecology* **15**, 281–289.

Prime Ministerial Task Group on Emissions Trading (2007). 'Report of the Task Group on Emissions Trading.' The Department of Prime Minister and Cabinet, Canberra.

Raison RJ (2006). Opportunities and impediments to the expansion of forest bioenergy in Australia. CSIRO, Canberra.

Raison RJ, Eamus D, Gifford R, McGrath J (2007). The feasibility of forest free air CO_2 enrichment (FACE) experimentation in Australia. Australian Greenhouse Office, Canberra.

Rawson HM (1992). Plant responses to temperature under conditions of elevated CO_2. *Australian Journal of Botany* **40**, 473–490.

Roden JS, Egerton JJG, Ball MC (1999). Effect of elevated CO_2 on photosynthesis and growth of snow gum (Eucalyptus pauciflora) seedlings during winter and spring. *Australian Journal of Plant Physiology* **26**, 37–46.

Rogers A, Ellsworth DS (2002). Photosynthetic acclimation of *Pinus taeda* (loblolly pine) to long-term growth in elevated pCO_2 (FACE). *Plant, Cell and Environment* **25**, 851–858.

Rundgren M, Björck S (2003). Late-glacial and early Holocene variations in atmospheric CO_2 concentration indicated by high-resolution stomatal index data. *Earth and Planetary Science Letters* **213**, 191–204.

Sharma M, Burkhart HE, Amateis RL (2002). Modelling the effect of density on the growth of loblolly pine trees. *Southern Journal of Applied Forestry* **26**, 124–133.

Smettem K, Crombie S, Harper R, Farrington P, Rivasborge M, Williamson D (1999). 'Defining soil constraints to blue gum water use, growth and survival for managing groundwater recharge'. Final Report for LWRRDC, Canberra.

Steffen W, Canadell JG (2005). Carbon Dioxide Fertilisation and Climate Change Policy. Australian Greenhouse Office, Canberra.

Sutherst RW, Maywald GF (1985). A computerised system for matching climates in ecology. *Agriculture, Ecosystems and Environment* **13**, 281–299.

Tissue DT, Griffin KL, Turnbull MH, Whitehead D (2001). Canopy position and needle age affect photosynthetic response in field-grown *Pinus radiata* after five years of exposure to elevated carbon dioxide partial pressure. *Tree Physiology* **12–13**, 915–923.

Turnbull MH, Tissue DT, Griffin KL, Rodgers GND, Whitehead D (1998). Photosynthetic acclimation to long-term exposure to elevated CO_2 concentration in *Pinus radiata* D. Don. is related to age of needles. *Plant, Cell and Environment* **21**, 1019–1028.

Woodward FI (1987). Stomatal numbers are sensitive to increases in CO_2 from pre-industrial levels. *Nature* **327**, 617–618.

10

BROADACRE GRAZING

CJ Stokes, S Crimp, R Gifford, AJ Ash and SM Howden

KEY MESSAGES:

- The main projected challenges for the grazing industry under climate change are declines in pasture productivity, reduced forage quality, livestock heat stress, greater problems with some pests and weeds, more frequent droughts, more intense rainfall events, and greater risks of soil degradation. These challenges partly reflect the projected drying trend across the rangelands, which is uncertain, and, at least in some locations, climate conditions could become more favourable for pastoralism, presenting new opportunities. Uncertainties over which locations will fare better under climate change and which will be worst affected present a major challenge for adaptation at regional and property scales.

- Increased adoption of climate forecasting-based strategies for coping with climate variability will assist graziers to incrementally adjust management to 'track' the early stages of climate change (but these strategies need to incorporate considerations of long-term climate change trends).

- The adaptation challenge and opportunity need to be clearly defined by quantifying the range of plausible impacts that uncertain climate change could have on the grazing industry and framing adaptation options in relation to existing management pressures. Likely responses of graziers and policy-makers to these impacts need to be determined and comprehensively evaluated.

- The most arid and least productive rangelands may be the most severely impacted by climate change, while the more productive eastern and northern grazing lands may provide some opportunities for slight increases in production. However, a rigorous analysis of the regional variation in impacts of projected climate change on grazing lands still needs to be conducted, allowing for both the heterogeneous nature of the grazing resources and expected regional variations in climate changes.

- Participatory research approaches that incorporate producer knowledge will assist in assessing vulnerability of the grazing industry to climate change, indentifying practical adaption options, and determining the limits of adaptations for coping with climate change.

Introduction

Extensive grazing is by far the most widespread agricultural land use in Australia. Most of the continent is unsuitable for intensive agricultural production and is used instead for low-intensity production of beef and sheep (meat and wool). The grazing industry has been an important contributor to the overall economic growth of Australia, but has also been susceptible to the influence of variable climate (Campbell 1958; Anderson 1991). The sensitivity of agricultural production to climatic fluctuations has been identified as an important contributor to volatility in Australia's economy (White 2000). For example, modelling studies by ABARE determined that

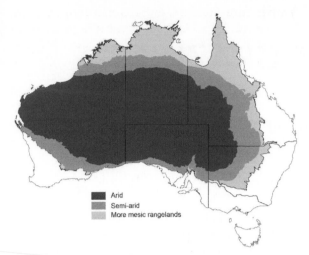

Figure 10.1: Classification of Australian rangelands based on a moisture index: Arid (rainfall <20% of potential evaporation [≈250 mm per year]), Semi-arid (rainfall = 20 to 40% of potential evaporation [≈250–350 mm per year]), remainder (rainfall >40% potential evaporation [≈350 mm per year]).

the drought event of 1994–95 reduced the gross value of farm production by as much as 9.6% or $2.4 billion (Hogan *et al.* 1995).

The native and improved pastures that form the rangelands cover over three-quarters of the continent. This broad geographic spread is reflected in their great diversity. However, for simplicity, we classify the rangelands into just three categories: (1) the arid rangelands and deserts of the centre, (2) the adjacent semi-arid region, which includes the productive Mitchell Grass Downs, and (3) the remaining more mesic rangelands consisting of the tropical savannas to the north and productive, highly modified pasture in the subtropical woodlands to the east (Figure 10.1). In addition, more intensive broadacre grazing occurs in some higher rainfall regions on the southern margins of the rangelands. The variety of ecological characteristics within these grazing lands means that there are likely to be regional and local differences in the effects of climate change.

Climate change is likely to add to and exacerbate existing pastoral management challenges such as undesirable grass species, shrub invasions, soil erosion, salinisation, soil acidification, and problems with animal nutrition and health. While technological advances have been developed to address some of these issues (Quirk 2002), with climate change it will become even more important to ensure widespread adoption

of these practices. We review the potential impacts of climate change on Australia's grazing industries and options for adapting to these changes. Some regions are likely to be better off and others worse off under climate change. But, even if the grazing industry as a whole is able to maintain its current level of productivity, significant management changes will be required (1) to cope with the uncertainty around which changes will occur in which locations, and (2) to make the necessary adjustments to adapt to local changes in climate. We focus primarily on the challenges that climate change may bring, since these will require the most effort, assistance and preparation to address.

Impacts of projected climate change

Changes in atmospheric concentrations of greenhouse gases will affect grazing systems through complex interactions involving effects on plant growth, rising temperatures, changing climate, grazing management and a potentially wide range of indirect impacts that may affect vegetation, natural resources and animal production. But the most direct influences will be through the ways that changes in the quantity and quality of forage affect livestock performance. In the longer term, if graziers do not adjust their land management practices to suit the changed climate conditions, then risks of land degradation may increase and/or new opportunities may be lost (McKeon *et al.* 2004). We consider some of the key management issues that graziers will have to tackle to ensure that pastoral enterprises remain productive and viable: (1) forage productivity and quality; (2) vegetation composition (including weeds and fire management); (3) soil degradation, and (4) animal husbandry and health (including pests and disease).

Forage productivity and quality

Livestock performance is strongly related to the availability of young, digestible plant material (Mannetje 1974; Ash *et al.* 1982), which in turn is strongly influenced by the frequency of climate conditions suitable for plant growth. McKeon *et al.* (2008) have shown that a climate-sensitive index of the number of pasture growth days is strongly correlated with animal performance and livestock carrying capacities. Projected changes in rainfall regimes (Chapter 2), rising temperatures and higher levels of atmospheric CO_2 will all alter patterns of plant growth and pasture productivity.

The dominant influence of climate change on pasture growth is expected to be through changes in the amount of rainfall (Hall *et al.* 1998; Crimp *et al.* 2002). In addition, changes in the temporal distribution of rainfall may reduce the effectiveness of rainfall in regionally specific ways, through projected increased variation within seasons (fewer, more intense rainfall events) and from year to year (more frequent droughts). Modelling studies suggest that pasture growth will amplify changes in rainfall, so that the magnitude of change in forage production will exceed the percentage change in rainfall. For example, in a simulation study of the sensitivity of pasture growth across Australian rangelands to changes in rainfall, a 10% decline in rainfall reduced pasture growth by 11 to 15% (depending on soil type) while a 10% increase in rainfall produced a 12 to 18% increase in pasture growth (Crimp *et al.* 2002). This amplification is greatest in the arid and semi-arid areas, and is least in the tropical north. Several other studies have demonstrated the sensitivity of pasture production to small changes in climate (Scanlan *et al.* 1994; Johnston *et al.* 1996) and non-linear responses to rainfall (Hall *et al.* 1998). In addition, changes in river flow regimes and beneficial flooding may alter the production of locally important ephemeral pastures on floodplains (White 2001) and these changes too are likely to magnify any changes in rainfall. For example, in arid regions a 10% change in rainfall can lead to a 30 to 40% change in runoff (Chiew *et al.* 1995).

Rising temperatures could benefit pastures in cooler, southern climates by increasing the length of the growing season and reducing frost damage. However, increased plant growth in the cooler months could deplete soil moisture at the expense of subsequent pasture growth in the spring with the net effect highly situation-dependent. In warmer climates, increased heat stress, and possibly increased evaporative demand, would likely have negative effects on pastures.

The most certain aspect of climate change is that rising atmospheric CO_2 is affecting rangeland ecosystems even in the absence of rainfall and temperature changes (and this effect will increase in strength as CO_2 levels continue to rise). In water-limited rangelands, the dominant means by which CO_2 will affect plant growth will be through changing patterns of plant water use. Experiments on the C_3 (temperate) pasture component responses to CO_2 have demonstrated that stimulated plant production was almost entirely attributable to indirect effects of moisture savings, rather than directly stimulated photosynthesis (Volk *et al.* 2000; Niklaus *et al.* 1998). Correspondingly there is growing evidence that C_4 (tropical) rangeland grasses may not be substantially less responsive to CO_2 than C_3 grasses, particularly under field conditions (Morgan *et al.* 2004; Owensby *et al.* 1993). The effect of CO_2 in stimulating pasture growth may be greatest in ecosystems receiving intermediate amounts of rainfall (about 500–1000 mm per year, depending on latitude) (Nowak *et al.* 2004; Stokes and Ash 2006) where water is limiting during most periods of active plant growth. There are strong interactions of pasture responses to CO_2 with other variables such as temperature, soil moisture and soil nutrient availability, especially nitrogen (Fischer *et al.* 1997; Suter *et al.* 2002). Consequently, the influence of CO_2 on plant growth under field conditions, while still present in both moist temperate (Campbell *et al.* 1997) and arid (Smith *et al.* 2000) conditions, can be quite variable (Nowak *et al.* 2004), and could become more so as vegetation composition and species interactions change over time (Edwards *et al.* 2001; Smith *et al.* 2000).

The effects of CO_2 have yet to be properly incorporated and validated in grazing models, but some initial attempts, using a conservative representation of CO_2 effects, have been made to simulate how combined changes in temperature, rainfall and CO_2 will influence pasture production and livestock carrying capacities (McKeon *et al.* 2009). These sensitivity analyses showed that the combined effects of a 10% *increase* in rainfall, warmer temperatures (+3°C) and doubling of CO_2, increased pasture growth by 15–30% over most of Australia's rangelands, with the weakest effects (0–10%) in the dry monsoonal regions of northern Australia (where soil fertility would be expected to limit pasture growth). When rainfall was instead *reduced* by 10%, there was a slightly greater than 10% decline in pasture production over much of inland Australia (indicating that the effects of increased temperature and declining rainfall outweighed the conservatively represented benefits of increasing CO_2). The simulations suggest that rising temperatures amplify the effects of declining rainfall while rising CO_2 may have some ameliorating effect, emphasising the importance of being able to correctly estimate the impact of these opposing effects.

Animal nutrition depends strongly on three aspects of forage quality: energy supply (non-structural carbohydrate), protein (nitrogen) content and digestibility. All of these forage attributes are influenced by climate directly, as well as indirectly through altered phenology of pasture growth (since forage quality declines as leaves age and cure). Stimulation of forage production under elevated levels of atmospheric CO_2 is associated with changes in forage quality through declines in forage protein content (Wand et al. 1999), and increased forage non-structural carbohydrates in C_3 species (Wand et al. 1999; Lilley et al. 2001) but not C_4 (Wand and Midgley 2004). There is also decreased digestibility of tropical grasses, although there may be little change in digestibility of other species (Lilley et al. 2001). Warmer conditions tend to significantly decrease non-structural carbohydrate concentrations (and digestibility in tropical species) while also slightly reducing leaf protein content (Wilson 1982). In addition there will be changes to seasonal patterns of variation in forage quality. Growth seasons could be extended either by increased growth into the winter period in temperate areas (Cobon and Toombs 2007) or prolonged availability of soil moisture (and delayed grass senescence) as a result of CO_2-induced conservation of water by grasses. In contrast, declining spring and autumn rainfall will tend to shorten growing seasons. Altered seasonal patterns of variation in forage protein and utilisable energy will have consequences for ruminant nutrition (Beever 1993), but the combined effects of these interacting climate change influences has yet to be determined. It will be difficult to generalise how the nutritive quality of forage will change in any one area because in some situations the increase in digestible energy content will dominate while elsewhere the effect of decreased protein content will dominate (Howden et al. 2008).

Vegetation change

Climate change, in combination with management (such as grazing, burning, tree regrowth control or legislated lack thereof, applied nutrients, and herbicide use), can have major impacts on botanical species composition. Past shifts in vegetation composition in response to climate variability indicate the responsiveness of pastures to climate changes (McKeon et al. 2004; Bisset 1962). As vegetation changes it will be essential to maintain perennial grasses and shrubs to provide dry season and drought feed, fuel for fires where appropriate, and surface cover to protect soils. Recent drought episodes have demonstrated the potential of climate extremes to cause widespread plant mortality, even for long-lived trees (Fensham 1998). It is not known how most existing species will respond to unprecedented extremes of temperature and desiccation, so changes in vegetation will have to be monitored in order to adapt appropriately.

Further changes in rangeland vegetation are expected in response to rising atmospheric CO_2 concentration (Warwick et al. 1998; Howden et al. 1999b; Howden et al. 2001b). Rising CO_2 will affect pastures by changing patterns of soil moisture availability, increasing the availability of moisture deeper in the soil profile (Gifford et al. 1996). Differences in species responses to CO_2 are therefore likely to be strongly influenced by differences in rooting patterns and the ability of plants to rapidly exploit conserved soil moisture from reduced transpiration. Deep-rooted woody plants and legumes are likely to be advantaged over grasses at higher CO_2 levels. In southern pastures with mixed C_3 and C_4 grasses, rising temperatures may favour an increase in C_4 species (Cullen et al. 2008), which generally provide a less nutritious forage than C_3 grasses.

Changing fire regimes could also influence vegetation, particularly the balance between woody plants and grasses. Climate change could affect fire regimes in several ways. First, any changes in pasture production (discussed above) will affect fuel loads, unless utilisation levels by livestock are adjusted to match changes in grass growth. Second, pastures could cure earlier under warmer climates, shifting the timing of fires to earlier in the season, and increasing the potential for more intense fires later in the season. Third, hotter and drier conditions could increase the risk of wild fires and make prescribed burns more difficult to manage. Fourth, if increases in year-to-year rainfall variability are accompanied by similar increases in variation of fuel loads, then fire regimes could shift toward fewer, more intense fires, particularly in the northern tropical pastures. Finally, human responses to these changes could alter how fire is used in rangelands. From a management perspective, there may be a growing requirement to use fire as a tool to control increasing woody vegetation and other invasive species (Howden *et al.* 2001b).

Climate change will influence both pasture species and invasive weedy species (particularly woody weeds and legumes). The 'desirability' of species and vegetation change will need to be viewed and re-evaluated in the context of the potential species that are most suitably adapted to the emerging climate (rather than trying to maintain familiar species that used to grow in particular locations before). It will be more productive to recognise, facilitate and direct climate-induced changes in species distributions rather than trying to completely prevent them. Accepting and managing vegetation change may require substantial shifts in attitudes, particularly regarding the definition and roles of 'invasive' (needing control) and 'useful' (in terms of production, environmental, biodiversity and aesthetic values) species.

Soil degradation

Changing climate, through its influence on hydrological processes, will strongly influence soil degradation processes such as erosion and salinisation. Rainfall intensity is projected to increase in the rangelands while the number of dry days is projected to increase (CSIRO and Australian Bureau of Meteorology 2007; Tebaldi *et al.* 2006; Chapter 2 in this volume). These changes in rainfall regimes are likely to increase soil erosion through the combined effects of increasing runoff while reducing protective vegetation cover (McKeon *et al.* 2004). Erosion risks will likely be further exacerbated by increased year-to-year variability in rainfall, creating a greater chance of erosion events where a wet year (high runoff) follows a drought (when plant cover is low and soils become highly susceptible to erosion). Even slight erosion of surface soils can reduce infiltration and remove a large proportion of important soil nutrients, thus markedly reducing potential pasture productivity (McKeon and Hall 2000). Soil erosion may become an increasingly challenging management consideration.

Both changing rainfall patterns and altered plant water use (with rising CO_2 levels) will influence salinity risk, but the combined effects of these influences on hydrological processes in rangelands has yet to be evaluated. On the positive side, a growing interest in biosequestration of carbon in rangeland soils (particularly moderately degraded soils where it may be most feasible to restore depleted carbon stores; see Chapter 15) is raising awareness of the multiple benefits of managing grazing lands to maintain and improve soil health.

Animal husbandry and health

Climate change will substantially increase the frequency of heat stress days, particularly in northern Australia. This will reduce livestock productivity, decrease reproductive rates and increase concerns about animal welfare in locations where grazing populations are concentrated, such as feedlots (Howden and Turnpenny 1997; Howden *et al.* 1999a; Mader and Davis 2004; Amundson *et al.* 2006). The long-term upward trend in cattle numbers in Australia has been made possible by continued breeding improvement in livestock, especially resistance to drought, pests and diseases (Lloyd and Burrows 1988). As conditions in southern regions become harsher, pastoralists will have to rely on hardier northern cattle breeds (and crosses), but these hardier traits are associated with lower productivity, fecundity and meat quality.

Livestock water requirements will increase as temperatures warm, e.g. a 13% increase in demand for 2.7°C warming (Howden and Turnpenny 1997). This will also mean that livestock will be unable to travel as far from watering points, limiting use of the grazing resource in extensive grazing operations and tending to increase grazing pressure and risks of soil degradation near watering points.

Pests and diseases are other important considerations for animal health and performance. The grazing industry is already susceptible to pests and disease under current climate conditions (Mcleod 1995; Sutherst 1990; Sutherst *et al.* 1996). For example, cattle ticks cost the northern tropical beef industry $41 million annually in control measures and $91 million in productivity losses (Hall *et al.* 1998; Mcleod 1995), while roundworms, lice and blowflies cost the Australian sheep industry $552 million annually in control measures and production losses (Mcleod 1995). The potential expansion in the geographic range of pests and diseases, particularly a poleward range shift of tropical species (Sutherst 2001), represents a major risk to the grazing industry. For example, projections indicate a southward expansion in distribution of the insect vector of blue-tongue disease, *Culicoides wadia* (Sutherst 2001). Cattle ticks (*Boophilus microplus*) may also spread further south with projected losses in live weight gain of 21 600 tonnes per year by 2100, compared to present estimated losses of 6600 tonnes per year (White *et al.* 2003).

Broad-scale indirect impacts

In addition to direct climate change impacts, broader-scale indirect effects such as changes in local and international markets, economic returns (price vs. costs), policy and land use (Box 10.1) are also likely to shape the future of the grazing industry. For example, meat prices are strongly influenced by production of overseas competitors (influenced in turn by world grain production) and hence global variation in climate change impacts will also influence the financial performance of local grazing enterprises (White 1972; Herne 1998). The availability of grain for livestock production may also become restricted by growing demand from the emerging biofuel industry. Internationally, many trading competitors in the livestock industry are likely to be affected in similar ways to Australia. The key exceptions appear to be New Zealand, northern Europe and eastern USA and Canada. The strong trend to increased consumption of livestock products such as meat and dairy in Asia (Dalton and Keogh 2007) seems unlikely to be affected by climate change concerns (except perhaps for issues related to greenhouse gas emissions, discussed below). In contrast, climate changes seem likely to have a negative impact on both wool demand and wool supply in that warming climates may reduce demand for woollen clothes while countries like China may have improved conditions for grazing (Keeling *et al.* 1996; Myneni *et al.* 1997; Harle *et al.* 2007).

Climate-related changes in policy will also have important implications for the grazing industry, particularly policies aimed at reducing agricultural greenhouse gas emissions (Chapter 15). Government and consumer concerns over methane emissions from ruminant livestock could see the imposition of financial disincentives (e.g. through the proposed Carbon Pollution Reduction

Box 10.1: Example of a land management response to changing rainfall in the rangelands

Enterprises operating at the extreme of their natural range may be forced to change land use under climate change conditions. For example, dryland cropping is an important industry for the Fitzroy Basin, but it is located at the northern margin of the wheat cropping region of Australia. Prior to the 1970s, the Emerald region was primarily used for grazing beef cattle despite the potential for higher gross margins in cropping. Subsequently in the next 30 years cropping developed in importance and it is possible that the relative suitability of cropping versus grazing is an artefact of recent climate (Howden *et al.* 2001a). If the increase in cropping was due to long-term multi-decadal climate variability then cropping is likely to decline in the region as conditions return to those experienced earlier in the weather record. If the increase in cropping was related to climate change (a persistent upward or downward trend in climate) then cropping in the region is likely to persist. A further consequence of this shift in land use could be greater sediment loads in rivers (Cobon *et al.* 2007), which would eventually be deposited into the water around the southern part of the Great Barrier Reef.

Scheme) and reduced demand for meat. There could be several other consequences of incentives to reduce greenhouse gas emissions from grazing lands including: (1) loss of productive grazing lands to reforestation; (2) changes in land management to improve soil carbon stores, and (3) modification of burning practices in rangelands, particularly the northern savannas, to reduce emissions from fires.

Regional variation in impacts

Regional variation in climate change projections (Chapter 2) combined with geographic diversity in the sensitivity of rangelands to these changes (McKeon *et al.* 2009) means that responses of rangelands to climate change are likely to differ from region to region. Anticipating this variation and engaging with pastoralists directly will assist in preparing an appropriate range of adaptation strategies for land managers. A comprehensive analysis of regional impacts of climate change, combining current climate projections with eco-system models, has yet to be conducted. Accepting the additional caveat that regional projections of climate change involve a high degree of uncertainty, particularly for rainfall, we suggest some possible regional differences in climate change impacts across the rangelands (using the three rangeland regions mapped in Figure 10.1).

Based on the climate models presented in Chapter 2, reductions in rainfall and increases in temperature are both projected to be greatest in the arid rangelands. Rising temperatures are likely to be greatest towards the centre of Western Australia, while drying trends are projected to be greatest across the central to south-western parts of the country. The semi-arid mixed sheep and cattle grazing areas in the south-west of the rangelands are also projected to be adversely affected by declining rainfall. These climate changes are likely to be amplified by the high sensitivity of these pastures to climate change (McKeon et al. 2009).

The more productive northern and eastern rangelands are likely to be the least adversely affected by climate change (Crimp et al. 2002). The northern savannas are the only area of the country where rainfall is not projected to decline with climate change, and the proximity to the coast should moderate increases in temperature (Chapter 2). However, monsoon rains are projected to increase in intensity, increasing the risks of soil erosion and potentially reducing the effectiveness of rainfall (by increasing runoff and reducing available soil moisture for plant growth). In the eastern woodlands there is projected to be some drying and increases in year-to-year variability in rainfall may increase the frequency of droughts (Cai 2003; Nicholls 2003; Whetton and Suppiah 2003). Simulation models suggest that moderate amounts of warming may benefit pasture growth by extending the growing season, so that there may be a slight net increase in pasture productivity (Howden et al. 1999c; Crimp et al. 2002), but small increases in the length of the growing season and greater growth efficiency due to higher CO_2 failed to compensate for reduced moisture levels in central Queensland (Cobon et al. 2005a). Rising CO_2 may also benefit these pastures (Stokes and Ash 2006), but this effect may be partially offset in the longer term if tree biomass increases.

The general pattern of climate change impacts across the rangelands may therefore be that the least productive arid rangelands are the most negatively affected. Marginal pastoral enterprises in arid rangelands may be at the greatest risk of becoming non-viable under climate change. In contrast, the more productive eastern and northern rangelands may be the most likely to provide opportunities for slight increases in productivity (Heyhoe et al. 2007). Rising temperatures are most likely to cause heat stress problems in the northern rangelands, but in the cooler, southern areas, some warming may benefit pasture growth.

Options for dealing with climate variability

Australia's rangelands are characterised by high year-to-year variability in rainfall that, in turn, drives high variability in plant growth, nutrients available to livestock and availability of land management options (e.g. use of fire and spelling) (McKeon et al. 2004). Analysis of historic relationships between climate variability and changes in pastoral land condition have shown that episodes of major degradation in Australian rangelands have often been associated with inappropriate management responses to drought, that land can be degraded very rapidly, and that recovery of degraded lands is very difficult (McKeon et al. 2004). Climate variability therefore presents not only an economic challenge for managing variable cash flows, but also a serious challenge to maintaining the long-term productivity of rangelands. This challenge is likely to increase with climate change if climate variability increases and conditions become warmer and drier.

In many parts of Australia, particularly in the east, much of the year-to-year variability in rainfall is associated with the El Niño-Southern Oscillation (ENSO) and the Inter-decadal Pacific Oscillation (IPO) (McKeon et al. 2004). Based on these phenomena, operational seasonal forecasts have been developed using changing patterns in sea surface temperatures (Day et al. 2000). These forecasts have been used to formulate tactical strategies whereby stocking rates could be pre-emptively adjusted according to the likelihood of favourable or unfavourable seasonal conditions (Ash et al. 2007; McKeon et al. 2000; Stafford Smith et al. 2000; Cobon et al. 2005b). Such climate-responsive management strategies also build in automatic tracking of longer term climate cycles and trends, and therefore provide the ability to adapt to the early stages of climate change (McKeon *et al.* 1993) providing that climate change does not fundamentally alter broad-scale processes behind ENSO and the IPO. These approaches should appeal to graziers and their advisors who, faced with the continuing uncertainty of climate change projections, may prefer to focus primarily on responding to immediate rainfall variability (however, see 'Maladaptations' below, for cautions in using this approach.)

Aside from seasonal forecasts, there are several other strategies for coping with climate variability. These include using a conservative set stocking rate; diversifying sources of income, and diversifying climate risk geographically across multiple pastoral properties in regions with non-synchronous patterns of climatic variability. While these strategies would not necessarily provide the automatic benefit of tracking climate change, they may still be of adaptive value in increasing the capacity of pastoral enterprises to cope with future uncertainty.

Options for dealing with projected climate change

In order to adapt to climate change, pastoralists will need to prepare management responses to take full advantage of new opportunities and minimise any negative impacts (as outlined above). If graziers monitor the changes taking place in their enterprises, including their management responses to them, they will be in a better position to understand and adapt to climate change. At a national industry level, there are likely to be winners and losers, but significant changes in management will have to take place in adapting to climate change, even if the industry as whole is ultimately no worse off.

Grazing and pasture management

In those extensively managed rangelands that are negatively impacted by climate change, there may be few options to compensate for declining pasture productivity. Past efforts to increase pasture production in more humid rangelands have often relied on removing trees and shrubs to increase the availability of water, nutrients and light for grass growth (Burrows *et al.* 1988). However, this has been controversial and restricted by legislation because of the impacts on biodiversity,

greenhouse gas emissions and catchment hydrology. Nonetheless, to counteract the trend towards woodier vegetation (Burrows *et al.* 2002), it may become more desirable to use fire and selective thinning to maintain current tree levels and pasture productivity. In temperate and Mediterranean rangelands the need to restore populations of native woody species, which have declined under past management, will compound the challenge of declining pasture production where the climate becomes drier (Dorrough *et al.* 2006; Pettet and Froend 2000).

Current management, and particularly rehabilitation, of pastures requires careful grazing management including conservative stocking rates, strategic spelling and responsive adjustments to stocking rates based on seasonal climate forecasts (McKeon and Howden 1992; Johnston *et al.* 2000; Cobon and Clewett 1999). These practices are likely to become more important with climate change and will be necessary to ensure desirable pasture species establish and are maintained as species ranges shift under climate change. Similarly, careful grazing management will be required to facilitate the establishment of any introduced species. With shifts to rainfall regimes that increase the risk of soil erosion, it will become increasingly important to ensure that ground cover is maintained in rangelands. It will also be necessary to redefine safe carrying capacities, pasture utilisation levels and grazing management practices, and to continually review and adjust these in accord with the changing climate (McKeon *et al.* 2009).

In more intensively managed pastures there will be some additional options for maintaining forage production and quality. In temperate pastures, there is generally insufficient metabolisable energy in fodder for protein to be fully utilised. Breeding grass varieties with high levels of non-structural carbohydrates, the source of energy for the rumen (Evans *et al.* 1996), could therefore improve forage quality, even if forage protein declines under climate change. Introduced legumes and fertiliser could also be used to increase nitrogen input to, and productivity of both pastures and livestock (although soil acidification and increased N_2O emissions need to be considered). It may also be possible to breed and sow new pastures that are better adapted to warmer temperatures and higher CO_2. In extensively managed rangelands where these options will be less viable, pastoralists will likely have to rely on increased use of feed supplements (N, P and energy) and rumen modifiers to compensate for declining forage quality. In rangelands that are close to grain-producing areas it may be possible to concentrate on utilising pasture growth earlier in the season and destock earlier, to make greater use of feedlots to finish livestock.

Managing pests, diseases and weeds

Several existing methods will be suitable for combating the spread of rangeland pests and disease under climate change including applications of pesticide and chemicals to respond to outbreaks; strategic use of fire to control weeds; biological weed control; vaccinations to enhance resistance to existing pests and disease, and selection of tick-resistant cattle (*Bos indicus*) in northern Australia. Pesticides are likely to become less effective options in the future because of rising costs and resistance, so alternative options will increasingly need to be considered. Developing improved predictive tools and indicators may provide opportunities to reduce reliance on pesticides. Quantitative modelling has proved particularly useful in managing cattle ticks in northern Australia by identifying areas and periods of greatest risk. Other options that could be developed to improve management of pests and diseases include identifying opportunities to introduce more species of dung fauna (to eradicate buffalo fly larvae), encouraging greater use of traps (buffalo fly and sheep blowfly) and vaccines (cattle ticks and worms). It will also be important to improve monitoring and border surveillance to (1) restrict the southward expansion of pests and diseases whose ranges are currently limited by cold temperatures (e.g. flies and ticks), and (2) prevent the establishment of new exotic pests.

Woody weeds, particularly legumes in tropical rangelands, are likely to require more attention with climate change. Chemical and mechanical control will likely be more economically viable under such conditions (Burrows *et al.* 1990). Where pasture productivity increases with climate change, there may also be opportunities for more frequent use of fire to control woody weeds (e.g. Howden *et al.* 2001a).

Livestock management

The grazing industry will be vulnerable to an increasing incidence of extreme temperatures and declining water availability under climate change. Howden and Turnpenny (1997) and Howden *et al.*

(1999a) showed that the incidence of heat stress has increased significantly since 1957 across large areas of Australia. This suggests that the practice of selecting cattle lines with effective thermoregulatory controls or adaptive characteristics within breeds, such as feed conversion efficiency and coat colour (Finch *et al.* 1984), would need to continue if current levels of productivity are to be maintained. This practice may need to become more common in more southerly regions as the frequency of heat stress days increases. Additional adaptation strategies such as modifying the timing of mating could also serve to match nutritional requirements of cow and calf to periods with favourable seasonal conditions. This means that the animal production system (cow/calf, steer trading, finishing for market) would have to become more flexible in order to accommodate potential changes to seasonal variability (McKeon *et al.* 2004), including changes in timing of supplementation and weaning (Fordyce *et al.* 1990).

For some livestock operations such as stockyards and feedlots, the construction of shading and spraying facilities may represent an economically feasible adaptation measure (see also Chapter 11). It may also be necessary to plant plots of suitable shade trees, and increase the number of water points. In areas that become more prone to flooding it will be important to provide livestock with access to areas of higher ground.

Broader-scale considerations

As confidence in climate change projections grows and trends in observed weather data become more apparent, the motivation for adaptation in the grazing industry will be likely to increase. However, the adoption of new property management practices will also require demonstration of the benefits of new adaptation options; buffering against establishment failure of new practices during less favourable climate periods; alteration of transport and market infrastructure to support altered production; continuous monitoring of climate change impacts and management responses to adjust actions and ensure effective and appropriate adaptation, and development and modification of government policies and institutions to support implementation of the required changes (McKeon *et al.* 1993).

Government, at all levels, is continually developing and modifying strategies, initiatives and policies to deal with environmental issues such as land condition, biodiversity, greenhouse gas emissions, drought, salinity and water quality. Similarly, management practices in the grazing industry are constantly being adapted to an ever-changing operating environment, such as shifting rural population densities, reduced on-farm profitability, changes in government legislation for drought relief, and enforcement of legislation on resource management and animal welfare (McKeon *et al.* 1993). It will be important to identify and promote any synergies with existing initiatives and address any conflicts that create barriers to adaptation. Of particular importance will be understanding how climate change will influence the incidence of 'exceptional circumstance' ('drought') conditions (Hennessy *et al.* 2008), and ways in which drought policy can be altered to encourage climate change adaptation. Similarly, initiatives and strategies that encourage reductions of greenhouse gas emissions, carbon sequestration and potential trading of carbon may present synergies for exploring changes to on-farm management (Chapter 15).

Risks of maladaptation

There are several situations in which responses aimed at dealing with specific aspects of climate change could have unintended negative consequences when viewed in the broader context of land management and climate change adaptation. For example, in locations where climate change is non-linear (e.g. if rainfall were to increase for several decades before ultimately declining), the recommended strategy of adjusting management to track seasonal forecasts of climate variability could result in tracking short-term fluctuations that are the reverse of long-term climate change trends. Under such conditions, management changes that have been made to take advantage of the short-term favourable conditions may leave pastoralists more poorly prepared for longer term drying of the climate. Once climate models are able to provide a suitably reliable indicator of long-term trends, these should be incorporated into future tools for dealing with climate variability so that decisions to take advantage of temporarily favourable opportunities can be balanced against preparing for longer term impacts of climate change.

The benefits of introducing legumes to improve forage quality will need to be carefully considered against the risks of soil acidification and pasture degradation from overgrazing of native pasture species. More generally, caution needs to be exercised in any proposed solutions to improve pastures by introducing species/varieties that are hoped to be superior under altered climate conditions. The past history of introducing 'desirable' species to pastures has not always been successful, with some species becoming weeds, reducing biodiversity or otherwise negatively affecting ecosystem health (e.g. Lonsdale 1994). Likewise, efforts to sequester carbon in rangelands will also have to be carefully considered against the long-term costs and benefits of maintaining the enhanced vegetation and soil carbon stores (Chapter 15) often in the face of climate changes which will tend to reduce them. In developing sound climate change adaptation strategies it will be important to consider not only the direct, intended benefits of proposed actions, but to also take a broader systems view of the possible unintended consequences (both positive and negative) of those actions.

Costs and benefits

To date there have not been any comprehensive analyses of the costs and benefits of climate change impacts to the pastoral industry, nor of the costs and benefits of adaptation measures to offset these impacts. There have been some modelling studies to determine the sensitivity of pasture production to climate change in different rangeland regions (Crimp *et al.* 2002; McKeon *et al.* 2008) and Howden *et al.* (2003) assessed the economic trade-offs between beef production and carbon sequestration (if a carbon trading market including agriculture were to emerge). But there has been no assessment of the costs of developing and implementing adaptation strategies and the benefits these could yield in reducing the impacts of climate change or exploiting new opportunities.

Discussion

Climate change will have a variety of location-specific impacts because of regional differences in (1) the types of climate change that occur; (2) the sensitivity of grazing lands to climate change, and (3) the capacity of graziers to respond to these changes. A range of adaptation options will therefore be required to accommodate this diversity of situations. We have discussed some of these options in this chapter (Table 10.1).

Table 10.1: Summary of climate change adaptation options for the grazing industry. Priority 1 (high), 2 (medium) and 3 (low).

Adaptation option	Priority
Broad-scale adaptation	
Modify existing Federal and State Drought Schemes to encourage adaptation	1
'Mainstream' climate change considerations into existing government policies and initiatives, e.g. Greenhouse Challenge, salinity, water quality and Landcare activities	1
Work with the pastoral industry to evaluate potential adaptive responses to the system-wide impacts of a range of plausible climate change scenarios	1
Continuously monitor climate change impacts and adaptation responses, adjusting actions to support and ensure effective and appropriate adoption	2
Grazing and pasture management	
Introduce responsive stocking rate strategies based on seasonal climate forecasting (and which include consideration of climate change trends)	1
Progressively recalculate and adjust safe stocking rates and pasture utilisation levels taking into account observed and projected climate change	1
Accept climate-induced changes in vegetation and modify management accordingly	2
Make greater use of strategic spelling	2
Improve on-property water management, particularly for pasture irrigation	2
Improve nutrient management using sown legumes and phosphate fertilisation where appropriate	2
Develop software to assist proactive decision-making at the on-farm scale	2
Expand routine record keeping of weather, pests and diseases, weed invasions, inputs and outputs	2
Diversify on-farm production and consider alternate land uses	3
Managing pests, diseases and weeds	
Improve predictive tools and indicators to monitor, model and control pests	2
Increase the use of biological controls (with caution)	2
Incorporate greater use of fire and alternative chemical and mechanical methods for controlling weeds and woody thickening	2
Livestock management	
Select animal lines that are resistant to higher temperatures but maintain production	2
Modify timing of mating, weaning and supplementation based on seasonal conditions	2
Provide extra shade using trees and constructed shelters	2

Over the short term (next 20 years), possible increases in the occurrence of extreme events and uncertainty over climate change trends will make risk-based adaptation approaches such as seasonal climate forecasting appealing. This will facilitate the incremental changes in management practices that land managers are likely to incorporate anyway as part of their usual business of adjusting to changing operating conditions. Such responses are likely to occur with little policy intervention and may include adaptations such as greater use of feed supplements, shifts to hardier animal breeds and modified herd management. Many of these options could be promoted through existing initiatives that encourage improved management of natural resources in grazing lands, such as Landcare and extension activities by State agencies.

Over the longer term (20–100 years), greater government intervention will probably be required to support adaptation in grazing lands, particularly in those locations where the rate or magnitude of climate change overwhelms the existing capacity of local communities to respond. This will require scientists, affected grazier communities and supporting institutions (1) working together to anticipate climate change impacts; (2) planning ahead to develop and evaluate appropriate adaptation options, and (3) building capacity within communities (Chapter 15) to implement these strategies. Such actions will require policy support to encourage adoption and assist communities through the risky transition periods. It will be essential to build climate adaptation considerations into existing policy, particularly those relating to drought, and to address policy conflicts where they exist.

Given the uncertainties involved, it will also be important to monitor the outcomes of changing climate, land management and policy so that inappropriate assumptions and actions can be detected early, learned from and corrected (Howden *et al.* 2007).

Grazing will continue to play an important role in shaping Australia's economy and land use over the coming decades. Significant contributions to sustainable management have resulted from past efforts to better understand grazing ecology, grazing practices and productivity (Quirk 2002). Climate change is adding another dimension to the existing challenges facing grazing enterprises. By developing and implementing adaptation measures that improve the sustainability of pastoral enterprises and their capacity to adapt to change, the grazing industry will be better prepared for the challenges that lie ahead.

Acknowledgements

We are grateful to Greg McKeon, John Carter, Grant Stone and David Cobon for their contributions to an earlier version of this chapter, and Mick Quirk, Beverly Henry and Pauline Simonetti for comments that helped improve this chapter.

References

Amundson JL, Mader TL, Rasby RJ, Hu QS (2006). Environmental effects on pregnancy rate in beef cattle. *Journal of Animal Science* **84,** 3415–3420.

Anderson JR (1991). A framework for examining the impacts of climate variability. In *Climatic Risk in Crop Production: Models and Management for the Semi-arid Tropics and Sub-tropics*. (Eds RC Muchow, JA Bellamy) pp. 3–17. CAB International, Wallingford, UK.

Ash AJ, Prinsen JH, Myles DJ, Hendricksen RE (1982). Short-term effects of burning native pasture in spring on herbage and animal production in south-east Queensland. *Proceedings of the Australian Society of Animal Production* **14,** 377–380.

Ash A, McIntosh P, Cullen B, Carberry P, Smith MS (2007). Constraints and opportunities in applying seasonal climate forecasts in agriculture. *Australian Journal of Agricultural Research* **58,** 952–965.

Beever DE (1993). Ruminant animal production from forages: present position and future opportunities. In *Proceedings of the XVII International Grassland Congress*. pp. 535–542. New Zealand Grassland Association, Palmerston North.

Bisset WJ (1962). The black spear grass (*Heteropogon contortus*) problem of the sheep country in central western Queensland. *Queensland Journal of Agricultural Science* **19,** 189–207.

Burrows WH, Scanlan JC, Anderson ER (1988). Plant ecological relations in open forests woodlands and shrub lands. In *Native Pastures in Queensland: The Resources and Their Management*. (Eds WH Burrows, JC Scanlan, MT Rutherford) pp. 72–90. Queensland Department of Primary Industries QI87023, Brisbane.

Burrows WH, Carter JO, Scanlan JC, Anderson ER (1990). Management of savannas for livestock production in north-east Australia – contrasts across the tree grass continuum. *Journal of Biogeography* **17,** 503–512.

Burrows WH, Henry BK, Back PV, Hoffmann MB, Tait LJ, Anderson ER, Menke N, Danaher T, Carter JO, McKeon GM (2002). Growth and carbon stock change in eucalypt woodlands in northeast Australia: ecological and greenhouse sink implications. *Global Change Biology* **8,** 769–784.

Cai W (2003). Australian droughts, climate variability and climate change: insights from CSIRO's climate models. In *Science for Drought. Proceedings of the National Drought Forum, QC 0300*. (Eds R Stone, I Partridge) pp. 21–23. Queensland Department of Primary Industries, Brisbane.

Campbell BD, Smith DMS, McKeon GM (1997). Elevated CO_2 and water supply interactions in grasslands: a pastures and rangelands management perspective. *Global Change Biology* **3**, 177–187.

Campbell KO (1958). The challenge of production instability in Australian agriculture. *Australian Journal of Agricultural Economics* **2**, 23.

Chiew FHS, Whetton PH, McMahon TA, Pittock AB (1995). Simulation of the impacts of climate change on runoff and soil moisture in Australian catchments. *Journal of Hydrology* **167**, 121–147.

Cobon DH, Clewett JF (1999). 'DroughtPlan CD. A compilation of software, workshops, case studies, reports and resource material to help manage climate variability in northern Australia. QZ90002'. Queensland Department of Primary Industries, Brisbane.

Cobon DH, Bell KL, McKeon GM, Clewett JF, Crimp S (2005a). Potential climate-change impacts on beef production systems in Australia. In *Proceedings XX International Grassland Congress*. p. 557. Dublin.

Cobon DH, Park JN, Bell KL, Watson IW, Fletcher W, Young M (2005b). Targeted seasonal climate forecasts offer more to pastoralists. In *20th International Grassland Congress*. p. 556. Dublin.

Cobon DH, Toombs NR (2007). 'Practical adaptation to climate change in regional natural resource management. Qld Case Studies – Fitzroy Basin. Production and natural resource indicators in beef systems under climate change conditions. Final report to the Australian Greenhouse Office'. Queensland Department of Natural Resources and Mines, Brisbane.

Cobon DH, Toombs NR, Zhang X (2007). Climate change impacts on the sediment load for the Nogoa catchment of the Fitzroy Basin. In *Proceedings of MODSIM Conference*. pp. 853–859. Christchurch, New Zealand.

Crimp SJ, Flood NR, Carter JO, Conroy JP, McKeon GM (2002). 'Evaluation of the potential impacts of climate change on native pasture production: implications for livestock carrying capacity. Final Report to the Australian Greenhouse Office'. Queensland Department of Natural Resources and Mines, Brisbane.

CSIRO and Australian Bureau of Meteorology (2007). 'Climate change in Australia: technical report 2007'. CSIRO and Australian Bureau of Meteorology, Melbourne.

Cullen B, Eckard R, Johnson I, Lodge G, Walker R, Rawnsley R, Dassanayake K, Christie K, McCaskill M, Clark S, Sanford P, Browne N, Sinclair K, Chapman D, Leiffering M, Snow V, Hovenden M, Perring M (2008). 'Whole farm systems analysis and tools for the Australian and New Zealand grazing industries: project report'. University of Melbourne, Melbourne.

Dalton G, Keogh M (2007). 'The implications for Australian agriculture of changing demand for animal protein in Asia. Research Report Overview'. Australian Farm Institute, Surry Hills NSW.

Day KA, Ahrens DG, Peacock A, Rickert KG, McKeon GM (2000). Climate tools for northern grassy landscapes. In *Proceedings of the Northern Grassy Landscapes Conference, 29–31 August 2000*. pp. 93–97. CRC for the Sustainable Development of Tropical Savannas, Katherine NT.

Dorrough J, Moxham C, Turner V, Sutter G (2006). Soil phosphorus and tree cover modify the effects of livestock grazing on plant species richness in Australian grassy woodland. *Biological Conservation* **130**, 394–405.

Edwards GR, Clark H, Newton PCD (2001). The effects of elevated CO_2 on seed production and seedling recruitment in a sheep-grazed pasture. *Oecologia* **127**, 383–394.

Evans DR, Humphreys MO, Williams TA (1996). Forage yield and quality interactions between white clover and contrasting ryegrass varieties in grazed swards. *Journal of Agricultural Science* **126**, 295–299.

Fensham RJ (1998). The influence of cattle grazing on tree mortality after drought in savanna woodland in north Queensland. *Australian Journal of Ecology* **23**, 405–407.

Finch VA, Bennett IL, Holmes CR (1984). Coat color in cattle – effect on thermal balance, behavior and growth, and relationship with coat type. *Journal of Agricultural Science* **102**, 141–147.

Fischer BU, Frehner M, Hebeisen T, Zanetti S, Stadelmann F, Luscher A, Hartwig UA, Hendrey GR, Blum H, Nosberger J (1997). Source-sink relations in *Lolium perenne* L. as reflected by carbohydrate concentrations in leaves and pseudo-stems during regrowth in a free air carbon dioxide enrichment (FACE) experiment. *Plant Cell and Environment* **20**, 945–952.

Fordyce G, Tyler R, Anderson VJ (1990). Effect of reproductive status, body condition and age of *Bos indicus* cross cows early in a drought on survival and subsequent reproductive performance. *Australian Journal of Experimental Agriculture* **30**, 315–322.

Gifford R, Campbell BD, Howden SM (1996). Options for adapting agriculture to climate change: Australian and New Zealand examples. In *Greenhouse: coping with climate change.* (Eds W Bouma, GI Pearman, MR Manning) pp. 399–416. CSIRO, Melbourne.

Hall WB, McKeon GM, Carter JO, Day KA, Howden SM, Scanlan JC, Johnston PW, Burrows WH (1998). Climate change in Queensland's grazing lands: II. An assessment of the impact on animal production from native pastures. *The Rangeland Journal* **20**, 177–205.

Harle KJ, Howden SM, Hunt LP, Dunlop M (2007). The potential impact of climate change on the Australian wool industry by 2030. *Agricultural Systems* **93**, 61–89.

Hennessy KJ, Fawcett R, Kirono D, Mpelasoka F, Jones D, Bathols J, Whetton P, Stafford Smith M, Howden M, Mitchell C, Plummer N (2008). 'An assessment of the impact of climate change on the nature and frequency of exceptional climatic events'. CSIRO and the Australian Bureau of Meteorology: Melbourne. http://www.bom.gov.au/climate/droughtec/

Herne B (1998). U.S. cattle cycle is the key to rising prices. *Brigaletter* **32**, 1–2.

Heyhoe E, Kim Y, Kokic P, Levantis.C., Ahammad H, Schneider K, Crimps S, Nelson R, Flood N, Carter J (2007). Adapting to climate change. *Australian Commodities* **14**, 167–178.

Hogan L, Woffendon K, Honslow K, Zheng S (1995). The impact of the 1994–95 drought on the Australian economy. In *Coping with Drought, Occasional Paper No. 87.* pp. 21–33. Australian Institute of Agricultural Science, Melbourne.

Howden SM, Turnpenny J (1997). Modelling heat stress and water loss of beef cattle in subtropical Queensland under current climates and climate change. In *Modsim '97 International Congress on Modelling and Simulation Proceedings, 8–11 December, University of Tasmania, Hobart.* (Eds DA McDonald, M McAleer) pp. 1103–1108. Modelling and Simulation Society of Australia, Canberra.

Howden SM, Hall WB, Bruget D (1999a). Heat stress and beef cattle in Australian rangelands: recent trends and climate change. In *People and Rangelands: Building the Future. Proceedings of the VI International Rangeland Congress.* (Eds D Eldridge, D Freudenberger) pp. 43–45. Townsville.

Howden SM, McKeon GM, Carter JO, Beswick A (1999b). Potential global change impacts on C_3–C_4 grass distribution in eastern Australian rangelands. In *People and Rangelands: Building the Future. Proceedings of the VI International Rangeland Congress.* (Eds D Eldridge, D Freudenberger) pp. 41–43. Townsville.

Howden SM, McKeon GM, Walker L, Carter JO, Conroy JP, Day KA, Hall WB, Ash AJ, Ghannoum O (1999c). Global change impacts on native pastures in south-east Queensland, Australia. *Environmental Modelling & Software* **14**, 307–316.

Howden SM, McKeon GM, Meinke H, Entel M, Flood N (2001a). Impacts of climate change and climate variability on the competitiveness of wheat and beef cattle production in Emerald, north-east Australia. *Environment International* **27**, 155–160.

Howden SM, Moore JL, McKeon GM, Carter JO (2001b). Global change and the mulga woodlands of southwest Queensland: greenhouse gas emissions, impacts, and adaptation. *Environment International* **27**, 161–166.

Howden SM, Stokes CJ, Ash AJ, MacLeod ND (2003). Reducing net greenhouse gas emissions from a tropical rangeland in Australia. In *26 July – 1 August 2003, Durban, South Africa*. (Eds N Allsopp, AR Palmer, SJ Milton, KP Kirkman, GIH Kerley, CR Hurt, CJ Brown) pp. 1080–1082. Document Transformation Technologies, Irene, South Africa.

Howden SM, Soussana JF, Tubiello FN, Chhetri N, Dunlop M, Meinke H (2007). Adapting agriculture to climate change. *Proceedings of the National Academy of Sciences of the United States of America* **104**, 19691–19696.

Howden SM, Crimp SJ, Stokes CJ (2008). Climate change and Australian livestock systems: impacts, research and policy issues. *Australian Journal of Experimental Agriculture* **48**, 780–788.

Johnston PW, McKeon GM, Day KA (1996). Objective 'safe' grazing capacities for south-west Queensland Australia: development of a model for individual properties. *The Rangeland Journal* **18**, 244–258.

Johnston PW, Buxton R, Carter JO, Cobon DH, Day KA, Hall WB, Holmes WB, McKeon GM, Quirk MF, Scanlan JC (2000). Managing climate variability in Queenslands grazing lands – new approaches. In *Applications of Seasonal Climate Forecasting in Agricultural and Natural Ecosystems – The Australian Experience*. (Eds GL Hammer, N Nicholls, C Mitchell) pp. 197–226. Kluwer Academic Press, Amsterdam.

Keeling CD, Chin JFS, Whorf TP (1996). Increased activity of northern vegetation inferred from atmospheric CO_2 measurements. *Nature* **382**, 146–149.

Lilley JM, Bolger TP, Peoples MB, Gifford RM (2001). Nutritive value and the nitrogen dynamics of *Trifolium subterraneum* and *Phalaris aquatica* under warmer, high CO_2 conditions. *New Phytologist* **150**, 385–395.

Lloyd PL, Burrows WH (1988). The importance and economic value of native pastures to Queensland. In *Native Pastures in Queensland: The Resources and Their Management*. (Eds WH Burrows, JC Scanlan, MT Rutherford) pp. 1–12. Queensland Department of Primary Industries, Brisbane.

Lonsdale WM (1994). Inviting trouble: introduced pasture species in Northern Australia. *Australian Journal of Ecology* **19**, 345–354.

Mader TL, Davis MS (2004) Effect of management strategies on reducing heat stress of feedlot cattle: feed and water intake. *Journal of Animal Science* **82**, 3077–3087.

Mannetje L (1974). Relations between pasture attributes and liveweight gains on a subtropical pasture. *Proceedings of the 12th International Grasslands Congress*. pp. 299–304. Moscow.

McKeon GM, Howden SM (1992). Adapting the management of Queensland's grazing systems to climate change. In *Climate Change: Implications for Natural Resource Conservation. University of Western Sydney Occasional Papers in Biological Sciences No 1*. (Ed. S Burgin) pp. 123–140. University of Western Sydney, Sydney.

McKeon GM, Howden SM, Abel NOJ, King JM (1993). Climate change: adapting tropical and subtropical grasslands. In *Grasslands for Our World*. (Ed. MJ Baker) pp. 426–435. SIR Publishing, Wellington, New Zealand.

McKeon G, Hall W (2000). 'Learning from history: preventing land and pasture degradation under climate change. Final Report to the AGO.' Queensland Department of Natural Resources, Mines and Energy, Brisbane.

McKeon GM, Ash AJ, Hall WB, Stafford Smith DM (2000). Simulation of grazing strategies for beef production in north-east Queensland. In *Applications of Seasonal Climate Forecasting in Agricultural and Natural Ecosystems – The Australian Experience*. (Eds GL Hammer, N Nicholls, C Mitchell) pp. 227–252. Kluwer Academic Press, Amsterdam.

McKeon G, Hall W, Henry B, Stone G, Watson I (2004). 'Pasture degradation and recovery in Australia's rangelands: learning from history.' Queensland Department of Natural Resources, Mines and Energy, Brisbane.

McKeon GM, Flood N, Carter JO, Stone GS, Crimp JS, Howden SM (2008). 'Simulation of climate change impacts on livestock carrying capacity and production: report for the Garnaut Climate Change Review.' Queensland Environmental Protection Agency, Brisbane.

McKeon GM, Stone GS, Syktus JI, Carter JO, Flood N, Fraser GW, Crimp SJ, Cowley R, Johnston PW, Stokes CJ, Cobon D, Ryan JG, Howden SM (2009). Climate change impacts on rangeland livestock carrying capacity: more questions than answers. *The Rangeland Journal* **31**, 1–29.

Mcleod RS (1995). Costs of major parasites to the Australian livestock industries. *International Journal for Parasitology* **25**, 1363–1367.

Morgan JA, Mosier AR, Milchunas DG, Lecain DR, Nelson JA, Parton WJ (2004). CO_2 enhances productivity, alters species composition, and reduces digestibility of shortgrass steppe vegetation. *Ecological Applications* **14**, 208–219.

Myneni RB, Keeling CD, Tucker CJ, Asrar G, Nemani RR (1997). Increased plant growth in the northern high latitudes from 1981 to 1991. *Nature* **386**, 698–702.

Nicholls N (2003). Climate change: are droughts becoming drier and more frequent? Science for drought. In *Proceedings of the National Drought Forum QC 03004*. (Eds R Stone, I Partridge) pp. 1–7. Queensland Department of Primary Industries, Brisbane.

Niklaus PA, Spinnler D, Korner C (1998). Soil moisture dynamics of calcareous grassland under elevated CO_2. *Oecologia* **117**, 201–208.

Nowak RS, Ellsworth DS, Smith SD (2004). Functional responses of plants to elevated atmospheric CO_2 – do photosynthetic and productivity data from FACE experiments support early predictions? *New Phytologist* **162**, 253–280.

Owensby CE, Coyne PI, Ham JM, Auen LM, Knapp AK (1993). Biomass production in a tallgrass prairie ecosystem exposed to ambient and elevated CO_2. *Ecological Applications* **3**, 644–653.

Pettet NE, Froend RH (2000). Regeneration of degraded woodland remnants after relief from livestck grazing. *Journal of the Royal Society of Western Australia* **83**, 65–74.

Quirk M (2002). Managing grazing. In *Global Rangelands: Progress and Prospects*. (Eds AC Grice and KC Hodgkinson) pp. 131–145. CAB International, Wallingford, UK.

Scanlan JC, Hinton AW, McKeon GM, Day KA, Mott JJ (1994). Estimating safe carrying capacities of extensive cattle-grazing properties within tropical, semi-arid woodlands of north-eastern Australia. *The Rangeland Journal* **16**, 76.

Smith SD, Huxman TE, Zitzer SF, Charlet TN, Housman DC, Coleman JS, Fenstermaker LK, Seemann JR, Nowak RS (2000). Elevated CO_2 increases productivity and invasive species success in an arid ecosystem. *Nature* **408**, 79–82.

Stafford Smith DM, Buxton R, McKeon GM, Ash AJ (2000). Seasonal climate forecasting and the management of rangelands: do production benefits translate into enterprise profits? In *Applications of Seasonal Climate Forecasting in Agricultural and Natural Ecosystems – The Australian Experience.* (Eds GL Hammer, N Nicholls, C Mitchell) pp. 271–290. Kluwer Academic Press, Amsterdam.

Stokes CJ, Ash AJ (2006). Special example 1: impacts of climate change on marginal tropical animal production systems. In *Agroecosystems in a Changing Climate.* (Eds PCD Newton, RA Carran, GR Edwards, PA Niklaus) pp. 323–328. CRC Press, London.

Suter D, Frehner M, Fischer BU, Nosberger J, Luscher A (2002). Elevated CO_2 increases carbon allocation to the roots of *Lolium perenne* under free-air CO_2 enrichment but not in a controlled environment. *New Phytologist* **154**, 65–75.

Sutherst RW, Yonow T, Chakraborty S, O'Donnell C, White N (1996). A generic approach to defining impacts of climate change on pests, weeds and diseases in Australasia. In *Greenhouse: Coping with Climate Change.* (Eds W Bouma, GI Pearman, MR Manning) pp. 190–204. CSIRO Publishing, Melbourne.

Sutherst RW (1990). Impact of climate change on pests and diseases in Australia. *Search* **21**, 232.

Sutherst RW (2001). The vulnerability of animal and human health to parasites under global change. *International Journal for Parasitology* **31**, 933–948.

Tebaldi C, Hayhoe K, Arblaster JM, Meehl GA (2006). Going to the extremes. *Climatic Change* **79**, 185–211.

Volk M, Niklaus PA, Korner C (2000). Soil moisture effects determine CO_2 responses of grassland species. *Oecologia* **125**, 380–388.

Wand SJE, Midgley GF, Jones MH, Curtis PS (1999). Responses of wild C_4 and C_3 grass (Poaceae) species to elevated atmospheric CO_2 concentration: a meta-analytic test of current theories and perceptions. *Global Change Biology* **5**, 723–741.

Wand SJE, Midgley GF (2004). Effects of atmospheric CO_2 concentration and defoliation on the growth of *Themeda triandra*. *Grass and Forage Science* **59**, 215–226.

Warwick KR, Taylor G, Blum H (1998). Biomass and compositional changes occur in chalk grassland turves exposed to elevated CO_2 for two seasons in FACE. *Global Change Biology* **4**, 375–385.

Whetton P, Suppiah R (2003). Climate change projections and drought. In *Science for Drought. Proceedings of the National Drought Forum QC 03004.* (Eds R Stone, I Partridge) pp. 130–136. Queensland Department of Primary Industries, Brisbane.

White BJ (1972). Supply projections for the Australian beef industry. *Review of Marketing and Agricultural Economics* **40**, 1–12.

White BJ (2000). The importance of climatic variability and seasonal forecasting to the Australian economy. In *Applications of Seasonal Climate Forecasting in Agricultural and Natural Ecosystems – The Australian Experience.* (Eds GL Hammer, N Nicholls, C Mitchell) pp. 271–289. Kluwer Academic Press, Amsterdam.

White IA (2001). 'With reference to the channel country – review of available information. QI01068.' Queensland Department of Primary Industries, Brisbane.

White N, Sutherst RW, Hall N, Whish-Wilson P (2003). The vulnerability of the Australian beef industry to impacts of the cattle tick (*Boophilus microplus*) under climate change. *Climatic Change* **61**, 157–190.

Wilson JR (1982). Environmental and nutritional factors affecting herbage quality. In *Nutritional Limits to Animal Production from Pastures*. (Ed. JB Hacker) pp. 111–131. CAB International, Farnham Royal, UK.

11

INTENSIVE LIVESTOCK INDUSTRIES

CJ Miller, SM Howden and RN Jones

KEY MESSAGES:

▪ Climate change, including changes in the intensity and frequency of extreme events, will challenge traditional intensive livestock farming systems.

▪ Warmer and drier conditions are projected for most intensive livestock-producing regions, raising the likelihood and incidence of heat stress in stock. Traditional high energy and water use options for improving the environment of livestock under heat stress conditions are likely to be maladaptive. Low energy and low emission options should be identified and evaluated.

▪ Suppliers and consumers in global commodity markets will be affected by climate change and associated issues such as international food security, and governmental policy responses. The costs of inputs required to maintain productivity are likely to increase. Farmers and producers need to have a greater awareness of environmental, economic and social conditions beyond their farm gates than ever before.

▪ Livestock enterprises must have the flexibility to rapidly change management systems in response to dynamic environmental, economic and social conditions. Proactive adaptation is about risk management and creating opportunities for prosperity under dynamic and challenging conditions.

Introduction

Dairy

The Australian dairy industry is the third highest value rural industry behind beef and wheat, valued at $3.2 billion in 2006–07, and the fifth most important agricultural export, valued at $2.5 billion (Australian Bureau of Statistics). The bulk of milk production occurs in Victoria (approximately 65%), although all States have viable dairy industries that supply fresh milk to nearby cities and towns. The industry occurs within eight administrative dairy regions, established for the purposes of targeting research, development and extension (Figure 11.1):

• Sub Tropical Dairy; extending from Kempsey in NSW to the Atherton Tablelands in Far North Queensland.

• Dairy Industry Development Company (DIDCO); covers the eastern fringe of New South Wales.

• Murray Dairy; the largest dairying region in Australia, straddling the Murray River from the Alps to Swan Hill. It covers the Northern Irrigation and north-east regions of Victoria and the Riverina and Upper Murray regions of New South Wales.

• Gipps Dairy; covers the Gippsland region, Victoria.

• West VIC Dairy; covers the south-west of Victoria.

• Dairy TAS; Tasmania.

• Dairy SA; South Australia.

• Western Dairy; the majority of farms occur in coastal south-west Western Australia.

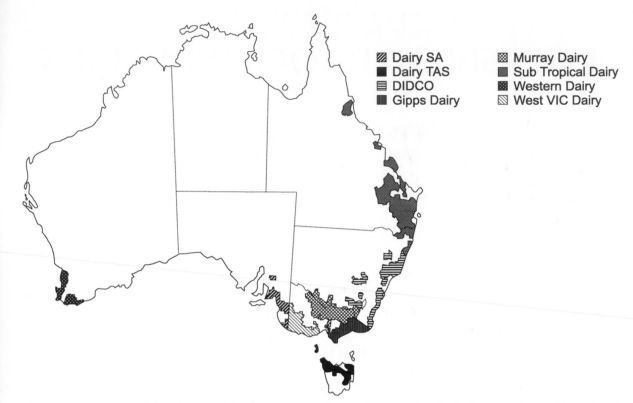

Figure 11.1: Location and distribution of the dairy regions of Australia. Note that the highest number and intensity of dairy farms is in Victoria. Source: Dairy Australia

Milk production is largely based on pasture systems that are affected by seasonal rainfall and temperature patterns, although a significant proportion of milk production is from irrigated pasture. Farms range from low input to high input across the country, with calving occurring both seasonally and year-round. An increasing number of farmers are supplementing pasture feed with other feedstock, such as grain, conserved forages (hay and silage) and by-products such as citrus pulp. Dairy feedlots are becoming more common as a consequence of the current drought. Farms in the Murray Dairy region have been highly productive, given the Mediterranean climate, the proximity to grain growing regions, and access to a reliable supply of water for irrigation, although this reliable water supply is already challenged and may change further (e.g. Cai and Cowan 2008).

Pigs

Pig production occurs in approximately 2800 farms spread across all States of Australia, with the highest proportion of producers located around the grain, sorghum or maize-growing regions. For example, 30% of Queensland's producers are located on the Darling Downs, an extensive grain-producing region (QPIF 2009). Intensive piggeries usually house their animals in specialised sheds for the duration of their life. For example, approximately 50% of Australia's pigs are raised in deep litter housing systems. These sheds tend to use passive end-to-end ventilation systems, with some having cross-flow ventilation options. Free range piggeries run their animals in paddocks that have rooting areas, wallows, and huts for shelter. Climatic and soil conditions limit the suitability of many areas for free range farming, as ambient temperatures cannot be kept below 27°C.

Poultry

The Australian poultry industry is primarily focussed on chicken meat (broiler) production and egg production. There are a small number of turkey and other fowl producers; the issues and options raised here, while not explicitly considered, are likely to be similar to chicken production.

Chicken meat production is becoming increasingly regionalised, following initial development near the major capital cities (Poultry CRC 2009):

- New South Wales – outer metropolitan Sydney, Central Coast, Newcastle, Tamworth, Griffith, Byron Bay, Canberra

- Queensland – south-east Queensland

- Victoria – Mornington Peninsula, east Melbourne, Geelong, Bendigo

- South Australia – outer metropolitan Adelaide, Two Wells

- Western Australia – outer metropolitan Perth

- Tasmania – outer metropolitan areas

- Northern Territory – No commercial farms

Breeding farms owned by the integrated meat companies are located away from traditional poultry-rearing areas to reduce disease risks. Grow-out farms, where chickens grow from day-olds until they are ready for processing, are generally within 100 km of the processing plant, and require a guaranteed feed source, guaranteed water supply and guaranteed three phase electric power. Most vertically integrated companies own feed mills, with their location driven by transport costs of feed ingredients and proximity to the farms.

The majority of commercial grow-out farms are intensive and highly mechanised, with the chickens raised in large open sheds. These sheds tend to be large, with three to 10 sheds per farm, holding 40 000 to 60 000 chickens per shed. Shed temperature, humidity and air quality are carefully controlled and regulated.

Egg production and supply to the Australian market are largely met by 423 companies (ABS 2008), located around major metropolitan or regional centres and with easy access to feed stock. Farms range in size, with the largest having between 100 000 and 500 000 hens contained in multiple-level sheds, although most have fewer than 20 000 birds. There is an increasing number of free-range or barn raised egg farms due to consumer preferences or perceptions around animal welfare. As with chicken meat farms, shed temperature, humidity and air quality are carefully controlled and regulated, and a guaranteed supply of water is required (Poultry CRC 2009).

Feedlots

Beef feedlots are concentrated in the major agricultural regions of Australia where they have adequate access to cattle, water, grain and other feedstuffs. The majority are in southern Queensland and New South Wales. At the last available estimate there were around 600 accredited feedlots, with capacity for some 860 000 head of cattle (Australian Lot Feeders' Association 2002). Regional and annual climatic variability, and product quality assurance, have driven the use of feedlots for growing beef to market specifications, in preference to pasture or rangelands.

An increasing number of dairy farmers are using feedlots to supplement or replace their reliance on pasture, particularly in south-east Queensland, but also in New South Wales, Northern Victoria and parts of South Australia and Western Australia. Some farmers are also buying up grain-producing properties in order to help ensure a reliable feed supply.

Impacts of projected climate change

Climate projections for Australia in 2030 indicate a warming of about 1°C relative to the average temperature from 1980–99. The degree of warming is expected to be less in coastal areas and higher inland (Chapter 2). The majority of climate models indicate decreased rainfall over the next 20 years, with decreased annual average and winter rainfall in southern areas, decreased spring rainfall in southern, western, and eastern areas, and autumn decreases along the west coast. The change in rainfall is highly uncertain in the far north, with essentially equal probability of increases or decreases. Specific projections for the intensive livestock-producing regions of Australia in 2030 indicate that climatic conditions will be warmer and drier, resulting in increased evaporation and reduced runoff (Hennessy 2007). Temperatures are projected to increase between 0.4 and 1.5°C by 2030, with increased maximum temperatures in southern Australia and increased minimum temperatures in northern Australia. The greatest declines in rainfall are projected to occur in spring and winter.

The most significant biophysical impacts of climate change for intensive livestock farming are likely to be due to increased climatic variability and the increased frequency and intensity of extreme events, rather than the change in average conditions *per se* with the exception of reductions in mean rainfall which could have far reaching consequences. Agricultural production is inherently sensitive to climate and seasonal weather patterns, with the growth of pastures and feed crops, and the welfare and productive capacity of livestock, directly linked to environmental conditions. The vulnerability of intensive livestock farming to climate change will be as a consequence of the exposure of the biophysical elements of the farm and farm system to acute weather events or chronic climatic conditions, and to the range of social and economic conditions that facilitate or constrain adaptation of those farming systems (Belliveau *et al.* 2006). The vulnerability of an agricultural system is therefore reduced or exacerbated by the level of adaptive capacity in the system.

Animal welfare and husbandry

A potentially significant direct impact of climate change on intensive livestock is likely to be heat stress (Howden and Turnpenny 1997). Heat stress is a function of air temperature, relative humidity, air movement, solar radiation as well as metabolic heat load, coat colour and other animal attributes. An animal suffers heat stress when it is unable to cool itself, or thermoregulate, to within its thermal tolerance levels. This is a major issue for livestock production in the warmer parts of the world. Below a lower critical temperature an animal must increase its metabolic rate and activity (e.g. shivering) to maintain body temperature, while at the upper critical level the animal must expend energy (e.g. through ear flapping and panting) and water in order to cool themselves. Animals respond to heat stress in a number of ways including 1) reduced feed intake; 2) increased water intake; 3) changed metabolic rate and maintenance requirements; 4) increased evaporative water loss; 5) increased respiration rate; 6) changed blood hormone content; 7) increased body temperature, and 8) behavioural change (e.g. shade seeking).

Heat stress results in significant economic and production losses for dairy operations throughout the world since temperatures above 26°C are outside the comfort zone for high-producing cows, regardless of humidity (Aharoni *et al.* 2005). The optimal atmospheric temperature range for lactating dairy cows is 5°C to 25°C (Roenfeldt 1998). Above 26°C cows enter a heat-stress zone, where they can no longer adequately cool themselves, and their milk composition changes and production declines. One of the unintended consequences of achieving improvements in milk yield through selective breeding is that it is more difficult for high-producing cows to thermoregulate, even in temperate areas (Hansen 2007; Kadzere *et al.* 2002). This results from high productivity generating large amounts of metabolic heat which needs to be shed so that the animal can maintain body temperature at acceptable levels. Highly productive cows in early lactation may be most susceptible, given their energy requirements and metabolism, and this is particularly so in warmer months (Igono and Johnson 1990). Cool night-time temperatures may partly ameliorate the effect of high daytime temperatures (Igono and Johnson 1990).

Along with reduced productivity, heat stress also reduces the reproductive success of livestock, including cattle (García-Ispierto *et al.* 2007), pigs (Wettemann and Bezar 1985) and poultry (Cooper and Washburn 1998). Consequently intensive livestock operations that house stock indoors or in close quarters (e.g. feedlots) aim to maintain optimal air temperature and humidity (e.g. Hahn and Oosburn 1969).

One commonly used measure of heat stress is the Temperature–Humidity Index (THI; Johnson *et al.* 1963); a robust predictor of heat stress that is used operationally for heat stress assessment in dairy cattle in South Africa (Du Preez *et al.* 1990). The THI also defines the distribution of beef cattle varieties in Africa (King 1983). A THI reading over 72 indicates the potential for heat stress in cattle and higher readings are associated with progressively more negative impacts. While actual humidity may decrease in some situations under climate change, increases in temperature are likely to override this effect and lead to increases in heat stress and the frequency of heat stress in almost all instances across Australia (Howden *et al.* 1999; Howden and Turnpenny 1997).

Pasture productivity and quality

Pasture requires adequate sunlight, water and nutrient, and appropriate soil and air temperatures in order to grow and meet the metabolic requirements of lactating cows. Perennial exotic pastures, comprising ryegrass (*Lolium perenne*) and clover (e.g. *Trifolium repens*) or paspalum (*Paspalum dilatatum*), are the dominant pastures of the temperate and Mediterranean climatic regions of Australia. Some farms also include annual pasture species, with nearly a quarter of farmed areas in the Murray Dairy region consisting of irrigated annual pasture (Armstrong *et al.* 2000). This is rapidly increasing in northern Victoria as annual pasture gives the best response per ML of water. Dairy farms in the subtropical dairy region may have a higher ratio of native grasses or introduced legumes in their pastures, although perennial ryegrass and kikuyu still have a significant pasture role.

It is likely that the effects of climate change on pasture will be most apparent through observed changes in pasture growth and quality, and greater inter-annual variability in pasture production (Harle *et al.* 2007). Problems with the persistence and quality of ryegrass-dominant pastures during dry conditions are already an issue for Australian farmers (Callow *et al.* 2003; Callow *et al.* 2006; Waller and Sale 2001). Heat stress and moisture deficits during summer conditions reduce growth, herbage quality and dry matter content of these pastures (Waller and Sale 2001). Perennial pasture requires the equivalent of 1200–1300 mm of water in the Murray Dairy region (Austin 1998), and pasture can only be sustained at a productive level throughout the year in this region by applying irrigation water when evaporation exceeds rainfall (Bethune and Armstrong 2004). The issue of pasture persistence is due to the fact that the majority of ryegrass cultivars used in Australia are from northern Europe, where they grow from spring to autumn under mild conditions, and are unsuited to hot dry summers. The optimal ambient temperature for ryegrass growth is 17–21°C with growth ceasing above 30–35°C. New heat-tolerant cultivars (e.g. Dobson) from the Mediterranean region have been developed, but their uptake is not yet widespread (Fulkerson *et al.* 2003).

One upside of the projected temperature increase in southern Australia is that pasture growth during autumn or winter may be enhanced, or at least start earlier in the season provided that water is available. This would obviously benefit farmers through the provision of fodder during seasons that normally require supplementary feeding (due to cold temperatures), and/or would allow for the harvesting of excess growth for hay or silage.

Atmospheric CO_2 enrichment, a causal factor for climate change, has the potential to increase plant biomass, crop yield, and water use efficiency, although there may also be decreases in protein content and increased non-structural carbohydrate content. CO_2 enrichment also favours C_3 grasses over C_4 grasses, a compositional change that is generally positive for dairy production. However, in terms of pasture composition, the positive effects of CO_2 enrichment are likely to be counteracted through increased temperatures and reduced rainfall. C_4 grasses are likely to be favoured over C_3 grasses in a warming environment, with the increased water efficiency potentially reducing the impact of drier conditions to a small degree. The balance between these two contrasting processes is likely to be situation-dependent and is not well understood at present (e.g. Howden *et al.* 1999). More research is required given the degree of uncertainty around the relationship between CO_2 enrichment, temperature and rainfall and the impact on pasture growth and productivity under Australian conditions (Harle *et al.* 2007).

Water availability

Water is essential for intensive livestock production, being used for irrigation, stock drinking and evaporative cooling (e.g. Howden and Turnpenny 1997), and consumption increases with temperature. Climate projections indicate that there will be reduced rainfall for most of southern Australia which will have an effect on rain-fed pasture growth and crops used for supplementary feed, runoff into rivers and storages, and aquifer recharge. This is likely to result in regulatory reductions in the availability of water for agriculture, an increased cost for water, as well as increased competition for water from other agriculture or horticulture, and for urban consumption.

Climate change policy

Government policy in response to climate change is an evolving entity, however the signals indicate that the economy will be subject to a major structural reform that will favour, if not explicitly require, low greenhouse gas emitting enterprises, and will give an economic advantage to energy generated or sourced from no or low greenhouse gas emitting technologies. Consequently, energy sourced from coal burning technology will become more expensive, while energy sourced from low emitting technology will become increasingly competitive. This is relevant to intensive production units that have a high energy demand for 'air conditioning', where livestock are farmed in indoor climate-controlled conditions. As part of an emissions trading market, farmers may have to account for the greenhouse gasses emitted through the production of their agricultural produce, as well as any income generated as a consequence of being in the emissions trading market, for taxation purposes.

The use of nitrogen-based fertilisers has increased dramatically over the last 15 years, and this is largely responsible for the dramatic increases in milk productivity (Eckard et al. 2007). However, nitrogenous fertilisers are implicated in the emission of nitrous oxide (N_2O), a greenhouse gas, to the atmosphere. Government policy instruments have been recommended to cap these emissions, including restrictions or certification on use, or the required use of nitrification inhibitors (Hatfield-Dodds et al. 2007). Such policy instruments are likely to increase costs for farmers, although many farmers are currently inefficiently using nitrogenous fertilisers and there are opportunities to reduce or target their use based on weather conditions and pasture requirements (Eckard et al. 2003; Eckard et al. 2006; Eckard et al. 2007; Eckard and Franks 1998).

Options for dealing with climate variability

Climate variability is a centuries-old challenge for Australian farmers, and attempts to farm under these conditions have resulted in both economic prosperity and economic hardship. The most successful farmers have been those able to adapt their farm management practices tactically or strategically to deal with the environmental and economic challenges of farming in Australian conditions. There are a number of adaptation options available to farmers to deal with climate variability and climate change, and over the last decade many intensive livestock producers have already made tactical changes in their farm and financial management practices to survive drought conditions.

While tactical adaptation allows intensive livestock farmers to address within-season climatic variability, it is important that farmers also make strategic adaptations to their farm management practices to ensure that they and their enterprises are both robust and resilient to the future effects of climate change, and are poised to take advantage of new opportunities. Strategic adaptations are the structural changes made in farm management or land use that will apply in future years, rather than in the current production season (Smit and Skinner 2002). Strategic adaptations could simply be the result of a farmer or producer deciding to continue with a management practice that resulted from an operational decision, e.g. use of supplementary fodder to increase stock numbers and total farm productivity in good pasture years, when the initial objective was to maintain productivity of existing stock in a poor pasture growing year. Conversely, the farmer or producer may choose to adapt by completely transforming their production system, e.g. from pasture-based to feedlot dairies. Successful tactical adaptation to climate change can be assisted by farmers understanding the implications of seasonal forecasts for their farming system, and having the flexibility or capacity to make changes in advance of or during the seasonal conditions. It also requires that farmers actively monitor and respond appropriately to issues like pasture growth and supplementary feed or water markets, while being efficient (i.e. not wasteful) in the application of inputs such as irrigation and fertiliser. While this may seem evident, a significant proportion of farmers does not use, or uses inadequately, the tools available to help them make such management decisions (Kerr 2004; Parminter et al. 2004).

The primary factor in adapting to climate variability is incorporating seasonal climate forecasts into the farm decision-making process. There have been significant advances in seasonal weather forecasting in the last few decades, initially based on the El Niño-Southern Oscillation (ENSO) (Hammer et al. 2000), with recent advances in General Circulation or Global Climate models (GCM) likely to make seasonal forecasts more accurate. Combining climate projections with the many decision-support tools available, such as GrazPlan (Donnelly et al. 2002) or Yield Prophet (www.yieldprophet.com.au), allows farmers to assess risk and make informed decisions.

Howden *et al.* (2007) identify a number of enter-prise-level or tactical management options for adapting pastoral farm management practices to climate variability and climate change. Options include proactive feed budgeting through matching stocking rates with variable pasture production and supplementary fodder, timing of artificial insemination to avoid hot periods, and ensuring adequate storage of water. Most of these options require flexible operating systems and a degree of forward planning, although many of them would be expected to be best practice.

Other options for dealing with climate variability in the intensive livestock industry include:

- maximising the efficiency of irrigation systems and water use efficiency of farm systems (Armstrong 2004; Armstrong *et al.* 2000; Watson and Drysdale 2006);

- selecting heat- or drought-tolerant phenotypes (stock, pasture or feedstock) (e.g. Franca *et al.* 1998; Fulkerson *et al.* 2003);

- shifting from perennial pastures to a mix of annual and perennial pastures or incorporate fodder cropping (e.g. Chapman *et al.* 2008a; b);

- climate-controlled production sheds through mechanical or natural air conditioning and ventilation (e.g. Hahn 1981; 1985);

- utilising water-based cooling mechanisms in feedlots, such as misting (e.g. Gaughan *et al.* 2004), and ensuring adequate shade, e.g. through hard structures or shelterbelts (e.g. Gaughan *et al.* 1998; Hahn 1981; 1985; Hahn and Oosburn 1969);

- agistment of stock outside the region when and where necessary (McAllister *et al.* 2006);

- feeding stock raised in feedlots during cool periods, e.g. at night (Holt *et al.* 2004), and

- ration-balancing and the use of supplementary osmolytes such as betaine to reduce heat stress and water loss.

Options for dealing with projected climate change

The adaptation options identified here are more strategic than tactical, and may in fact be trans-formational for some farmers or producers. They are not intended to be a comprehensive review of all the options available, nor will they be applicable to all situations.

Protecting stock from sun and heat

It is possible that the frequency of heat stress conditions in some regions may affect productivity to the degree that it is no longer profitable to run dairy cows on pasture, or to free-range pigs or poultry, particularly during summer. In many cold regions of the world, cattle are housed indoors during winter where the climatic conditions can be controlled. The converse, e.g. housing dairy cattle indoors with a controlled climate during summer, may become necessary in some areas to ensure sustained productivity. This feedlot model for dairy is not necessarily a desirable or practical adaptation and, as with other options, would require a comprehensive cost–benefit analysis on a case-by-case basis.

The use of shelter and shade where dairy cows are farmed in feedlots is essential, and techniques such as using sprinklers, misters, or shade cloth to keep the animals cool are often used. However humidity may be increased by the use of evaporative cooling techniques, which adds stress to the animals, so there is a need to ensure adequate air velocity through the area (Berman 2006). It has been suggested that night feeding of cattle in feedlots may reduce the energy expenditure associated with foraging during the heat of day and result in sustained milk yield over time (Aharoni *et al.* 2005).

It may be necessary to build structures in paddocks, pastures, or ranging areas, e.g. with shade cloth, where it is not possible to immediately provide shelter through other means, e.g. shelter belts.

Box 11.1: Example of the dairy industry managing livestock heat stress

Heat stress in livestock is a real issue for Australian farmers. Dairy Australia has developed an early warning system for notifying dairy farmers of impending heat stress conditions, with an email and mobile phone text alert system. Farmers have been provided with information via a booklet, and an interactive website (www.coolcows.com.au) that enables them to make informed decisions about immediate and ongoing stock management practices under heat stress conditions.

Develop new pasture or fodder crop varieties that are tolerant of climate variability

Ryegrass-clover pastures are the mainstay for dairy farming in cool–temperate climates, and are used in drier Australian dairy regions to provide pasture during the cool seasons, or in the warm seasons in conjunction with irrigation. During hot dry conditions, the cultivars used in Australia tend to become unthrifty or die. Consequently, scientists have been selecting ryegrass ecotypes that are adapted to more Mediterranean conditions and these cultivars will become increasingly available. However, there is a need to provide pasture and fodder crop types, cultivars, and hybrids that are robust to a variable climate, rather than just to prevailing or idealised climatic conditions (Smit and Skinner 2002).

One 'blue sky' option is to genetically engineer C_3 pasture grasses to express C_4 metabolic traits. Plants with C_4 photosynthesis tend to be more water efficient and competitive under hot sunny conditions than C_3 perennial species, while C_3 species are more viable under cooler conditions. Such a macro-evolutionary process is naturally occurring in wildflowers of the genus *Flaveria* (Asteraceae), which have C_3-C_4 intermediates, as well as distinct C_3 and C_4 species (Kutschera and Niklas 2007). C_4 ryegrass is not currently an option for farmers, but it is not beyond the bounds of possibility that such a pasture species could be developed in the near future.

Selecting for thermotolerance or capacity for thermoregulation

Australia's intensive livestock industry has for many years invested in selective breeding to increase livestock productivity and quality, be it for increased meat, milk or fibre yield. The likely increases in mean and maximum temperatures suggest that the industry should also be selecting for increased capacity for thermoregulation or thermotolerance. The Australian beef industry has done this by crossing or totally replacing their traditional cattle *Bos taurus* with *Bos indicus*, which has superior capacity for body temperature regulation under heat stress (Hansen 2004). However, there are likely to be trade offs between heat tolerance/regulation and productivity, given that these traits are likely to be achieved by reducing animal size or metabolic activity. Selection for coat or feather characteristics may contribute to a small degree, but is considered that the biggest gains will be made at the metabolic level.

Modifying livestock feeding system

Manipulating the diet and feeding times of livestock can help in reducing heat stress during hot conditions. Metabolic heat load increases following food consumption, and it is easier for an animal to thermoregulate in cool evening or night-time conditions rather than hot daytime conditions (Aharoni *et al.* 2005; Brosh *et al.* 1998). Similarly feeding livestock with low metabolisable energy (ME) fodder in the morning and high ME fodder in the evening will also reduce the likelihood of heat stress (Aharoni *et al.* 2005; Brosh *et al.* 1998).

Redesign buildings for passive or energy efficient cooling

The energy demand for cooling in production sheds can be reduced by applying new building designs or materials (e.g. Raman *et al.* 2001). This can be a capital-intensive option, but the energy cost savings achieved through retrofitting existing sheds may offset the capital outlay. New sheds could be designed and built with passive cooling and heating as the key driver, to be supplemented by powered air conditioning where necessary. Energy can also be generated onsite (see below) to power the air conditioning system.

Supplementary or complete power generation onsite

Advances in photovoltaic technology are bringing down the cost of solar-generated electricity and significantly increasing its efficiency and capacity to generate electricity (Green 2006). Production sheds and other buildings often provide suitable sites for photovoltaic cells which can supplement or completely supply the electricity needs for livestock production. This type of power source will also provide peak power at the time of peak demand for cooling, a time when potential competition for energy and hence energy prices are greatest.

Farmers or producers also have other options to produce low emission energy, such as wind power, or biomass boilers or anaerobic digestion to produce methane using effluent, waste feed stock, or bedding litter for onsite generation of electricity. Many of Australia's sugar cane farmers now consider that they are in the business of producing energy, in the form of sugar or from ethanol production, and this change in mindset has enabled them to create new business models of production and income generation. Taking such an approach would allow livestock producers to benefit from the new carbon economy and emissions trading.

Clustering of compatible industries

Industrial agricultural parks that co-locate industries involved in waste processing, energy generation, water capture and recycling, feedstock and foodstuff manufacture, etc., with livestock production have the option to reduce energy demand from fossil fuels and increase value in the supply chain. This form of industrialisation can close the resource – waste loop, ensuring efficient supply and reducing the output of pollution to the environment. Pollution often represents wasted resources and wasted profit.

The siting of these agricultural industrial parks should be determined after considering the potential for increased exposure of the site to climate change as well as the more traditional social, environmental or economic issues.

Increasing landscape resilience or robustness through revegetation, rehydration, and soil organic carbon

Agricultural development has relied heavily on clearing native woody vegetation and draining wetlands. While this has resulted in significant economic returns nationally, it has reduced the capacity of agricultural land to cope with and recover from events such as droughts and floods. The natural capital of the land, its soils and water-holding capacity, has been degraded and the impacts of this for agriculture are likely to be exacerbated under projected climate change. While research is needed to identify the most appropriate configurations and techniques, the resilience of agricultural landscapes to climate change can be enhanced by strategic revegetation and the recreation of wetland systems (e.g. Ryan 2007). Such revegetation can also provide immediate shelter from solar radiation for stock and equable microclimates (Langston *et al.* 2003), while at larger scales may also affect regional climates and precipitation regimes (Avissar and Pielke 1991; Pielke 2001; Pielke *et al.* 2007; Pitman *et al.* 2004).

Soil organic carbon (SOC) is an essential contributor to soil health, productivity and water-holding capacity, yet many contemporary agricultural practices result in the loss of SOC (e.g. Lal *et al.* 2007; Smith 2008). Grazing management to stop overgrazing or soil compaction, use of legumes and perennial pasture species, and spreading of manure are all ways to increase SOC and increase agricultural resilience to climate change.

Use of seasonal climate forecasts and decision-support tools in risk management

Decision-making in the face of uncertainty requires that risk is assessed and managed. Seasonal climate forecasts are becoming increasingly useful and intensive livestock producers can use these to inform risk assessments and decision-making, particularly if forecasts are incorporated into the many decision-support tools available. For example, they can be used to make strategic feedbase decisions based on likely on-farm conditions as well as projecting the potential impacts of seasonal climate on feedbase supplies from other regions.

Risks of maladaptation

There are a number of risks of maladaptation to climate change in intensive livestock farming. The greatest risk is increasing reliance on fossil fuel-based energy and water-intensive solutions. This is likely to reduce the resilience of the industry to energy or water shocks, as well as adding to the problem of emissions and reduced water availability. Other adaptation options are likely to impact on the financial viability of the enterprise. For example, genetic selection for highly productive dairy cattle has reduced their capacity to thermoregulate and cope with increased temperatures; conversely selecting for heat tolerant phenotypes can reduce productivity (Hansen 2007).

Reliance on supplementary fodder in the dairy industry places farmers at risk of fluctuating prices and competition from international markets keen on ensuring their national food security. To date, increased input costs have been sustained because the milk commodity price was high, however recent decreases in milk price have highlighted the vulnerability of this system. Such a strategy can be maladaptive if it means that farmers cannot survive financially in years where the return on milk is near or lower than their input costs and operating capital is low. Dairy farmers with water rights in the Murray Dairy region have been able to survive the recent drought by selling their water at a high price to more efficient users of that water. However, this becomes a risky and potentially maladaptive strategy if, hypothetically, the national interest determined that those rights be relinquished to urban supply and more efficient water users. Financial compensation might ensure the immediate livelihood of the farmers, however the dairy industry is likely to collapse in the region.

At an individual farmer or producer level, adaptation is critical to reducing immediate vulnerability and maintaining productivity. However it is important to ensure that short-term tactical adaptations do not reduce the long-term adaptive capacity or profitability of the producer or the enterprise.

It is important that adaptation options, including policy options, be tested through multi-scale systems analysis so that the negative and positive feedbacks resulting from the proposed adaptation options can be explicitly identified. For example, dairy deregulation was economically rational at the time and has resulted in the bulk of Australia's dairy production moving to Victoria, particularly in irrigated regions. Under climate change scenarios it could be argued that this policy has resulted in a maladaptive industry response, increasing vulnerability. This argument would suggest that the dairy industry would have been better positioned to focus on regions likely to retain reasonable rainfall and pasture growth, such as north-eastern New South Wales or the Atherton Tablelands in Queensland.

Consequently it may be useful for industry bodies involved in intensive livestock production to engage in adaptation policy analysis in order to contribute positively to policy debates and to ensure that farmers and producers are provided with the best possible advice on adaptation options.

Costs and benefits

To our knowledge there are few cost–benefit analyses of adaptation options for dealing with climate change and intensive agriculture. The most relevant cost–benefit analyses relating to intensive livestock farming that we could find focused on the issue of heat stress management. Davison *et al.* (1996) used cost–benefit analysis to show that net savings could be made by providing shelter for dairy cattle over much of Australia. Jones and Hennessy (2000) conducted a risk assessment of heat stress on dairy cattle in the Hunter Valley, New South Wales, which farmers could use in conjunction with a cost–benefit analysis for their own situation. Their provisional cost–benefit analysis, using the method of Davison *et al.* (1996) demonstrated that the capital costs of $30 per cow to install shade and sprinklers, in order to adapt to increased heat and humidity, resulted in a gross return of $14 per cow per annum. While this analysis didn't take all factors into account, it did suggest that heat stress management in the upper Hunter Valley was cost effective.

We consider that cost–benefit analysis is an integral component of assessing the value or utility of adaptation options, particularly those recommended at an industry level. A cost–benefit analysis was recently conducted to evaluate salinity mitigation measures for the Mary River, Northern Territory (McInnes 2004), and the method used has been proposed as the benchmark method for future cost–benefit evaluations of projects involving environmental change.

Discussion

The challenge of successfully adapting intensive livestock production to climate change and ensuring a prosperous future requires action at three levels: the farm, industry and government. The role of industry and government is to enable adaptation through policy and service provision, while farmers carry it out through their farm management systems. We have discussed a variety of possible adaptation options in this chapter and these are summarised below (Table 11.1). At a policy level, the intensive livestock industry will benefit its suppliers by engaging with government in conducting a comprehensive policy analysis and review across agricultural, energy, taxation and trade portfolios (Howden *et al.* 2007), particularly in light of the structural changes being made to the economy in response to climate change by the Australian Government. Policy analysis at a whole system level is necessary to reduce the risks of maladaptation or counteracting policy and regulation inherent in portfolios with different objectives.

Industry and government can assist adaptation by supporting research into traditional areas such as heat- or variable climate-tolerant fodder crops,

Table 11.1: Potential climate change adaptation options for the intensive livestock industry. Note that the options identified here are not comprehensive and those chosen will depend on individual circumstances. Priority 1 (high) and 2 (medium).

Adaptation options	Priority
Feedbase management	
Selection of drought-tolerant pasture species	1
Use of perennial and annual pasture species	1
Fodder conservation and conserved fodder use strategies	1
Forward contracting supply of supplementary feedstock	1
Livestock management	
Selection for thermotolerance and thermoregulation capacity	2
Provision of shade and shelter through infrastructure or tree planting and protection	1
Feeding of feedlot-grown stock in cool periods of the day	1
Agist stock during unsuitable conditions	1
Farm management	
Improve water use efficiency of irrigation	1
Increase groundwater recharge and soil moisture maintenance through revegetation and soil organic carbon management	2
Use decision-support tools with seasonal climate forecasts to make proactive decisions	1
Infrastructure and energy	
Climate control in buildings through natural air conditioning	2
On-site power generation	2
Use alternative energy sources such as biodiesel	2

and new stock breeds, but also into whole systems analysis. This form of analysis is necessary to understand, for example, the risks posed to feed supplies due to reduced or more variable grain yields and/or increased competition from use of feed stock for biofuels or international food markets. Similar analysis is required to assess the vulnerability of the irrigated dairy industry to potential reductions in irrigation supply, and placing this within a regional assessment framework.

Industry will have a role in developing guidelines for building or retrofitting livestock production buildings to be energy and water efficient, incorporating options for energy generation and passive or environmental air conditioning. They will also have a role in communicating the need to take into account climate change in the design of new infrastructure to allow for increased capacity to keep animals cool but also to reduce greenhouse gas emissions in both the construction and maintenance phases. These options may also be built into planning codes and building regulations for future sustainable developments.

At the level of the farm, adaptations to climate change or climate variability can take at least two forms (Smit and Skinner 2002). Adaptations can be tactical, i.e. changes made in the production system in reaction to prevailing conditions, such as drought. Tactical adaptation usually occurs within a production season, and may cease when conditions return to 'normal', although they can become fixed in the production system if they offer longer term benefits. Strategic adaptations, on the other hand, are changes made to the production system in order to address prevailing and future conditions, and are usually made to ensure the robustness or resilience of the enterprise to the projected conditions.

It has become apparent that livestock farmers must have a greater situational awareness of issues beyond their farm gate than they ever have before, in order to adapt to, and prosper in, a changing economic and climatic environment. Farm management systems are no longer just about growing grass and producing milk, or raising chickens or pigs. These systems need to incorporate such things as an understanding of seasonal and out-year forecasts, the dynamics of global commodity and financial markets, the dynamics of Australian horticulture, and consumer demands. The farmer will then make more informed and effective decisions about investing in a changing and uncertain future for the intensive livestock industry.

References

ABS (2008). Value of agricultural commodities produced, 2005–06. Australian Bureau of Statistics, Canberra. www.abs.gov.au/AUSSTATS/abs@.nsf/DetailsPage/7503.02005-06?OpenDocument (Accessed 16 November 2009).

Aharoni Y, Brosh A, Harari Y (2005). Night feeding for high-yielding dairy cows in hot weather: effects on intake, milk yield and energy expenditure. *Livestock Production Science* **92**, 207–219.

Armstrong DP (2004). Water use efficiency and profitability on an irrigated dairy farm in northern Victoria: a case study. *Australian Journal of Experimental Agriculture* **44**, 137–144.

Armstrong DP, Knee JE, Doyle PT, Pritchard KE, Gyles OA (2000b). Water-use efficiency on irrigated dairy farms in northern Victoria and southern New South Wales. *Australian Journal of Experimental Agriculture* **40**, 643–653.

Austin N (1998). Sustainable use of water and fertiliser in the irrigated dairy industry of Australia. PhD thesis, University of Melbourne.

Australian Lot Feeders' Association (2002). Australian Lot Feeders' Association Overview. Sydney.

Avissar R, Pielke RA (1991). The impact of plant stomatal control on mesoscale atmospheric circulations. *Agricultural and Forest Meteorology* **54**, 353–372.

Belliveau S, Smit B, Bradshaw B (2006). Multiple exposures and dynamic vulnerability: evidence from the grape industry in the Okanagan Valley, Canada. *Global Environmental Change* **16**, 364–378.

Berman A (2006). Extending the potential of evaporative cooling for heat-stress relief. *Journal of Dairy Science* **89**, 3817–3825.

Bethune M, Armstrong DP (2004). Overview of the irrigated dairy industry in Australia. *Australian Journal of Experimental Agriculture* **44**, 127–129.

Brosh A, Aharoni Y, Degen AA, Wright D, Young BA (1998). Effects of solar radiation, dietary energy, and time of feeding on thermoregulatory responses and energy balance in cattle in a hot environment. *Journal of Animal Science* **76**, 2671–2677.

Cai WJ, Cowan T (2008). Evidence of impacts from rising temperature on inflows to the Murray-Darling Basin. *Geophysical Research Letters* **35**, L07701, doi:10.1029/2008GL033390.

Callow MN, Fulkerson WJ, Donaghy DJ, Morris RJ, Sweeney G, Upjohn B (2006). Response of perennial ryegrass (Lolium perenne) to renovation in Australian dairy pastures. *Australian Journal of Experimental Agriculture* **45**, 1559–1565.

Callow MN, Lowe KF, Bowdler TM, Lowe SA, Gobius NR (2003). Dry matter yield, forage quality and persistence of tall fescue (*Festuca arundinacea*) cultivars compared with perennial ryegrass (*Lolium perenne*) in a subtropical environment. *Australian Journal of Experimental Agriculture* **43**, 1093–1099.

Chapman DF, Kenny SN, Beca D, Johnson IR (2008a). Pasture and forage crop systems for non-irrigated dairy farms in southern Australia. 1. Physical production and economic performance. *Agricultural Systems* **97**, 108–125.

Chapman DF, Kenny SN, Beca D, Johnson IR (2008b). Pasture and forage crop systems for non-irrigated dairy farms in southern Australia. 2. Inter-annual variation in forage supply, and business risk. *Agricultural Systems* **97**, 126–138.

Cooper MA, Washburn KW (1998). The relationship of body temperature to weight gain, feed consumption, and feed utilization in broilers under heat stress. *Poultry Science* **77**, 237–242.

Davison TM, McGowan M, Mayer DG, Young B, Jonsson N, Hall A, Matschoss A, Goodwin P, Goughan J, Lake M (1996). Managing hot cows in Australia. Queensland Department of Primary Industry, Brisbane.

Donnelly JR, Freer M, Salmon L, Moore AD, Simpson RJ, Dove H, Bolger TP (2002). Evolution of the GRAZPLAN decision support tools and adoption by the grazing industry in temperate Australia. *Agricultural Systems* **74**, 115–139.

Du Preez JH, Hattingh PJ, Giesecke WH, Eisenberg BE (1990). Heat stress in dairy cattle and other livestock under southern African conditions. III. Monthly temperature-humidity index mean values and their significance in the performance of dairy cattle. *Onderstepoort Journal of Veterinary Research* **57**, 243–248.

Eckard R, Johnson I, Chapman D (2006). Modelling nitrous oxide abatement strategies in intensive pasture systems. *International Congress Series* **1293**, 76–85.

Eckard RJ, Chapman DF, White RE (2007). Nitrogen balances in temperate perennial grass and clover dairy pastures in south-eastern Australia. *Australian Journal of Agricultural Research* **58**, 1167–1173.

Eckard RJ, Chen D, White RE, Chapman DF (2003). Gaseous nitrogen loss from temperate perennial grass and clover dairy pastures in south-eastern Australia. *Australian Journal of Agricultural Research* **54**, 561–570.

Eckard RJ, Franks DR (1998). Strategic nitrogen fertiliser use on perennial ryegrass and white clover pasture in north-western Tasmania. *Australian Journal of Experimental Agriculture* **38**, 155–160.

Franca A, Loi A, Davies WJ (1998). Selection of annual ryegrass for adaptation to semi-arid conditions. *European Journal of Agronomy* **9**, 71–78.

Fulkerson WJ, Slack K, Bryant R, Wilson F (2003). Selection for more persistent perennial ryegrass (*Lolium perenne*) cultivars for subtropical/warm temperate dairy regions of Australia. *Australian Journal of Experimental Agriculture* **43**, 1083–1091.

García-Ispierto I, López-Gatius F, Bech-Sabat G, Santolaria P, Yániz JL, Nogareda C, De Rensis F, López-Béjar M (2007). Climate factors affecting conception rate of high producing dairy cows in northeastern Spain. *Theriogenology* **67**, 1379–1385.

Gaughan JB, Davis MS, Mader TL (2004). Wetting and the physiological responses of grain-fed cattle in a heated environment. *Australian Journal of Agricultural Research* **55**, 253–260.

Gaughan JB, Goodwin PJ, Schoorl TA, Young BA, Imbeah M, Mader TL, Hall A (1998). Shade preferences of lactating Holstein-Friesian cows. *Australian Journal of Experimental Agriculture* **38**, 17–21.

Green MA (2006). *Third Generation Photovoltaics: Advanced Solar Energy Conservation*. Springer, Berlin.

Hahn GL (1981). Housing and management to reduce climatic impacts on livestock. *Journal of Animal Science* **52**, 175–186.

Hahn GL (1985). Management and housing of farm animals in hot environments. In *Stress Physiology in Livestock. Ungulates*. (Ed. MK Yousef) pp. 151–176. CRC Press, Boca Raton.

Hahn GL, Oosburn DD (1969). Feasibility of summer environmental control for dairy cattle based on expected production losses. *American Society of Agricultural Engineering* **12**, 448–451.

Hammer GL, Nicholls N, Mitchell C (Eds) (2000). *Applications of Seasonal Climate Forecasting in Agricultural and Natural Ecosystems. The Australian Experience.* Kluwer Academic Publishers, Dordrecht, The Netherlands.

Hansen PJ (2004). Physiological and cellular adaptations of zebu cattle to thermal stress. *Animal Reproduction Science* **82–83**, 349–360.

Hansen PJ (2007). Exploitation of genetic and physiological determinants of embryonic resistance to elevated temperature to improve embryonic survival in dairy cattle during heat stress. *Theriogenology* **68S**, S242–S249.

Harle KJ, Howden SM, Hunt LP, Dunlop M (2007). The potential impact of climate change on the Australian wool industry by 2030. *Agricultural Systems* **93**, 61–89.

Hatfield-Dodds S, Carwardine J, Dunlop M, Graham P, Klein C (2007). 'Rural Australia providing climate solutions. Preliminary report to the Agricultural Alliance on Climate Change'. CSIRO, Canberra.

Hennessy KJ (2007). *Climate change in Australian dairy regions.* CSIRO Marine and Atmospheric Research, Aspendale, Victoria.

Holt SM, Gaughan JB, Mader TL (2004). Feeding strategies for grain-fed cattle in a hot environment. *Australian Journal of Agricultural Research* **55**, 719–725.

Howden SM, McKeon GM, Walker L, Carter JO, Conroy JP, Day KA, Hall WB, Ash AJ, Ghannoum O (1999). Global change impacts on native pastures in south-east Queensland, Australia. *Environmental Modelling and Software with Environment Data News* **14**, 307–316.

Howden SM, Soussana J-F, Tubiello FN, Chhetri N, Dunlop M, Meinke H (2007). Climate change and food security special feature: adapting agriculture to climate change. *Proceedings of the National Academy of Sciences* **104**, 19691–19696.

Howden SM, Turnpenny J (1997). Modelling heat stress and water loss of beef cattle in subtropical Queensland under current climates and climate change. In *Modsim '97. International Congress on Modelling and Simulation Proceedings, 8–11 December, University of Tasmania, Hobart.* (Eds DA McDonald, M McAleer) pp. 1103–1108. Modelling and Simulation Society of Australia.

Igono MO, Johnson HD (1990). Physiological stress index of lactating dairy cows based on diurnal pattern of rectal temperature. *Journal of Interdisciplinary Cycle Research* **21**, 303–320.

Johnson HD, Ragsdale AC, Berry IL, Shanklinm MD (1963). 'Temperature-humidity effects including influences of acclimation in feed and water consumption of Holstein cattle.' Research Bulletin of Missouri Agricultural Experimentation Station, No. 846.

Jones RN, Hennessy KJ (2000). Climate change impacts in the Hunter Valley. A risk assessment of heat stress affecting dairy cattle. CSIRO Atmospheric Research, Aspendale, Victoria.

Kadzere CT, Murphy MR, Silanikove N, Maltz E (2002). Heat stress in lactating dairy cows: a review. *Livestock Production Science* **77**, 59–91.

Kerr D (2004). Factors influencing the development and adoption of knowledge based decision support systems for small, owner-operated rural business. *Artificial Intelligence Review* **22**, 127–147.

King JM (1983). 'Livestock water needs in pastoral Africa in relation to climate and forage'. ILCA Research Report No 7, Addis Ababa.

Kutschera U, Niklas KJ (2007). Photosynthesis research on yellowtops: macroevolution in progress. *Theory in Biosciences* **125**, 81–92.

Lal R, Follett F, Stewart BA, Kimble JM (2007). Soil carbon sequestration to mitigate climate change and advance food security. *Soil Science* **172**, 943–956.

Langston A, Ryan P, Langridge J, Abel N, Plant R, Ive J (2003). Ecosystem services from a dryland catchment. In *Natural Values. Exploring Options for Enhancing Ecosystem Services in the Goulburn Broken Catchment*. (Ed. Nea Abel) pp. 54–82. CSIRO Sustainable Ecosystems, Canberra.

McAllister RJ, Gordon IJ, Janssen MA, Abel N (2006). Pastoralists' responses to variation of rangeland resources in time and space. *Ecological Applications* **16**, 572–583.

McInnes R (2004). Cost-benefit analysis of Mary River salinity mitigation. Australian Greenhouse Office, Canberra.

Parminter TG, Botha CAJ, Smeaton D (2004). 'Needs analysis for dairy industry grazing management research and extension'. AgResearch Australia, Melbourne.

Pielke RA (2001). Influence of the spatial distribution of vegetation and soils on the prediction of cumulus convective rainfall. *Reviews of Geophysics* **39**, 151–177.

Pielke RA, Adegoke JO, Chase TN, Marshall CH, Matsui T, Niyogi D (2007). A new paradigm for assessing the role of agriculture in the climate system and in climate change. *Agricultural and Forest Meteorology* **142**, 234–254.

Pitman AJ, Narisma GT, Pielke RA, Holbrook NJ (2004). Impact of land cover change on the climate of southwest Western Australia. *Journal of Geophysical Research – Atmospheres* **109** **(D18)**.

Poultry CRC (2009). Poultry hub. Poultry Cooperative Research Centre, Armidale, New South Wales. www.poultryhub.org (Accessed 16 November 2009).

QPIF (2009). Pig production. Queensland Primary Industries and Fisheries, Brisbane. www.dpi.qld.gov.au/27_122.htm (Accessed 16 November 2009).

Raman P, Mande S, Kishore VVN (2001). A passive solar system for thermal comfort conditioning of buildings in composite climates. *Solar Energy* **70**, 319–329.

Roenfeldt S (1998). You can't afford to ignore heat stress. *Dairy Management* **35**, 6–12.

Ryan J (2007). *Combining farmer decision making with systems models for restoring multi-functional ecohydrological systems in degraded catchments*. University of Queensland, Brisbane.

Smit B, Skinner MW (2002). Adaptation options in agriculture to climate change: a typology. *Mitigation and Adaptation Strategies for Global Change* **7**, 85–114.

Smith P (2008) Land use change and soil organic carbon dynamics. *Nutrient Cycling in Agroecosystems* **81**, 169–178.

Waller RA, Sale PWG (2001). Persistence and productivity of perennial ryegrass in sheep pastures in south-western Victoria: a review. *Australian Journal of Experimental Agriculture* **41**, 117–144.

Watson DJ, Drysdale G (2006). Irrigation practices on north-east Victorian dairy farms: a survey. *Australian Journal of Experimental Agriculture* **45**, 1539–1549.

Wettemann RP, Bezar FW (1985). Influence of environmental temperature on prolificacy of pigs. *Journal of Reproductive Fertility Supplement* **33**, 199–208.

12

WATER RESOURCES

RN Jones

KEY MESSAGES:

- Recent climate change in southern and eastern Australia has resulted in catchment yields as low as the worst-case model projections for 2030 to 2050. These changes have not always been gradual: step changes in rainfall and streamflow were observed in south-western Australia in the mid 1970s and in southern and eastern Australia in 1997, shifting many water systems beyond their historical operating limits.

- Changes in some climate processes linked to rainfall decreases have been attributed to anthropogenic climate change, as have warming temperatures over most of Australia. Natural climate variability may also be contributing to recently observed rainfall changes.

- The use of historical climate to construct the likely range of operating conditions for water resource management and to provide a baseline from which to measure potential future change is no longer sufficient. A 'whole of climate' approach to operational and strategic decision-making is recommended, combining the analysis of past and recent climate with model projections of future change.

- For eastern and southern Australia, the use of a 'whole of climate' approach recognises that the observed decreased rainfall occurring over the past decade is a significant and persistent departure. A 'new normal' or operating baseline for rainfall for this area is required. Most of Australia is projected to warm at a rate of 0.2°C or more per decade for the next few decades. Rainfall over southern Australia is projected as very likely to decrease in future, so further declines may be anticipated in line with continued global warming. Changes in other regions are less clear.

- The challenge for agriculture is to continue to improve its productivity during a period of historically unprecedented low water supply. Continuing water shortages over the southern and eastern parts of the continent can be anticipated.

- Agriculture in Australia will also need to play its part in the water reform process by aiming to achieve sustainable management during a period of resource constraint and increased competition between water users.

Introduction

Australian agriculture has been largely successful in coping with some of the highest moisture variability in the world, which affects rainfall, soil moisture, stream water, stored water and groundwater. However, measures that have coped effectively with historical climate are being tested by recent climate change, and will be tested further by changes to come.

Until recently, climate change was thought of as tomorrow's problem rather than today's. Recent changes to rainfall and temperature over much of southern and eastern Australia have moved many river systems beyond their historical climate

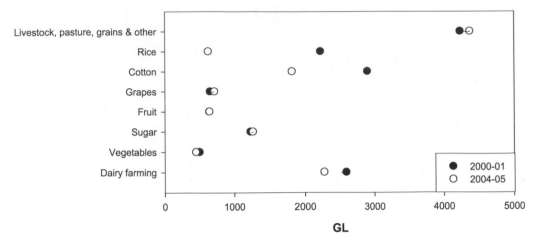

Figure 12.1: Total water use by agricultural activity, 2000–01 and 2004–05. ABS 2006

envelope. Current streamflow is similar to the worst-case projections accounting for anthropogenic climate change from 2030 to 2050.

The climate of the past is no longer an adequate reference for planning Australia's agriculture. The ability to develop sustainable management practices to cope with future moisture variability, and to address the much broader range of challenges faced by agriculture, will depend on understanding the risks to water supply posed by natural and human-induced climate change.

Yet the past has seen large-scale responses to a variable climate, significant resources are being invested in the national water reform process and the recent water shortages are provoking responses on all scales from national policy-making to enterprise level. This chapter summarises the challenges Australian agriculture faces in adapting to rapidly changing water supply and use.

Water use in agriculture

Most of Australian rainfall is evaporated or transpired. About 10% or so ends up as runoff, with <1% (about 15 000 GL annually) contributing to groundwater (Dunlop *et al.* 2001).

Australia has one of the highest *per capita* water consumption rates in the world (1.31 ML/person/ year; NLWRA 2001). Water consumption was 21 700 GL in 2000–01 (ABS 2004) falling to 18 800 GL in 2004–05 (ABS 2006). About two-thirds (15 000 GL or 69%) was used by the agricultural sector in 2000–01 (ABS 2004) falling to 12 200 GL (65%) in 2004–05 (ABS 2006). Under the current regime of capped allocations of general

and low security water and declining water supply overall, agriculture is comparatively more exposed to fluctuations in supply than the urban and industrial sectors.

Agricultural water use includes water for livestock and irrigation. The National Water Accounts (ABS 2004, 2006) are not clear as to what proportion of water use, especially self-extracted water (water extracted by the user and not delivered through a distribution system), is supplied to livestock, compared to the irrigation of crops and pastures. Slightly more than half the water consumed by the agriculture industry in 2004–05 was self-extracted (6582 GL or 54%), with distributed water (5329 GL or 44%) and reuse water (280 GL or 2%) accounting for the remainder. Surface water in rivers, dams and lakes accounted for 74% of self-extracted water, while groundwater accounted for 23%. The largest proportions of self-extracted surface water by state were Tasmania (92%), Victoria (84%) and Queensland (76%). Proportions of self-extracted groundwater were: Northern Territory (82%), South Australia (46%), Western Australia (26%), New South Wales (25%) and Queensland (23%).

The largest agricultural users of water in 2004–05 nationally were dairy farming (2276 GL or 19%), pasture other than for dairy (1928 GL), cotton (1822 GL or 15%), sugar (1269 GL or 10%) and grain crops (1162 GL) (ABS 2006). Agricultural water uses in 2000–01 and 2004–05 are shown in Figure 12.1. The majority of intensive crop and pasture irrigation occurs in the Murray–Darling drainage division. New South Wales contains the largest irrigation area with 910 000 hectares or 38% of the national total.

Irrigated land decreased by 8% in area from 2.6 Mha in 2000–01 to 2.4 Mha in 2004–05. Areas irrigated for livestock, sugar, fruit and grapes increased and areas irrigated for dairy farming, vegetables, cotton and rice decreased. Cotton experienced the greatest decrease from 437 378 ha in 2000–01 to 269 677 ha in 2004–05. The largest decrease in percentage terms was a 71% decrease in irrigated rice, from 178 965 ha to 51 216 ha.

The total gross value of irrigated agricultural production in 2004–05 was $9 076 million compared to $9 618 million in 2000–01, the decrease occurring mostly for cotton and rice crops. Irrigated production contributed 23% to the total gross value of agricultural commodities produced in 2004–05. Fruit was the largest contributor to the value ($1777 million or 20%), followed by vegetables ($1761 million or 20%) and dairy farming ($1632 million or 18%).

Impacts of projected climate change

Baselines and reference conditions

Water resource management requires a whole of climate approach incorporating ongoing natural climate variability. To assess change, a baseline or reference is needed. Climate baselines may include climate trends and different modes of variability. More comprehensive reference conditions might include an operational and environmental history and responses to past changes.

Ideally the baselines and reference conditions for management would cover (1) climate trends and different modes of climate variability; (2) past climate and adaptive responses that serve as the building blocks for future adaptation; (3) a record of how climate risks interact with other processes such as other biophysical changes, socio-economic change and system performance; (4) a record of system performance that can serve as the basis for projecting possible future changes, and (5) information about how climate risks have been managed in the past, including which adaptations were successful and which were unsuccessful.

Management decisions are influenced by a range of planning horizons. Some decisions are ongoing, but others may have to be made years in advance. Design and operation for water supply infrastructure needs to consider changing climatic risks that range from short-term flooding to long-term fluctuations in rainfall and supply. Multiple agents (e.g. users, operators, or regulators) may be involved in decision-making, depending on the time-horizon of the activity in question (Table 12.1). For the next few decades, the main drivers of change include natural- and human-induced climate change, socio-economic trends affecting demand patterns, land-use change and population growth.

Climate baselines

Two considerations are particularly important in understanding the context for decisions and

Table 12.1: Management and planning activities in the water sector as they relate to climate over different time horizons.

Decision type	Time horizon	Relevant climate aspects	Activity	Agent
Day to day management (Tactical)	Intraseasonal	Seasonal availability of water, short-term rainfall events, short-term evaporation rates	Delivery of orders and allocations, flow regulation, flood control, maintenance of water quality, irrigation, drainage	Water operators and users at enterprise level (e.g. farmers, engineers)
Seasonal management (Tactical)	Seasonal	Inter–annual rainfall variability, seasonal soil moisture balance	Allocations, seasonal planning, cropping and stock returns, drought control	Water authorities, farmers, regulatory bodies
Mid-term planning (Strategic)	Multi-seasonal (2–15 years)	Frequency of dry and wet years, decadal variability	Policy (e.g. action plans for dryland and irrigation salinity), economic reform, whole farm planning, catchment strategies, landcare	Catchment managers, legislators, governments, individual farmers, professional and research organisations
Long-term planning (Strategic)	Decades	Decadal variability, climate change	Infrastructure planning, sustainability	Planning bodies, whole of government approach, visionaries

planning around water in Australia. Firstly, water management has traditionally assumed that historical climate is stationary, that is, it is stable over time, with variations assumed to be temporary deviations from the norm only. Climate change is often then imposed as a gradual trend on top of this. We now characterise climate in dynamic terms with multiple modes of variability being possible. Secondly, most of Australia's water supply and distribution systems were developed during the latter half of the 20th century, a period of generally favourable rainfall, leading many water managers and users to believe that their systems were largely 'climate proof'. While most water delivery systems were managed conservatively, with allocations based on historical extremes often set according to the worst drought of record or similar criteria (Long and McMahon 1996), the reality has changed.

Recent climate change has produced record low accumulations of rainfall in parts of southern Australia. Potential evaporation rates have increased in areas of rainfall decline (Jovanovic *et al.* 2007; Kirono *et al.* 2007). Many catchments are now operating outside historical conditions, therefore historical climate no longer provides a reliable basis for operational or strategic operations. Although the climatic thresholds for agricultural and hydrological drought are different, many areas of dryland agriculture are likely to have been similarly affected.

Applying a whole of climate approach, while desirable, is difficult. Short- and long-term climate variability interacting with anthropogenic climate change pose a difficult task for climate science. Attribution of changes to particular modes of natural variability can help diagnose types of risk and how long they may persist. Attribution of changes to human-induced warming indicate how climate-related risks may change in the future.

Reconstructed palaeoclimatic records reveal natural climate variability larger than shown by the instrumental climate record (Harle *et al.* 2005). The historical climate record itself shows several modes of variability important to water resources (Chapter 1, Table 1.1). The most significant are decadal modes of variability lasting from 20 to more than 50 years linked to ocean-atmospheric phenomena such as the Inter-decadal Pacific Oscillation (IPO; Power *et al.* 1999). Changes between different modes can be abrupt as

identified by statistical tests that detect rapid changes while ignoring gradual trends. Variables that show distinct decadal modes include heavy daily rainfall, decadal mean rainfall, El Niño-Southern Oscillation (ENSO) behaviour and tropical cyclones.

The two most identifiable modes of decadal variability affect (1) interannual rainfall and (2) decadal average rainfall. In the first mode, oscillating El Niño–La Niña dominated phases of ENSO influenced by the IPO affect the frequency and magnitude of floods and droughts. Changes in a cycle lasting roughly 22 years can be observed around 1895, 1923, 1946–8, 1976 and 1999 (Power *et al.* 1999; Kiem *et al.* 2003; Kiem and Franks 2004; Verdon *et al.* 2004; Power *et al.* 2005). In the second mode, oscillating drought and flood-dominated periods affect mean rainfall and levels of intensity for several decades or more (Warner 1987; Vivès and Jones 2005). Statistically significant changes from drought to flood-dominated modes have been detected in Australian rainfall records in eastern Australia in 1946–48 and 1972 and from flood to drought-dominated modes in eastern Australia in 1895 and south-west Western Australia in 1946 and 1965–67 (Vivès and Jones 2005; Li *et al.* 2005). The first mode is associated with seasonal risks, whereas the second is associated more with stresses that accumulate over time.

Two research programs have assessed recent climate shifts. The Indian Ocean Climate Initiative in south-western Australia (IOCI 2002) and the South-eastern Australian Climate Initiative (SEACI) in south-eastern Australia (see www.ioci.org.au; www.mdbc.gov.au/subs/seaci). While, statistically, recent reductions in rainfall resemble those seen in the past, analysis of synoptic and other changes suggests that anthropogenic climate change is influencing rainfall declines seen in both regions (Timbal *et al.* in press). Climate change is also being linked to observed increases in temperature and evaporative demand (Nicholls, 2004).

Attribution to specific modes of decadal variability and/or to anthropogenic climate change, suggests that these changes are not short term and will persist, with either partial recovery or continued deterioration being possible. The Water Corporation in Western Australia has already responded to recent decreases in rainfall and surface water supply by altering their working baseline, previously based on data from the mid 1970s, to data recorded from 1997.

Water resource baselines

Reliable data on water budgets incorporating the hydrologic cycle, and water management and use are required for proper planning of water resources. However, until recently, Australia's water budget has been poorly known.

Two phases of National Land and Water Resources Audit (NLWRA) created the Australian Natural Resources Atlas (http://audit.ea.gov.au/ANRA/atlas_home.cfm) that maintains all the data collated for the Audit, including the status of Surface Water Management Areas (SWMAs) and groundwater management units (GMUs). Other data included are catchment statistics on water supply, quality and use at the management area/unit scale, historical streamflow data and spatial data sets of hydroclimatic variables such as potential evaporation.

Water accounts are a vital component of baseline data but are in their early stages and need to be improved (ABS 2004; Lenzen 2004). The Australian Bureau of Statistics maintains the National Water Account (ABS 2004, 2006), a physical account that consists of supply and use tables, information on water stocks and related issues. The Water Account integrates data from different sources, making it possible to link physical data on water to economic data, following the System of Integrated Environmental and Economic Accounting (UN 2003). Foran et al. (2005) uses these methods to construct input–output relationships for a triple bottom line physical accounting of water applying economic, social and environmental measures.

In April 2008, The Australian Government announced the $12.9 billion 'Water for the Future' program that includes $450 million for the 'Improving Water Information Program' administered by the Bureau of Meteorology. This includes a set of new statutory water information responsibilities: issuing national water information standards; collecting and publishing water information; conducting regular national water resources assessments; publishing an annual National Water Account; providing regular water availability forecasts; giving advice on matters relating to water information, and enhancing understanding of Australia's water resources.

An accurate national water budget delineating different sources, longitudinal records that can link climate and water use over time and the removal of systematic errors is urgently needed. The 'Improving Water Information Program' and associated research programs will greatly improve this capacity. Physical accounts will be managed by the Bureau of Meteorology and financial accounts by the Australian Bureau of Statistics.

Recent streamflow changes

Over the past decade, accumulated streamflow totals in catchments within and south of the Murray–Darling Basin and in south-west Western Australia have reached historical lows. In south-eastern Australia, the decrease in streamflow over the past decade is equivalent to the 'worst case' derived from climate model projections for 2030–50. Inflows into the Thomson Dam in Gippsland decreased by almost 35% in the period 1997–2004, compared to 1984–97 (Marsden and Pickering 2006). This is as severe as the 5th percentile scenario (drier than 95% of other scenarios) projected for 2050 by the Melbourne Water Study (CSIRO and Melbourne Water 2005).

Averaged over 1895 to 2006, the mean annual rainfall and modelled runoff for the Murray–Darling Basin (MDB) are 457 mm and 27.3 mm, respectively. Over the decade 1997–2006, mean annual rainfall decreased to 440 mm, about 4% lower than the long-term mean. The mean annual runoff for the MDB in the decade 1997–2006 was 21.7 mm, about 21% lower than the long-term mean. The biggest reductions occurred in the southern half of the MDB, where runoff was >30% lower than the long-term mean, and up to 50% lower in the southernmost parts (CSIRO 2008). This reduction has produced unprecedented levels of stress in the major irrigation systems of the Murray River. Emergency plans have been drawn up to secure supplies of essential water in the event of continuing low rainfall (MDBC 2007).

Climate scenarios that extend the current dry conditions for the next few years to decades have been used to identify chronic deficits that would trigger long-term severe water restrictions (e.g. DSE 2007). New infrastructure plans are being proposed in southern and eastern Australia to secure supply in response to these historically unprecedented conditions.

Projected surface water supply

Runoff integrates the combined impacts of changes in rainfall, temperature, and evapotranspiration, but is most sensitive to changing rainfall. Sensitivity studies indicate a 1% change in mean annual rainfall will result in a 2–3% change in mean annual runoff in Australian catchments (Chiew and McMahon 2002; Chiew 2006; Jones *et al.* 2006). The runoff sensitivity to rainfall is greater in catchments with low runoff coefficients. Each 1% increase in potential evaporation is estimated to produce a 0.5–1% reduction in mean annual runoff (Chiew *et al.* 2005; Jones *et al.* 2006).

Early assessments of impacts on water supply in Australia were highly uncertain about the direction of change (e.g. Chiew *et al.* 1995). Since coupled atmosphere–ocean climate models came into use in the late 1990s, projections used in impact assessments result in reduced streamflow over most of the country.

The first study that sampled a broad range of climate change uncertainties was for the Macquarie River in New South Wales, concluding that reductions in streamflow were highly likely (Jones and Page 2001). Extending these results across the Murray River system using climate change projections for Australia (CSIRO 2001) showed an overwhelming tendency towards runoff decreases (Jones *et al.* 2002). Recent studies are summarised in Hennessy *et al.* (2007).

Comprehensive studies extending beyond single catchments have only recently become available. Two methods have been used: rainfall-runoff models optimised for individual catchments (e.g. CSIRO 2008) and simple climate-hydrology response relationships developed from sensitivity analysis conducted across a range of catchments (Chiew *et al.* 2005; Jones *et al.* 2005). The first is resource-intensive but more accurate, yielding a wider array of outputs (e.g. soil moisture, flow extremes). The second is more flexible but the results can be updated quickly to take in new climate information, or to assess economic impacts (e.g. Quiggin *et al.* 2008).

The CSIRO Murray–Darling Basin Sustainable Yields Project (CSIRO 2008) analysed four scenarios for water supply across the entire basin, using rainfall-runoff modelling. Two baseline scenarios estimated long-term historical runoff (1895–2006) and that of the past decade (1997–2006), and two projections for 2030 assessed climate only and climate with multiple land-use drivers.

The median estimated change in mean annual Basin runoff in 2030 relative to 1990 is 5–10% lower in the north-east and southern half and about 15% lower in the southern-most parts. Averaged across the entire MDB, the median estimate is a 9% decrease in mean annual runoff. The range of uncertainty in all estimates across the MDB ranges from a 33% decrease to a 16% increase in mean annual runoff (CSIRO 2008).

The simple methods, while fast and easy to use, are less accurate and restricted to estimating mean changes. Based on annual response to changes in rainfall and potential evaporation, Jones and Durack (2005) calculated the range of change in mean surface water supply for Victoria, developing a simple tool that was then used by water authorities to assess sustainable yield for strategic catchment planning (DSE 2007). This model was refined to account for seasonality of streamflow (Jones *et al.* 2005) and extended Australia-wide (Preston and Jones 2008a, b and c).

Averaged estimates of regional runoff for 2030, allowing for a range of climate models, emissions futures, and climate sensitivities, are presented in Figure 12.2 (see page 47). Except for some of the arid interior (where there is little effective annual runoff) and the far northern extent of Queensland and the Northern Territory, runoff is projected to decline. Declines are greatest along the west coast, western Victoria/eastern South Australia, and north-east New South Wales to south-eastern Queensland.

The uncertainty of these 2030 estimates is considerable. The 90% confidence limits (the range lying between the 5th and 95th percentiles) produce regional runoff uncertainties of up to ±45% (e.g. central coastal Western Australia). Of Australia's 325 Surface Water Management Areas (SWMAs), 84% tend towards runoff reductions, with average changes ranging from 25–99% for individual SWMAs. Most of Australia is exposed to the risk of runoff reductions, and for some catchments, only decreases are projected.

However, these estimates are based solely on the signal of anthropogenic climate change; natural climate variability also needs to be accounted for. Changes to streamflow in the Macquarie River catchment applying a range of climate scenarios to three modes of decadal rainfall regime (drought-dominated, normal and flood-dominated) showed that a combination of drought-dominated conditions and climate change produced the greatest risks (Jones and Page 2001). This may be the condition that much of southern and eastern Australia is currently in.

Groundwater supply

Groundwater use has increased 58% nationally since 1983–4 and by over 200% in New South Wales, Victoria and Western Australia. Current use represents about 10% of the total groundwater that could be extracted sustainably although about 30% of groundwater management units are either over- or near fully allocated (NLWRA 2001). Areas where surface water supplies have become fully allocated or reduced by drought are seeing increased groundwater use, placing these supplies at greater risk.

The method of accounting for surface and groundwater supplies by the NLWRA led to double-counting where groundwater contributed to streamflow, an anomaly that needs to be corrected (Evans 2007). The need to properly account for the exchange of water between surface water and groundwater has been recognised as part of the National Water Initiative.

The volume and sustainability of groundwater supply relates to the balance between the input of water to a groundwater system (recharge), the output of water (discharge) and their relation to the total storage volume (Cartwright and Simmonds 2008). Climate change can affect groundwater balance in a number of ways, with the response often being spatially variable across the same aquifer.

Groundwater recharge is affected by changing rainfall annual mean, seasonality and timing, changing land use over large areas and changing surface water management. Groundwater discharge is affected by climate-induced changes in the balance of surface water and groundwater use, changing water levels leading to changed evapotranspiration, changing water use patterns by vegetation in areas of shallow water tables, and changing surface water management. Changes in the balance between groundwater recharge and discharge may impact on the following management issues: groundwater allocation for irrigation and other uses; soil salinity; protection of groundwater-dependent ecosystems; stream depletion caused by groundwater extraction, and deteriorating groundwater quality.

Many of these changes will affect an aquifer's sustainable yield. Despite a theoretical understanding of the fundamental relationships between surface and groundwater, the level of knowledge regarding the status, the estimation of sustainable yields and of localised processes for individual aquifers in Australia remains poor (Crosbie 2007). Improved accounting is needed to produce a traceable budget of groundwater production and use and to provide a reliable baseline for assessing climate change impacts. It is also required to manage interactions between surface and groundwater where both become a common tradeable resource (Budd et al. 2004). Resources have been allocated by the 'Improving Water Information Program' to more accurate metering and monitoring of both surface and ground water use.

Water resources at the catchment scale

Until recently, a nation-wide picture of water resources under climate change has not been available. The first of a series of projects, the MDB Sustainable Yields Project (CSIRO 2008) assessed changes in water supply due to commercial plantation forestry, the expansion of farm dams and maximum allowable groundwater use by 2030. These factors would increase surface water use across the MDB by a total of 375 GL per year compared to the median projected reduction from anthropogenic climate change of 451 GL per year. These new water uses are outside the current entitlement scheme, but if incorporated, represent a small reduction over the median projection. However, the change in water use at the regional level could be sizeable, especially if combined with reduced diversions due to climate change (CSIRO 2008).

A preliminary risk assessment of water supply from climate and other change factors was assessed at the catchment level by calculating a simple qualitative metric, a 'catchment risk score' (Preston and Jones, 2008c). This risk score was constructed using five indicators: status of surface and groundwater resources representing baseline conditions, recent rainfall and population trends representing changing supply and demand, and projected runoff changes in 2030.

Data sources for the five indicators are the First National Land and Water Resources Audit (NLWRA, 2001) for surface and groundwater status, the Australian Bureau of Statistics for recent population data, the Bureau of Meteorology for recent rainfall trends and projections of future runoff in 2030 as presented in Figure 12.2 (see page 47).

Mapping of catchment risk scores illustrates the distribution of water resources risk across Australia (Figure 12.3, see page 48). 'High' to 'very high' risks are concentrated in the eastern third of the continent from northern Victoria through to southern Queensland, created by full to over-development of surface and groundwater resources; declining trends in rainfall over recent decades; population growth, and projected reductions in runoff in 2030. The catchments of south-west Western Australia are also assigned a high risk, although significant potential remains for future development of deep aquifers.

Some less populated areas have relatively high risk scores. For example, depletion of aquifers such as the Great Artesian Basin, combined with declining rainfall and projected reductions in future runoff, will likely affect aquatic ecosystems as well as small regional communities. Most of the areas indicating the lowest risk to supply are regions where runoff increases correspond with limited resource development, such as the arid interior or the far northern regions of the Northern Territory and Queensland.

This simple risk metric should be interpreted cautiously (Preston and Jones 2008c). The risk scores are influenced by the self-assessment conducted by each of the states and territories of sustainable yields of surface and groundwater resources and levels of development provided to the NLWRA (2001). Such assessments are of questionable quality, particularly for groundwater resources. Improved data and methods such as those applied within the Murray–Darling Basin Sustainable Yield Project (CSIRO 2008) will provide more reliable estimates of development status.

Constraints on water resources in fully to over-allocated catchments and groundwater systems limit resilience and the capacity to adapt to climate change. Such areas have limited flexibility of supply that can be used to address growing demand, long-term reductions in rainfall and runoff, or periodic shocks such as droughts. For regions with the potential for further resource development, climate and non-climate changes may occur faster than management is able to respond.

Dryland agriculture

Water resources for dryland agriculture are largely a measure of available soil moisture. Surplus soil moisture levels affect productivity and animal health through waterlogging and disease. Deficiency in soil moisture is measured as agricultural drought (e.g. AMS 1997), which is different to hydrological drought. Agricultural drought is less sensitive to rainfall variability because the soil moisture reservoir usually fills before large volumes of runoff are produced. Water-related risks to dryland agriculture will be largely determined by changes in the frequency and magnitude of rainfall as described by Hennessy, Whetton and Preston (Chapter 2). Benefits are possible in wetter regions that become drier, allowing a transition from high-rainfall grazing to cropping.

Changes to water resources can affect stock and domestic water supplies. On-farm water supplies can become limited in periods of prolonged drought and may need to be purchased off-farm. Water quality can also be affected during periods of low volume in dams or flow in streams, becoming unfit for use due to increased levels of salinity, nutrients or toxic algae.

Irrigated agriculture

Most of Australia's agricultural water use is for irrigation. The irrigation industry is undergoing great change, driven by climate change and the water reform process. Drier conditions across most irrigation regions, combined with caps on supply, create severe shortages decades before such changes were previously anticipated to arise out of climate change. The water reform process has been modified to respond to these shortages (CofA 2007, 2008).

Few hydrological studies have examined the direct impact of reductions on irrigation. Wang *et al.* (1999) investigated the Campaspe River water supply in Victoria. Irrigation allocations were based on a 'water right' with up to a further 120% of sales water in years when supply was available. The modelled reliability of the basic water right was reduced by 1% in 2030, 4% in 2070, and 16%

for 4.1°C global warming. Under rules applied in the model, irrigation security was maintained at the expense of downstream environmental flows.

Irrigation allocations and environmental flows in the Macquarie River basin in New South Wales using scenarios from a range of models suggested a 'most likely' reduction in inflows of 0 to -15% by 2030 within a total range of 0 to -30% (Jones and Page 2001). Quiggin *et al.* (2008) found that under a business-as-usual emission scenario with strong reductions in rainfall, production over the MDB declined by 100% by 2100. Under scenarios where mitigation policy was applied and with less severe rainfall reductions, losses were greatly constrained and could be limited further by adaptation.

The catchment risk scores described in Figure 12.3 (see page 48) placed 76% of Australia's irrigated land area within catchments with 'high' or 'very high' risk scores (Jones *et al.* 2009). Most of these areas have been exposed to recent supply reductions.

Although increasing atmospheric CO_2 can potentially increase plant water use efficiency and growth rates, the extent of this benefit in irrigation systems is poorly known. Despite projected reductions for allocations in the northern MDB by 2030, overall increases in cotton production due to higher CO_2 more than compensated for losses in irrigation supply (Hassall and Associates 2002). However, recent supply reductions have reduced the overall size of the cotton harvest, outstripping any productivity increases. Further research in this area is needed.

Options for dealing with climate variability

Much of the irrigation system and its operating rules were set up during the latter half of the 20th century, a period of generally favourable rainfall. Irrigation infrastructure was well adapted to inter-annual variability, with large carry-over storages and extensive distribution systems with defined water rights. Self-extraction from watercourses downstream from those storages also benefited.

Water management for agriculture is currently in a state of transition, where many of the long-standing options are being altered to both cope with recent changes in supply but also to take on the long-term aims of water reforms that include

full-cost pricing, water markets, improved provision for environmental flows, provisions for water quality and the integration of surface and groundwater into management systems.

A number of options are currently being implemented in irrigation systems to deal with restricted water availability (adapted from CSIRO 2007). Limiting seepage in the delivery system improves channel efficiency and controls evaporation, saving up to 10–30% of the water diverted, more in some regions. Efficient on-farm delivery systems (such as microspray and trickle/drip systems or replacing flood irrigation with pipes) can further save up to 20% in on-farm distribution and 60% in the paddock. Scheduling irrigation according to soil moisture, evaporation measurements, timers and sensors, growth phase of crop/pasture, partial root drying (e.g. viticulture), reduces water applied to crops by at least 10–15%. Water management can be further improved by more accurate metering, full monitoring of all extractions, staged metering (detecting system losses) and improved data collection for physical accounting. Precision farming systems provide additional options to improve water use by tailoring irrigation layout and application to local variations in soil characteristics.

Aside from measures to improve the efficiency of irrigation systems, there are some additional options that deal specifically with variability in water availability. Water management could be improved on a season-to-season basis if system-wide allocations were forecast earlier. Such forecasts could, for example, provide likelihoods of reaching particular levels of allocation based on climatic states (e.g. phase of ENSO, Indian Ocean Dipole), storage levels and catchment soil moisture. Cropping and grazing management could also become more opportunistic, taking advantage of greater water availability in wetter years in much the same the way dryland agriculture has operated for some time. These opportunistic approaches could be particularly appropriate for rice and cotton. Water markets also provide the opportunity to hedge risk by forward selling, trading seasonal allocations, and 'banking' carry-over water where permitted.

There are some additional options for dealing with climate variability that apply to both dryland and irrigated farming. For example, it might be possible to take advantage of price and market conditions brought on by long-term drought conditions. Where climate conditions change, farm management could be improved by updating on-farm plans to better match the altered operating conditions. Furthermore, improvements to on-farm water harvesting and storage can help secure essential supplies for stock and domestic purposes. Finally, long-term commitments to improve soils and ecological function (e.g. revegetation and habitat for wildlife) can increase water holding and use capacity, thus improving productivity.

Options for dealing with projected climate change

Farms, catchments and water distribution systems operate at different scales, each with a corresponding set of adaptation options. Specific actions and combinations of those adaptation actions have inherent benefits and considerations for implementation that can be assessed. The planning process needs to apply sound risk assessment and management methods to prioritise and encourage preferred adaptation options while involving appropriate representation from affected and interested stakeholders.

Planning future supply

Strategic planning of irrigation futures between stakeholders and research institutions have the potential to provide better security for the industry, improve environmental outcomes and identify further knowledge needs. This is illustrated by the following examples. In Western Australia, the South-West Water Futures Project is assessing the best land and water resource use options for the south-west irrigation district to ensure maximum economic, environmental and social benefits. In Queensland, the Great Barrier Reef floodplain renewal project is helping to change floodplain land management to improve water quality and protect the reef. In northern Australia, the Northern Australia Irrigation Futures project is providing new knowledge, tools and processes to support debate and decision-making. And the Murrumbidgee catchment in the Murray–Darling Basin is providing knowledge, skills and technology to the world's biggest and most intensive irrigation regions

under UNESCO's global HELP (Hydrology, Environment, Life and Policy) program.

Improved seasonal prediction systems

Although seasonal allocations to holders of surface and some groundwater rights will continue, continuing shortages and full water pricing requires more intensive management. Seasonal predictions of changes in allocations and crop water balance will allow better forward planning for crop selection and areas planted, irrigation scheduling and minimise price risk if supplementary water is needed.

A number of adaptation options could be considered to improve management responses to seasonal variability. For example, ENSO-related indices can be used to predict streamflow directly, providing an equivalent level of predictability to the relationship between ENSO indices and rainfall (Khan *et al.* 2004). Forecast systems could be further improved by combining medium-range weather forecasting and catchment soil moisture, using both remote sensing and modelling. Based on these systems water distributors could provide likelihoods of changes to water allocations. Seasonal crop models could be updated throughout the season to reduce uncertainty and better target production inputs. These methods, now being applied in dryland systems, can be expanded to irrigation systems (Howden *et al.* 2008). In addition, greater use could be made of water trading to provide income for sellers, and to achieve production goals and/or secure production and assets for buyers.

Improved risk management

Risk management has generally been restricted to a single season's outlook. A more strategic approach with longer planning horizons is needed. This should include a 'whole of climate' approach that combines observations and climate projections for use in operational and strategic decision-making. For example, in eastern and southern Australia the observed decline in rainfall over the past decade can be used to construct a 'new normal' climate. Warming of 0.2°C or more for the next few decades can be anticipated, with the potential for further rainfall reductions. Where relevant, risk-based decision-making should incorporate climate change into all levels of operations and long-term planning.

Box 12.1: Water markets as an adaptation to variable supply

Water trading and full-cost pricing is a central part of the water reform process: higher pricing was expected to reduce waste by increasing efficiency and shift water from low to high income and yield activities. Australian water markets trade allocations and entitlements. Allocations are set for use within a specific licence based on availability of supply and can be traded within a single season. Entitlements cover the right to irrigate, a defined property right that can be bought and sold. When implemented in the mid-1990s, the water market was considered able to manage any scarcity associated with changing climate. However, three circumstances combined to spoil this assumption: the entry of unexercised rights into the market, the shift in climate occurring in 1997 and patchy information about water budgets and flows.

Allocation caps were set during the late 1990s because increasing water consumption in the MDB and other highly allocated catchments was driving those systems towards scarcity. The granting of property rights to 'sleeper' and 'dozer' water licences granted market value to all licences, not just the water being used (Crase et al. 2004). Since 1997, climate-induced water shortages, large

fluctuations in price, and a lack of information informing risk management have hampered the establishment of a mature trading market (Bjornlund 2005; Bjornlund and Rossini 2007).

The success of water markets needs to consider trade in allocations and entitlements separately. Trade in allocated water has been identified as increasing the ability of irrigators to manage risk (Bjornlund 2005, 2008), but the benefits of trade in entitlements are less clear and are more controversial (Tisdall and Ward 2003; Crase et al. 2004).

Under supply caps and seasonal allocations tied to water availability, farmers have had increased difficulty acquiring sufficient water during droughts, especially important for long-lived resources such as livestock and tree and vine crops. Purchasing water has generally been of benefit in these circumstances, although affecting profitability in some industries (e.g. dairy; Bjornlund and Rossini 2007). Higher water prices have increased farmers' involvement in water trading, e.g. by selling water in times of shortage, rather than risking crop failure or having to purchase expensive supplies in order to harvest a crop (Bjornlund 2005, 2008).

Improvements in risk management at the farm level will help farmers gain added benefits from water markets (Bjornlund 2006). Climate variability will also change. In particular, changing flood risks should not be overlooked, particularly in regions dominated by summer rainfall systems.

The understanding of integrated catchment management also needs to be improved among different users; for example, the relationships between water quality, surface and groundwater extraction, waterway management and land use, incorporating both climate and non-climatic influences. Institutional arrangements will need to be developed to manage this. There is also a need to employ contingency-based decision-making and planning approaches that are responsive to the transition between different levels of risk, e.g. between Tiers 1, 2 and 3 of water security as written into the national Water Act (CofA 2008). Strategic risk-management approaches will require improving the assessment of management outcomes using

research, community-wide discussion and outreach in an iterative process to examine multiple understandings of water-related values and preferences.

Steps are being taken to ensure that water markets can operate more efficiently. The information base is being improved through formal monitoring of the market, improving water data, better understanding of climate and water linkages, and the development of seasonal forecasting systems. However, there are still fears of maladaptation within rural communities (Tisdell and Ward 2003), where some regions may lose water to others, where speculative water trading diverts water from productive agricultural uses, and where structural change may disadvantage individual owner-operators (Tonts and Black 2002). Despite positive assessments of the seasonal allocation markets in helping farmers to manage risk, there is no consensus as to whether the economic, social and environmental values of water, and the interests of agriculture generally, are being adequately addressed within the market reform process.

Risks of maladaptation

The importance of water for a diverse range of environmental, social and economic outcomes requires an integrated approach to avoid significant unintended consequences. There are several foreseeable situations where risks of maladaptation could arise. (1) Replacement of relatively inefficient water systems by more energy-intensive, water-efficient systems may increase water security but may also increase greenhouse emissions. Efforts to offset increased emissions from such systems may add an opportunity cost if the offsets themselves could have been used to offset less tractable emissions elsewhere. Pumping costs from base load coal-powered energy systems, particularly in Victoria, already carries a significant greenhouse penalty. (2) More efficient agriculture could reduce the production of 'waste water', which would otherwise be available for inflows into streams and wetlands. (3) Existing infrastructure could be stranded from the trading of water rights out of a district, perhaps hastened by chronic water shortage. The demise of such infrastructure, which would otherwise be efficient and productive, could have flow-on effects into local economies. (4) Market power could be used to control a significant proportion of the water resource disadvantaging smaller operators, with little gain in water-use efficiency. Profiteering from speculation and trading could divert water from more productive uses. Both fears are currently very active in farming communities (Tisdall and Ward 2003). (5) The uptake of new and cutting-edge knowledge may be limited because of the fear of new investments failing compared to proven but less efficient traditional methods. And, (6) new irrigation developments may be sub-optimal because of an unwillingness to bear the upfront capital costs and ongoing costs of managing for sound environmental outcomes.

Costs and benefits

The costs and benefits of many individual adaptation options, such as technological improvements to improve on-farm efficiency, can be assessed through conventional agricultural economics. Further research in this area is being pursued through bodies such the Co-operative Research Centre for Irrigation Futures (http://

www.irrigationfutures.org.au). At present, however, such assessments are complicated by rising prices for both water allocations and entitlements. A debate on whether water should be valued monetarily or more broadly, taking into account economic, social and environmental values (Tisdall and Ward 2003) is also taking place (e.g. Crase *et al.* 2004; Batten 2007). Within the broader community, water holds very strong social and environmental values. Important decisions on the future of water need to take these values into account. Although some work in this area has been undertaken, further research is needed to ensure that these values are fully reflected in the costs and benefits of adaptations affecting the future of water resources.

Climate change, combined with full-cost measures for water being pursued through the National Water Initiative have the potential to change the economics of irrigated agriculture substantially. Any adverse climate change that increases the scarcity of water will increase the value of water within a water market (Bjornlund and Rossini 2005). Ultimately, any increase in the unit value of water will be capped by the value of commodities that can be produced. Agriculture may also not be able to compete with urban and industrial demand when supplies are scarce and prices high.

Australian water supply authorities are moving to full cost pricing that includes the cost of environmental externalities where feasible and practical (CoAG 2004). One price impact of climate change may be on the cost of underutilised, prematurely abandoned or damaged infrastructure. If climate change is expected to result in excess capacity and/or the abandonment of infrastructure before its construction costs can be written off, water utilities may set higher prices to recover that cost. For example, projects transferring water from one region to another to secure supply may not be robust to climate change impacts. Water utilities expecting losses in infrastructure performance, may seek to pass those risks onto end users. Private (e.g. on-farm) infrastructure may also be at risk if climate stress leads to a water right being sold.

Discussion

Based on the evidence outlined in this chapter, the time to adapt water management for climate

Table 12.2: Summary of climate change adaptation options for water in agriculture. Priority 1 (high) and 2 (medium).

Adaptation options	Priority
Operational	
More efficient on-farm use of water through improved technology and scheduling	1
Develop and apply probabilistic forecasts of water allocation changes	1
Improve distribution system operation and delivery	1
Increase monitoring of the water cycle and water market performance	1
Increase use of water management tools (crop models, decision support tools)	2
Increase crop choice to maximise efficiency and profit	2
Minimise losses and maintain environmental values in channels and streams	2
Strategic	
Develop more equitable sharing of climate risks among different uses	1
Improve water trading rules to remove perverse incentives and reduce the transfer of risk, especially during drought	1
Improve understanding of sustainable yield	1
Build climate change risks into caps/bulk allocation arrangements	2
Prepare for altered flood risks	2
Exercise control over private water storages and land use affecting supply	2
Improve income spreading strategies to manage risk	2
Build flexibility into allocation choices between agricultural, environmental, urban and industrial uses	2
Improve understanding of groundwater-surface water-climate interactions	2
Long-term planning	
Build climate change into integrated catchment management and strategic planning	1
Build adaptation to climate change into new infrastructure	1
Develop understanding of critical thresholds and limits within water collection delivery and use systems	1
Develop groundwater storage options	2
Provide design guidelines for land use to maximise water yield and water quality within a framework of long-term sustainability	2
Institutional capacity	
Develop risk-based decision-making into all levels of operation and planning from tactical to long-term	1
Develop better understanding of integrated catchment management among different users	1
Develop contingency based decision-making instead of 'one action fits all circumstances'	1
Improve multiple understandings of water related through research, discussion and community consultation	1
Develop a 'whole of climate' approach to operational and strategic decision-making	1

change is now. Climate change, most likely a combination of natural variability and human-induced change, has already caused historically unprecedented shortages in Australia's major irrigation systems. The chance of seeing some recovery in conditions in coming decades depends largely on the size of the contribution that natural variability made to this change. However, the relative contributions of natural and human-induced influences cannot as yet be attributed. This chapter has covered a wide range of potential adaptation options for dealing with the changing availability of water resources.

These options are summarised in Table 12.2 according to whether measures are currently being used, are under development or require further research. Given the importance of water resources to agriculture, all are given moderate to high priority.

Australian agriculture has a history of adapting to high moisture variability and of managing resource limitations. This experience can be used as the basis for adapting to further changes. Two important developments have been occurring at the policy and scientific level. The water reform process has been adapted in response to recent changes and now contains provisions developed explicitly to manage a changing climate (CofA 2007, 2008). The 'Water for the Future' Program contains significant provision for data collection and research which will help to reduce the current large uncertainties as to the available resource and its future under climate change. Improvements to physical and financial accounting will contribute to better resource management and improve market efficiency. Research projects on sustainable yield are also moving beyond the MDB to assess south-western Australia, northern Australia and Tasmania.

The picture in northern Australia is less clear. Although increases in flow are projected as being more likely than in southern Australia (Preston and Jones 2008a, b), and supported by recent increased rainfall in north-western and central Australia, decreases are also possible. This possibility needs to be factored into the planning of new irrigation developments in the north.

Water markets have helped farmers adapt to recent water shortages (Bjornlund 2008), specifically through annual sales of allocation water. The markets for entitlements are less mature and prices are also being affected by increased water scarcity, being closely linked to the deseasonalised allocation price (Bjornlund and Rossini 2007). Allocations being made using criteria that do not properly protect hydrological integrity, also decrease market effectiveness and increase the likelihood of failure in times of stress (Young and McColl 2008).

In this dynamic environment of change to both the resource and the way it is managed, the risk of maladaptation remains high. An adaptive approach to management is needed, where changes are regularly reviewed to ensure they are having the intended impact. Integrated assessment across scales will also be required to determine whether adaptations made at one scale (e.g. at farm level) are also having a positive impact at other scales (e.g. at catchment level).

References

ABS (Australian Bureau of Statistics) (2004). Water Account Australia 2000–01. Commonwealth of Australia, Canberra.

ABS (2006). Water Account Australia 2004–05. Commonwealth of Australia, Canberra.

Adams PD, Horridge M, Masden JR, Wittwer G (2002). Drought regions and the Australian economy between 2001–02 and 2004–05. *Australian Bulletin of Labour* **28**, 233–249.

AFFA (Department of Agriculture, Fisheries, and Forestry) (2005). *Information Handbook: Exceptional Circumstances Assistance: Guide to the Policy and Assistance Provided Under Exceptional Circumstance Agreements*. Commonwealth of Australia, Canberra.

AMS (American Meteorological Society) (1997). Meteorological drought – policy statement. *Bulletin of the American Meteorological Society* **78**, 1847–849.

Batten DF (2007). Can economists value water's multiple benefits? *Water Policy* **9**, 345–362.

Bjornlund H (2005). Irrigators and the new policy paradigm – an Australian case study. *Water Policy* **7**, 581–596.

Bjornlund H (2006). Can water markets assist irrigators managing increased supply risk? Some Australian experiences. *Water International* **31**, 221–232.

Bjornlund H (2008). Monitoring the impacts of water trading some methods and results from an Australian water market. *Proceedings 13th IWRA World Water Congress*, Montpellier France, Article 141. International Water Resources Association.

Bjornlund H, Rossini P (2005). Fundamental determining prices and activities in markets for water allocations. *International Journal of Water Resources Development* **21**, 355–369.

Bjornlund H, Rossini P (2007). Fundamentals determining prices in the market for water entitlements: an Australian case study. *International Journal of Water Resources Development* **23**, 537–553.

Budd K, Hostetler S, Gilmour J, Chiles G, Bleys E, Brodie R (2004). 'Review of the Australian Water Resources Assessment 2000.' Report to the Executive Steering Committee for Australia's Water Resource Information (ESCAWRI), Bureau of Rural Sciences, Canberra.

Bureau of Meteorology (2006). 'Living with drought'. http://www.bom.gov.au/climate/drought/livedrought.shtml

Cartwright I, Simmonds I (2008). Impact of changing climate and land use on the hydrogeology of southeast Australia. *Australian Journal of Earth Sciences* **55**, 1009–1021.

Chiew FHS (2006). Estimation of rainfall elasticity of streamflow in Australia. *Hydrological Sciences Journal* **51**, 613–625.

Chiew FHS, McMahon TA (2002). Modelling the impacts of climate change on Australian streamflow. *Hydrological Processes* **16**, 1235–1245.

Chiew FHS, Jones RN, Boughton WR (2005). Modelling hydrologic sensitivity to climate conditions. In *Proceedings of the 29th Hydrology and Water Resources Symposium*. The Institution of Engineers Australia, Canberra.

Chiew FHS, Piechota TC, Dracup JA, McMahon TA (1998). El Niño/Southern Oscillation and Australian rainfall, streamflow and drought: links and potential for forecasting. *Journal of Hydrology* **204**, 138–149.

CoAG (Council of Australian Governments) (2004). Communique, 25th June 2004. Commonwealth of Australia, Canberra.

CofA (2007). *Water Act 2007*. Commonwealth of Australia, Canberra.

CofA (2008). *Water Amendment Bill 2008*. Commonwealth of Australia, Canberra.

Crase L, Pagan P and Dollery B (2004). Water markets as a vehicle for reforming water resource allocation in the Murray–Darling Basin of Australia. *Water Resources Research* **40**, W08S05, doi:10.1029/2003WR002786.

Crosbie R (2007). 'The hydrological impacts of climate change on groundwater'. In *Hydrological Consequences of Climate Change*. CSIRO Discovery Centre, Canberra.

CSIRO (2001). 'Regional projections of climate change for Australia'. CSIRO Atmospheric Research, Melbourne.

CSIRO (2007). Science to improve Australia's irrigation systems. http://www.csiro.au/org/IrrigationResearch.html (Accessed January 2008)

CSIRO (2008). 'Water availability in the Murray–Darling Basin'. A report to the Australian Government from the CSIRO Murray–Darling Basin Sustainable Yields Project. CSIRO, Australia.

CSIRO and Australian Bureau of Meteorology (2007). 'Climate change in Australia'. CSIRO, Melbourne.

CSIRO and Melbourne Water (2005). 'Implications of climate change for Melbourne's water resources'. Melbourne Water, Melbourne.

DSE (Department of Sustainability and Environment) (2007). 'Our water our future: the next stage of the government's water plan'. Victorian Department of Sustainability and Environment, Melbourne.

Dunlop M, Hall N, Watson B, Gordon L, Foran B (2001). 'Water use in Australia'. Working Paper 01/02, CSIRO Sustainable Ecosystems, Canberra.

Evans R (2007). 'The effects of groundwater pumping on stream flow in Australia.' Technical Report, Land and Water Australia, Canberra.

Foran B, Lenzen M, Dey C (2005). *Balancing Act: A Triple Bottom Line Analysis of the Australian Economy*. 4 Volumes. CSIRO and University of Sydney, Canberra.

Harle K, Etheridge DM, Barbetti M, Turney C, Jones RN, Brooke B, Whetton PH, van Ommen TD, Goodwin I, Haberle S (2005). 'How past climate change can help reveal Australia's future: a public research document for the Australian Greenhouse Office'. CSIRO Atmospheric Research, Melbourne.

Hassall and Associates, CSIRO, DLWC, QNR&M (2002). 'Cotton rivers and climate change'. Prepared for the Cotton Research and Development Corporation, Hassall & Associates Pty Ltd, Sydney.

Hennessy K, Fitzharris B, Bates BC, Harvey N, Howden SM, Hughes L, Salinger J, Warrick R (2007). Australia and New Zealand. In *Climate Change 2007: Impacts, Adaptation and Vulnerability: Contribution of Working Group II to the Fourth Assessment Report of the Intergovernmental Panel on Climate Change*. (Eds ML Parry, OF Canziani, JP Palutikof, PJ van der Linden, CE Hanson) pp. 507–540. Cambridge University Press, Cambridge.

Howden SM, Carberry P, Hayman P (2008). SEACI technical report for Milestone 3.1.6, South-eastern Climate Initiative, Murray–Darling Basin Commission, Canberra.

IOCI (Indian Ocean Climate Initiative) (2002). 'Climate variability and change in southwest Western Australia'. Department of Environment, Water and Catchment Protection, Perth.

IPCC (2007). *Climate Change 2007: The Scientific Basis. Contribution of Working Group I to the Fourth Assessment Report of the Intergovernmental Panel on Climate Change*. Cambridge University Press, Cambridge.

Jones RN, Durack PJ (2005). 'Estimating the impacts of climate change on Victoria's runoff using a hydrological sensitivity model'. CSIRO Atmospheric Research, Melbourne.

Jones RN, Page CM (2001). Assessing the risk of climate change on the water resources of the Macquarie river catchment. In *Integrating Models for Natural Resources Management Across Disciplines, Issues and Scales*. (Eds Ghassemi P, Whetton P, Little R, Littleboy M). Modelling and Simulation Society of Australia and New Zealand, Canberra.

Jones RN, Chiew FHS, Boughton WC, Zhang L (2006). Estimating the sensitivity of mean annual runoff to climate change using selected hydrological models. *Advances in Water Resources* **29**, 1419–1429.

Jones RN, Whetton PH, Walsh KJE, Page CM (2002). 'Future impacts of climate variability, climate change and land use change on water resources in the Murray Darling Basin: overview and draft program of research'. Murray–Darling Basin Commission, Canberra.

Jones R, Durack P, Page C, Ricketts J (2005). Climate change impacts on the water resources of the Fitzroy River Basin. In 'Climate change in Queensland under enhanced greenhouse conditions: final report 2004–2005'. (Eds W Cai, S Crimp, R Jones, K McInnes, P Durack, B Cechet, J Bathols, S Wilkinson) pp. 19–58. CSIRO Marine and Atmospheric Research, Melbourne.

Jones R, Preston B, Brooke C, Aryal S, Bates B, Benyon R, Blackmore J, Chiew F, Kirby M, Maheepala S, Oliver R, Polglase P, Prosser I, Walker G, Young B, Young M (2009). 'Climate change and Australian water resources: preliminary risk assessment.' CSIRO Water for a Healthy Country Flagship, Canberra.

Jovanovic B, Jones DA, Collins D (2007). A high quality pan evaporation dataset for Australia. *Climatic Change* **87**, 517–535.

Khan S, Robinson D, Beddek R, Wang B, Dassanayake D, Rana T (2004). 'Hydro-climatic and economic evaluation of seasonal climate forecasts for risk based irrigation management.' Technical Report 5/04, CSIRO Land and Water, Canberra.

Kiem AS, Franks SW (2004). Multidecadal variability of drought risk – eastern Australia. *Hydrological Processes* **18**, doi:10.1002/hyp.1460.

Kiem AS, Franks SW, Kuczera G (2003). Multi-decadal variability of flood risk. *Geophysical Research Letters* **30**, 1035.

Kirono DGC, Jones RN (2007). Bivariate test for detecting inhomogeneity in pan evaporation series. *Australian Meteorological Magazine* **56**, 93–103.

Lenzen M (2004). 'Nature, preparation and use of water accounts in Australia.' CRCCH Technical Paper 04/02. Cooperative Research Centre for Catchment Hydrology, Melbourne.

Lenzen M, Foran B (2001). An input-output analysis of Australian water usage. *Water Policy* **3**, 321–340.

Li Y, Cai W, Campbell EP (2005). Statistical modelling of extreme rainfall in southwest Western Australia. *Journal of Climate* **18**, 852–863.

Long AB, McMahon TA (1996). 'Review of research and development opportunities for using seasonal climate forecasts in the Australian water industry.' Occasional Paper No. CV02/96, Land and Water Resources Research and Development Corporation, Canberra.

Marsden J, Pickering P (2006). 'Securing Australia's urban water supply: research notes for selected case studies'. Report for Department of Prime Minister and Cabinet, Marsden Jacobs and Associates, Melbourne. http://www.dpmc.gov.au/water_reform/docs/urban_water_research.pdf

MDBC (2007). Drought contingency measures. Fact sheet. Murray–Darling Basin Commission, Canberra (May 2007). http://www.mdbc.gov.au/news/drought_contingency_measures_may_2007

Mpelasoka F, Hennessy K, Jones R, Bathols J (2008). Future droughts over Australia under global warming. *International Journal of Climatology* **28**, 1283–1292.

Murphy BF, Timbal B (2008). A review of recent climate variability and climate change in southeastern Australia. *International Journal of Climatology* **28**, 859–879.

Nicholls N (2004). The changing nature of Australian droughts. *Climatic Change* **63**, 323–336.

NLWRA (National Land and Water Resources Audit) (2001). Australian Water Resources Assessment 2000. Land and Water Australia, Canberra.

Power S, Casey T, Folland C, Colman A, Mehta V (1999). Inter-decadal modulation of the impact of ENSO on Australia. *Climate Dynamics* **15**, 319–324.

Power S, Haylock M, Colman R, Wang X (2005). 'Asymmetry in the Australian response to ENSO and the predictability of inter-decadal changes in ENSO teleconnections.' BMRC Research Report No. 113. Bureau of Meteorology, Melbourne.

Power S, Sadler B, Nicholls N (2005). The influence of climate science on water management in Western Australia: Lessons for climate scientists. *Bulletin of the American Meteorological Society* **86**, 839–844.

Preston BL, Jones RN (2008a). A national assessment of the sensitivity of Australian runoff to climate change. *Atmospheric Science Letters* **9**, 202–208.

Preston BL, Jones RN (2008b). Evaluating sources of uncertainty in Australian runoff projections. *Advances in Water Resources* **31**, 758–775.

Preston BL, Jones RN (2008c). Screening climatic and non-climatic risks to Australian catchments. *Geographical Research* **46**, 258–274.

Quiggin J, Adamson D, Schrobback P, Chambers S (2008). 'The implications for irrigation in the Murray–Darling Basin.' Report commissioned by the Garnaut Climate Change Review.

Smith IN (2004). An assessment of recent trends in Australian rainfall. *Australian Meteorological Magazine* **53**, 163–173.

Suppiah R, Hennessy KJ (1998). Trends in total rainfall, heavy-rain events and number of dry days in Australia, 1910–1990. *International Journal of Climatology* **10**, 1141–1164.

Timbal B, Wheeler M, Hope P (in press). On the relationship of the rainfall in the southwest and southeast of Australia. Part II: Possible causes of recent declines. *Journal of Climate*.

Tisdell JG, Ward JR (2003). Attitudes toward water markets: an Australian case study. *Society & Natural Resources* **16**, 61–75.

Tonts M, Black A (2002). The impact of changing farm business structures on rural communities. Rural Industries Research and Development Corporation, RIRDC Publications, Canberra.

UN (United Nations) (2003). 'System of Integrated Environmental and Economic Accounting'. United Nations, New York.

Verdon DC, Kiem AS, Franks SW (2004). Multi-decadal variability of forest fire risk – eastern Australia. *International Journal of Wildland Fire* **13**, 165–171.

Vivès B, Jones R (2005). 'Detection of abrupt changes in Australian decadal rainfall variability (1890–1989).' CSIRO Atmospheric Research Technical Paper No. 73.

Wang QJ, Nathan RJ, Moran RJ, James B (1999). Impact of climate changes on security of water supply of the Campaspe system. In *Proceedings, Water 99 Joint Congress, Brisbane, Australia*.

Warner RF (1987). The impacts of alternating flood- and drought-dominated regimes on channel morphology at Penrith, New South Wales, Australia. IAHS Publication No. 68.

Young MD, McColl JC (2008). Double trouble: the importance of accounting for and defining water entitlements consistent with hydrological realities. *Proceedings Australian Agricultural and Resource Economics Society 52nd Annual Conference*.

13

MARINE FISHERIES AND AQUACULTURE

AJ Hobday and ES Poloczanska

KEY MESSAGES:

▓ Climate change will impact marine fisheries and aquaculture. General ocean warming around Australia, particularly on the east coast due to strengthening of the East Australian Current, is predicted to change the distribution and abundance of species targeted by marine fisheries, and change the location of suitable environments for aquaculture species.

▓ Understanding changes in species' distributions will facilitate adaptation by fisheries management to climate change. Management and policy changes will be needed in some cases to allow appropriate responses by these marine industries to such climate-related change.

▓ In addition to changing distributions, climate change is expected to affect the dynamics and productivity of many exploited species. This has important consequences for fisheries assessments as most approaches assume that population dynamics are relatively constant over time – termed 'stationarity'.

▓ Focused regional studies on the relationship between the climate variables and the species of interest will improve understanding of the potential impacts of climate change.

▓ Selective breeding programs will adapt some aquaculture species for warmer conditions, but changes in farm location may be required for some businesses.

▓ Development of decision-support tools for stakeholders and managers, together with improved delivery of climate information, will enhance adaptation in these marine industries. Such tools must be robust in the face of the increased uncertainty as a result of climate change.

Introduction

Marine fisheries and aquaculture are economically and socially important industries in Australia. The gross value of commercial fisheries production was estimated at $2.12 billion in 2005–06, of which about 35% was from the aquaculture industry (ABARE 2007). Rock lobster, prawns, abalone and tuna are the most valuable fisheries, accounting for 55% of Australia's gross value of fisheries production in 2005–06. Recreational fisheries involve an estimated 20% of Australians spending nearly $2 billion per year (Henry and Lyle 2003).

The Australian fishing region (EEZ) is one of the largest in the world, ranging from Torres Strait in the far north to waters adjacent to continental Antarctica, and from Lord Howe Rise in the east to Christmas Island in the west. Australian fisheries are managed and regulated using a combination of spatial management (e.g. geographic regions), input controls (e.g. vessel numbers), output controls (such as quotas) and technical controls (e.g. gear types). As a result, many of the areas overlap in some way. For the purposes of this review we have used a simplified set of geographic regions that include multiple fishery methods (Figure 13.1). Aquaculture is considered

Table 13.1: Observed and projected changes in physical and chemical characteristics of Australia's marine realm including the Southern Ocean. Projections are derived from the CSIRO Mk 3.5 model, under greenhouse gas emissions scenario IS92a, which is a mid-range scenario most similar to A1B from AR4, compared to baseline of 1990 SST = sea surface temperature, MLD = mixed layer depth. Observations of change come from a variety of sources summarised in the text.

Physical variables	Observed changes	Projected changes	
		2030s	2070s
Temperature and solar radiation	SST: Warming recorded Maria Island, Tasmania of approximately 1.5°C since 1950s Warming of 1°C determined for tropical Australia since 1950s	SST: Warming of 1–2°C around Australia with the greatest warming off south-east Australia (2°C). Solar Radiation: There will generally be increases in incident solar radiation	SST: Warming of 2–3°C, around Australia with the greatest warming off south-east Australia (3°C). At a depth of 500 m: warming of 0.5–1°C. Solar Radiation: Increase in incident solar radiation between two and seven units W/m^2
Winds, ocean currents, MLD & ocean stratification	Increases in the strength of East Australian Current Little evidence for other changes around Australia to date	Winds: Increase of 0–0.5 metres per second (m/s) in surface winds Currents: Increased strength of the East Australian Current MLD/Stratification: Increased stratification of almost all offshore Australian waters	Winds: An increase of 0–1 m/s in surface winds Currents: Increased strength of the EAC, but a general decline in the strength of other surface currents of between 0–1.2 m/s MLD/Stratification: Almost all offshore areas of Australia will have greater stratification and a shallowing of the mixed layer by about 1 m, reducing nutrient inputs from deep waters. Some coastal regions may have enhanced mixing and nutrient inputs
Precipitation, extreme events, and terrestrial runoff	Precipitation Long-term declines in southern regions, such as south-east Qld and south-west WA Storms Increases in intense events noted for recent years	Precipitation Average annual decrease of 0 to 5% over most of Australia. Storms Intensity of storms expected to increase	Precipitation Continued decrease over most of Australia. Storms Intensity of storms expected to increase
Sea level	20th century rate of sea level rise of 1.7 ± 0.3 mm/yr	Rise of 0.3–0.5 m around Australia (not including any rise due to ice sheet melting)	A rise of 0.6 to 0.74 m, with greater increase on the east compared with west coast (not including any rise due to ice sheet melting)
Acidification (pH)	pH of surface oceans has dropped by 0.1 units since the industrial revolution	Decline in pH by ≈0.1 units	Decline in pH by 0.2–0.3 units

in a single section, although we discuss production in different geographic areas.

The key variables expected to drive climate change impacts on fisheries and aquaculture are changes in temperature, ocean currents, winds, nutrient supply, rainfall, ocean chemistry and extreme weather conditions. Significant warming of ocean temperatures has already been documented around Australia (Ridgway 2007; Pearce and Feng 2007; Lough 2008). It is highly likely that these changes will in turn impact marine ecosystems (Denman *et al.* 1996; Cox *et al.* 2000; Bopp *et al.* 2001; Boyd and Doney 2002; Sarmiento *et al.* 2004b), and consequently the distribution, growth, recruitment, and catch of exploited marine

species, their prey and predators (Hobday *et al.* 2007b; Poloczanska *et al.* 2007). Predictions of climate-related changes to the Australian marine environment are taken from the CSIRO Mk 3.5 climate model (Gordon *et al.* 2002). Although there are subtle differences between the CSIRO model and other international models, many of the general trends are similar (e.g. Suppiah *et al.* 2007). Climate projections for shorter (2030) and longer (2070) time scales are provided (Table 13.1).

Aquaculture occurs in most coastal waters around Australia (Figure 13.1). Salmonids (salmon and trout), southern bluefin tuna, pearl oysters, edible oysters and prawns are the most valuable aquaculture species accounting for 86%

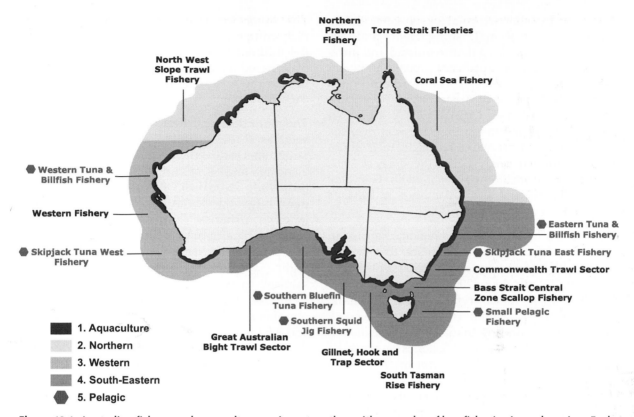

Figure 13.1: Australian fishery and aquaculture regions, together with examples of key fisheries in each region. Each geographic region is predicted to have generally coherent changes in the climate system. Aquaculture is widespread around the Australian continent. Pelagic fisheries, denoted by the hexagonal symbols, occur in all the regions and are treated as a separate section.

of the industry gross value but barramundi and abalone aquaculture industries are expanding rapidly (ABARE 2007). Barramundi, prawns and pearl oysters are grown mainly in Australian tropical and subtropical waters while salmon, trout, edible oysters and tuna are cultivated in the cooler southern waters (Table 13.2). The valuable salmon aquaculture industry is based almost entirely in Tasmania (Battaglene *et al.* 2008).

Northern fisheries target a wide variety of sea life including penaeid prawns and squid, shark, grey and Spanish mackerel, barramundi, threadfin salmon, Torres Strait lobsters, sea cucumbers, trochus shells, coral trout, red throat emperor, various other demersal finfish, live reef fish, and portunid crabs (Larcombe and McLoughlin 2007). A broad range of marine organisms is also taken by traditional (aboriginal) hunting, fishing and

Table 13.2: Aquaculture production value ($'000) for the most valuable species by state plus Northern Territory, 2005–06 (from ABARE 2007). A * indicates a farmed species where production figures are not available.

Species	NSW	Vic	Qld	WA	SA	Tas	NT
Salmon	0	0	0	0	0	221 013	0
Trout	1742	8624	0	0	447	0	0
Tuna	0	0	0	0	155 795	0	0
Barramundi	1238	0	13 900	0	2029	0	*
Pearl oysters	0	0	0	122 000	0	0	*
Edible oysters	34 093	0	570	0	32 480	16 720	0
Prawns	3387	0	46 500	0	0	0	*

gathering, by other artisanal (or subsistence) fishing and by recreational fishing (Henry and Lyle 2003). In addition, illegal, unregulated and unreported fishing is a significant problem in northern Australia (DEH 2004). The Northern Prawn Fishery (NPF) is one of Australia's most valuable commercial fisheries (ABARE 2007), averaging 8500 tonnes of prawn landings per year over the last decade and an estimated eight to 21 times that amount of bycatch species (Pender *et al.* 1992; Brewer *et al.* 1998; Stobutzki *et al.* 2000). The value of the NPF fishery was $73 million in 2004–05 (Stobutzki and McLoughlin 2007).

Fisheries in the south-eastern region comprise an extremely diverse set of activities that can be divided biologically by depth (coastal [0–50 m], shelf [50–200 m], slope [200–700 m] and deepwater [>700 m]), by fishery type (e.g. southern shark, demersal trawl, scallop dredge, abalone, rock lobster, squid), and by management jurisdiction and agency (State-managed fisheries versus those managed by the Federal Government). The region also has the longest European fishery history in Australia, has the highest numbers of commercial species, and is one of the most heavily exploited regions in Australia. In the 2006 annual report on the state of Australia's fisheries, 11 of the 19 overfished species occurred in the south-east fishery (Larcombe and McLoughlin 2007). Despite this, the south-east fisheries are still among the most valuable in Australia. In 2005–06 the combined value of the Tasmania, Victoria, New South Wales and Australian Government wild fisheries catch in the south-east region (other than pelagic fisheries) was about $700 million (ABARE 2007). The most valuable fisheries – those for molluscs ($240 million) and crustaceans ($100 million) – are mainly inshore and on the shelf.

Western fisheries harvest demersal, coastal and pelagic species (pelagic species are covered elsewhere). The level of fish (teleost) production is low on the west coast, primarily due to the poleward flowing Leeuwin Current, which carries warm, low-nutrient waters southward along the Western Australian continental shelf (Lenanton *et al.* 1991; Ridgeway and Condie 2004). The main species harvested, with respect to landings and value, are from demersal invertebrate fisheries. The western rock lobster fishery is Australia's most valuable single-species fishery ($250–350 million), with annual production averaging in excess of 11 000 t (ABARE 2007).

The other important species include prawns ($38 million), abalone ($12 million) and scallops ($9 million) (ABARE 2007). Overall, west coast State-based fisheries account for 29% (≈$542 million) of the Australian fisheries production value (ABARE 2007).

The main pelagic fisheries in Australia are managed as three separate fisheries, although the target species overlap. Tuna (yellowfin, bigeye, albacore and southern bluefin) and billfish (broadbill swordfish, striped marlin) are the main target species in the eastern and western longline fisheries (ETBF and WTBF), while southern bluefin tuna is the single target species in a purse-seine fishery in the Great Australian Bight (SBT fishery). A purse-seine fishery for skipjack tuna was recently separated from the ETBF for management purposes, although it remains small in value and tonnage (Larcombe and McLoughlin 2007). The gross value of production in the ETBF in 2005–06 was $28.7 million, down from $42.5 million in 2004–05 (ABARE 2007). This decline was due to a decrease in the catch of some species (e.g. swordfish), lower achieved prices because of a strengthening Australian dollar, as well as a shift to lower value species (e.g. albacore). Whether this pattern is related to changes in the regional oceanography or changes in fishing practices that altered to target deeper-living albacore, remains unclear. Declines in value for the other fisheries were relatively minor, a 3% decline (WTBF: 2005–06 $2.7 million) and 4% (SBT: 2005–06 $37.5 million, wild-caught value), and related to prices for the key species (ABARE 2007).

In the 2006 annual report on the state of Australia's fisheries, the Australian Bureau of Rural Sciences classified 19 of the 97 stocks considered as either over-fished and/or subject to overfishing, 27 as not over-fished, and 51 as status uncertain (Larcombe and McLoughlin 2007). The high proportion of stocks classified as uncertain reflects the addition of new stocks not previously classified and revised classification of some stocks for which assessments were previously thought to be more reliable. The high proportion of uncertain status, especially considering the addition of cumulative pressures associated with a changing climate, highlights the need for reliable assessment information and a growing understanding of complex relationships between fisheries stocks and climate.

Climate change will also impact marine species of interest to recreational and indigenous fishers. Most research to date has been directed at assessing the impacts of climate change on commercial fisheries, and little information has been gathered on the impacts to these other sectors. The dollar value of fisheries resources to the recreational sector is similar to that of the commercial sector. Significant investment is made by participants, with estimates of over 500 000 boats worth $3.3 billion used for recreational fishing (Henry and Lyle 2003). Approximately 41% of total recreational fishing effort occurs in coastal waters; estuarine waters account for 35%, offshore waters about 4%, while freshwater fishing represents 20% of total effort (Henry and Lyle 2003). A number of other commercial sectors depend on recreational fishing, including charter boat operators and associated tourism businesses with significant spending in some regional economies (Henry and Lyle 2003). Recreational fishers harvest significant numbers of finfish, small baitfish, crabs and lobsters, prawns and yabbies, cephalopods, molluscs and other taxa. The prominent finfish species in terms of numbers captured are whiting, flathead, Australian herring, bream, King George whiting, mullet, garfish, tailor, Australian salmon and pink snapper (Henry and Lyle 2003). There is significant overlap between commercial and recreational capture for many species, and for some the recreational catch is considerably higher than the commercial catch.

Fishing is a prevalent activity among indigenous Australians, with participation rates exceeding 90% reported in northern Australia (Henry and Lyle 2003). Capture of marine resources represents an important contribution to diet and to cultural events particularly in northern Australia. Target species for both recreational and indigenous fishers differ, and while specific knowledge about many groups is lacking, climate impacts on these species are expected. Indigenous fishers harvest millions of aquatic animals in northern Australia, including finfish (particularly mullet, catfish, sea perch/snappers, bream and barramundi), shellfish, prawns and yabbies, crabs and lobsters, and a range of other taxa. The most prominent non-fish species were mussels, cherabin, other bivalves, prawns, oysters and mud crabs (Henry and Lyle 2003). Indigenous fishers also harvest some commercial species (trochus, trepang), plus a number of species groups that have protected status for non-indigenous people, including dugong and turtles which have high cultural and spiritual significance to indigenous communities. Thus, although the indigenous population represents only 2.2% of the total Australian population (ABS 2002), the importance of the marine environment and the harvesting of marine resources by this group is nationally significant.

Impacts of projected climate change

Climate change is considered a potential threat to the sustainability of fisheries and aquaculture in Australia. Positive opportunities may also result from climate change, however considerable structural adjustment by the fishing sector (commercial or otherwise) may be needed to realise these benefits (Hobday et al. 2007b). The key variables expected to drive climate change impacts are changes in temperature, ocean currents, winds, nutrient supply, rainfall, ocean chemistry and extreme weather conditions. It is very likely that changes in any of these would also significantly change the marine ecosystems in Australian waters (Denman et al. 1996; Cox et al. 2000; Bopp et al. 2001; Boyd and Doney 2002; Sarmiento et al. 2004b), and consequently the distribution, growth, recruitment, and catch of exploited marine species, their prey and predators. Given the high percentage of marine species endemic to Australian temperate waters, it has been suggested that climate change would therefore have a greater impact on the biodiversity of Australia's temperate zones than in tropical waters (e.g. Pittock 2003; Poloczanska et al. 2007). However, a considerable number of marine species not endemic to Australia are confined to the Indo-West Pacific biodiversity hotspot, with Australia becoming a last refuge for some species affected by degrading environments in neighbouring tropical countries. This is worrying given that several lines of evidence indicate Australia's northern (tropical) ecosystems are vulnerable to climate change (e.g. Hill et al. 2002; Lough 2008) in addition to the effects of fishing. Thus, the impact of climate change on these ecosystems may have consequences for both fisheries and ecosystems (Hobday et al. 2007c; Poloczanska et al. 2007; Munday et al. 2008).

Aquaculture – around Australia

For most marine aquaculture species, growth, survival and the abundance of various life stages are sensitive to extreme temperatures and to shifts in temperature regimes. A change of only a few degrees might therefore mean the difference between a successful aquaculture venture and an unsuccessful one. It is expected that climate change will have adverse impacts on the production of species in Australia's cooler southern waters, particularly on Tasmania's valuable salmon aquaculture industry (Battaglene *et al.* 2008). The greatest warming of marine waters in the southern hemisphere is predicted for the Tasman Sea, and has been linked to a projected strengthening of the East Australian Current under global warming scenarios (Ridgway 2007). It is predicted that climate change will have particularly adverse impacts on the productivity of Tasmania's valuable salmon farming industry (Battaglene *et al.* 2008). The stocks of Tasmanian salmon (Atlantic salmon *Salmo salar*) originated from Canada and are farmed near the upper limits of their optimal growing temperature in Tasmanian waters during the summer months (Battaglene *et al.* 2008). The recent above average summer water temperatures in southern Tasmania have already increased mortality (Pittock 2003) and will necessitate action by the industry over the coming decades (Battaglene *et al.* 2008).

Climate change will also influence aquaculture ventures in tropical and subtropical regions (Preston and Poloczanska 2007). Analysis of intensively managed prawn farm ponds in Queensland demonstrated that variations in pond temperature had pronounced impacts on farm production, with maximum growth rates of tiger prawns (*Penaeus monodon*) during sustained periods of warmer pond water (Jackson and Wang 1998). This suggests that the production efficiency of tropical and subtropical prawns, such as tiger and banana prawns (*P. merguiensis*), might be increased by a rise in water temperature. Rising temperatures may not only enhance growth rates at existing sites, but also extend the area suitable for farming these species further south. On the other hand, an increase in pond water temperature might threaten the viability of farming cooler water species, such as the penaeid *P. japonicus*, whose production is restricted to a relatively narrow range of latitudes compared to the subtropical species (Preston *et al.* 2001). Disease outbreaks also may

become more prevalent for some species at higher water temperatures (Battaglene *et al.* 2008).

The projected increases in the intensity of storms and cyclones (Table 13.1) will increase flood risk, which is a threat to stock through overflows or damage to pond or dam walls. Any severe flooding event could result in mass mortalities of animals in aquaculture ponds, and lines or cages in coastal areas. For coastal and offshore aquaculture, more frequent and intense storms result in increased physical damage and stock losses, both of which are costly to operations. Many coastal processes, such as sediment transport, happen mostly during high-energy events (storms). An increase in storm activity may therefore change the direction of river flow, and initiate erosion. These and other effects can impact aquaculture facilities protected from direct exposure to increased wind and wave activity. For example, in 1996 Australian southern bluefin tuna (*Thunnus maccoyii*) farms at Port Lincoln in South Australia suffered losses of up to 75% of total production, which were attributed to asphyxiation of fish by sediments re-suspended during a severe storm (Preston *et al.* 1997).

Northern fisheries region

Penaeid prawn fisheries and other estuarine-dependent fisheries throughout northern Australia appear to be sensitive to climate related changes in rainfall and freshwater flow (see review by Robins *et al.* 2005). The sensitivity of northern Australia's penaeid prawns to freshwater flows is also well illustrated by the demonstrated relationships between the Southern Oscillation Index, which is strongly associated with regional rainfall patterns, and both banana prawn (positive) and tiger prawn (negative) catches (Love 1987; Catchpole and Auliciems 1999).

Changes in rainfall and freshwater flow patterns would also be likely to change nutrient runoff into northern Australia's coastal waters, which strongly influences the productivity of these otherwise low-nutrient tropical waters (Vance *et al.* 2003; McKinnon *et al.* 2007). Alteration of freshwater flows related to changes in climate patterns might also affect northern Australia fisheries. For instance, a loss of synchronicity between the timing of life history stages and environmental forces could disrupt reproductive stages of life cycles and community interactions, especially considering that other interacting species will be responding to these environmental changes in different ways. Seagrass beds and mangrove forests are considered critical nursery habitats for many marine species including commercially targeted prawns (Vance *et al.* 1990; Loneragan *et al.* 1994; Haywood *et al.* 1995; Sheaves 1998; Blaber 2000) and catches of tropical commercial species such as banana prawns (*Penaeus merguiensis*), mud crabs (*Scylla serrata*) and barramundi (*Lates calcarifer*) have been linked to mangrove abundance and extent (Lee 2004; Loneragan *et al.* 2005; Manson *et al.* 2005). These habitats are particularly vulnerable to cyclones, sea level rise, and their interactive effects. Projected sea level rise is expected to considerably reduce these habitats in the southern Gulf of Carpentaria (Hill *et al.* 2002), an area critical to much of northern Australia's prawn fisheries, while a projected increase in cyclone intensity will increase disturbance regimes in northern waters. Degradation of reef habitat from coral bleaching and ocean acidification is predicted to cause declines in the abundance of small reef fishes, which could potentially affect fishes at higher trophic levels, some of which are important to recreational and commercial fisheries (Munday *et al.* 2008; Pratchett *et al.* 2008). Range shifts of some commercially important species could occur in tropical waters. Munday *et al.* (2007) note that red throat emperor (*Lethrinus miniatus*), which is an important species to commercial and recreational fisheries, is mostly found south of 18°S and might be expected to contract further south as water temperatures increase.

South-east fisheries region

Warming in the south-eastern portion of Australia's ocean is projected to be the greatest in the southern hemisphere (Ridgway 2007). This warming will affect fished species in a variety of ways, including growth, distribution and abundance (Thresher *et al.* 2007b). Juvenile growth rates of shallow water, commercially exploited fish species in the south-east have increased significantly since the early 20th century, based on historical analysis of the width of annual increments in their otoliths (Thresher *et al.* 2007a). The changing growth rates do not appear to reflect changes in the abundance of the fish, but rather correlate significantly with the local water temperature (Thresher *et al.* 2007a, b). This growth increase is restricted to shallow water species;

among deeper (>1000 m) species examined, growth rates have been falling, mirroring declining water temperatures at intermediate depths. Consequently, increased growth rates could be widespread among shallow water marine species in south-east Australia. However, the consequences for exploited fish populations have not been examined but could be positive. This increased growth could be offset by reductions in productivity, if warmer waters have lower nutrient levels for plankton, or if stratification of the surface waters causes less mixing with deeper, nutrient rich, waters.

The projected warming of ocean waters will have profound effects on the distribution of many Australian species (Poloczanska *et al.* 2007; Munday *et al.* 2007; 2008). As a result, major changes in community composition and ecosystem function are expected as species' distributions shift poleward. Such changes are already being recorded in commercial and non-commercial temperate fish species in the northern hemisphere (Perry *et al.* 2005; Brodeur *et al.* 2006) and similar shifts are emerging from the sparse Australian data (Ling *et al.* 2008). Unlike northern hemisphere scenarios, however, for many species along Australia's southern coast, the east-west orientation of this temperate Australian coastline and the limited area of continental shelf to the south of Australia mean there are few opportunities for southern species to move further south as water temperatures increase (Poloczanska *et al.* 2007).

In the last decade, there have been conspicuous changes in the distribution of Tasmanian marine fish. Some 36 species in 22 families (≈10% of the inshore families of the region) have exhibited major distributional changes: some have become newly established south of Bass Strait; others have markedly increased in abundance in southern Tasmania by shifting their ranges south along the Tasmania coast, and others are totally new records for Tasmania (P. Last, in Lyne *et al.* 2005). Most of the species exhibiting a poleward shift in distribution are rocky-reef species normally found off the NSW coast, in habitats also associated with the long-spined sea urchin (*Centrostephanus rodgersii*). This urchin crossed Bass Strait in the mid-1960s and was first discovered on the east coast of Tasmania in 1978 (Johnson *et al.* 2005; Ling *et al.* 2008; Ling *et al.* 2009). The poleward shift in the distribution of this species has been associated with the decline of urchin barrens on the NSW coast, which adversely affected the local abalone fishery, while its arrival in Bass Strait and subsequent spread along the east coast of Tasmania has led to development of extensive urchin barrens in areas where they did not previously exist. The arrival of *C. rodgersii* off Tasmania appears to be disrupting the existing balance between macroalgae, abalone, rock lobsters and native urchins. It is predicted that without management intervention, *C. rodgersii* barrens will eventually cover 50% of the rocky reef habitat on Tasmania's east coast and have serious implications for the sustainability of rock lobster and abalone fisheries (Johnson *et al.* 2005).

Western fisheries region

The strength of the Leeuwin Current affects a range of exploited species along the west coast (Lenanton *et al.* 1991; Caputi *et al.* 1996; Gaughan 2007). The strength of the current has a significant positive influence on survival of the larval stage of the western rock lobster (*Panulirus cygnus)*, but a negative influence on the larval survival of the scallop (*Amusium balloti)* in Shark Bay and at the Abrolhos Islands (Pearce and Caputi 1994). For pelagic finfish species, the current has an adverse effect on the survival of pilchard larvae (*Sardinops sagax neopilchardus),* but a positive impact on whitebait (*Hyperlophus vittatus)* and also on recruitment of Western Australian salmon (*Arripis truttaceus)* and Australian herring (*Arripis georgianus)* to South Australia. The current appears to have a correspondingly negative impact on the recruitment of Australian herring in the south-west of Western Australia (Pearce and Caputi 1994).

Western Australian fisheries, such as the rock lobster fishery, also correlate (some positively, some negatively) with phases of the El Niño-Southern Oscillation (ENSO) as well as the strength of the Leeuwin Current through unknown mechanisms (Penn *et al.* 2005; Gaughan 2007). For example, pearl oyster (*Pinctada maxima*) catch rates are affected by a number of environmental variables including El Niño events; catch rates were enhanced two years after El Niño events (Hart *et al.* 1999). Explanations for the positive relationship between ENSO-Leeuwin Current strength and levels of rock lobster settlement have shifted from the direct effect of ocean currents on larval transportation to the indirect effects on their growth and mortality (Caputi *et al.* 2001; 2003). Despite

uncertainty about the mechanism, the resulting fluctuation in value of the rock lobster catch indicates that subsequent socio-economic impacts can be large. Future changes in the frequency of El Niño and the strength of the Leeuwin Current may thus directly impact the rock lobster fishery, perhaps through a systematic shift in the larval settlement-Leeuwin Current relationship (Matear *et al.* 2007; Gaughan 2007).

Pelagic fisheries

Ocean temperatures have a demonstrable effect on the distribution of the target species in Australia's pelagic fisheries (e.g. SBT: Reddy *et al.* 1995). Evidence is also strong in other parts of the world, and temperature is one of the strongest drivers of pelagic fish distribution (Laurs *et al.* 1984; Andrade and Garcia 1999; Schick *et al.* 2004; Kitagawa *et al.* 2006). The strongest environmental signal in the ocean, the ENSO phenomena, has been shown to have a major impact on the distribution of tropical tunas (Lehodey *et al.* 1997).

On the east coast, pelagic species are captured in the Coral Sea, East Australian Current, and Tasman Sea regions (Young *et al.* 2009). There are seasonal changes in the abundance of species such as yellowfin (*Thunnus albacares*) and bigeye tuna (*T. obesus*) captured by longline fishing that are positively linked to the expansion and contraction of the East Australian Current (Campbell 1999). At a finer scale the distribution of yellowfin tuna has been linked to the distribution of mesoscale environmental features such as eddies generated by the East Australian Current (Young *et al.* 2001). The known relationships between the distribution and abundance patterns of some pelagic species on the east coast suggest that changes in the strength of the East Australian Current would have dramatic effects on the availability of key pelagic species to the fishery (Hobday in press), although the mobility of the fishing fleets might reduce the immediate economic impact. Changes in productivity can also affect the pelagic ecosystem and the harvested species at the top of the food chain, but research in this area is in its infancy (Young *et al.* 2009). Preliminary analyses have found spatial differences in productivity and pelagic ecosystem structure, and it is believed that these regional differences could mimic the temporal changes that might occur as a result of climate change (Young *et al.* 2009).

In southern Australia, juvenile southern bluefin tuna (SBT) have been the subject of several studies investigating distribution and abundance relationships to mesoscale environmental variability (Hobday 2001; Cowling *et al.* 2003) or to prey (Young *et al.* 1996). In general, the environmental linkages to abundance are not strong at the mesoscale. A recent study did not find a link between an apparent decline in a fishery-independent abundance index of age-1 SBT and environmental conditions (SST, Leeuwin Current strength, winds) in south-western Australia (Hobday *et al.* 2004). This index of abundance has not yet been validated, and so environment-SBT relationships may in fact exist. In contrast, at a larger scale, seasonal changes in the abundance of juvenile SBT (ages one to five) in southern Australia are well documented. SBT are resident along the shelf during the austral summer (Cowling *et al.* 2003) and then migrate south during the winter (Bestley *et al.* 2008). Interannual variation in SBT abundance within the main fishing grounds in the Great Australian Bight has not been linked to the environment, although variation in the arrival time of schools has been attributed to unspecified environmental factors (Cowling *et al.* 2003). Variation in the availability of SBT prey (sardines and anchovies) as a result of changes in wind-driven upwelling (Dimmlich *et al.* 2004) are also likely if climate change affects the strength of upwelling-favourable winds (Hertzfeld and Tomczak 1997), which might ultimately affect the distribution and abundance of pelagic predators. Finally, the impact of climate change on the winter SBT feeding grounds in the Southern Ocean may be more dramatic (e.g. Sarmiento *et al.* 2004a); it remains an area for investigation.

The two pelagic longline fisheries on the east and west coast of Australia have a large number of by-product and by-catch species, which may also experience climate-related impacts, while two lower value fisheries (squid and small pelagic fisheries) target key species at intermediate trophic levels. These intermediate levels contain crucial species for the rest of the ecosystem and could be particularly sensitive to climate impacts (e.g. Cury *et al.* 2000; Rose 2005; Hunt and McKinnell 2006; Pecl and Jackson 2008).

Pelagic squid captured in the southern squid jig fishery (SSJF) have more flexible life history strategies and greater tolerances to environmental change than the fish making up the small pelagic fishery (SPF) (Pecl and Jackson 2008). Squid could benefit from climate-induced changes in the regional oceanography, possibly at the expense of the species of the SPF which feed mainly on zooplankton that are restricted to temperate waters (Young *et al.* 1993).

On the east coast, small pelagic fishery (SPF) changes in fishing method (purse-seine to midwater trawl) have confounded potential environmental relationships with small pelagic fish distribution and abundance (Lyle *et al.* 2000) that have been so clearly documented elsewhere in the world (e.g. Jacobson *et al.* 2001; Chavez *et al.* 2003). However, off the coast of Tasmania, declining growth rates of jack mackerel and a change in the age structure of the catch through the 1990s may have both an environmental and fishing-related component (Lyle *et al.* 2000; Browne 2005). Changes in the relative dominance of the East Australian Current and the sub-Antarctic water masses, and consequently the regional prey communities, have also been implicated in changes in local productivity off the east coast of Tasmania (Young *et al.* 1993, 1996). For example, the disappearance of krill (*Nyctiphanes australis*), from the shelf ecosystem of eastern Tasmania during a warm (La Niña) event in 1989 was linked to the simultaneous disappearance of their main predator, jack mackerel (*Trachurus declivis*) (Young *et al.* 1993). Given that *N. australis* is at the base of most Tasmanian shelf marine ecosystems, and that it is a cool-water species, any persistent warming of the regional oceanography would have a profound effect on krill-dependent food chains. These food chains include cephalopods (Pecl and Jackson 2008), seabirds (Bunce 2004), and small pelagic fish and tunas (Young *et al.* 2001).

Recreational and indigenous fishers

Some impacts of climate change discussed above for commercial fisheries will be similar for recreational and indigenous fisheries, including movement of target and bait species to more southern latitudes or to greater water depth. Continued ocean warming may result in an increase in tropical species for south-eastern Australia in particular. Changes in availability, due to changes in local abundance and/or growth rates may interact to reduce recreational catch where size limits are employed. Species reliant on freshwater flows (estuarine) may be particularly impacted by terrestrial and marine impacts.

Environmental changes may affect both recreational and indigenous fishers more severely than commercial fishers who generally use larger boats and more sophisticated and robust gear. Reductions in fish catch may be due to increased storm and wave activity which may reduce the number of suitable days for fishing, or access to certain waters by both vessel and shore-based fishers. There may also be climate-related changes in seasonal availability of fish, which may lead to a reduction in the time period when fishing activity can occur, particularly if open seasons are not amended to allow for seasonality changes. With regard to the species targeted, recreational fishers may be more flexible in adapting as fish distributions change.

Options for dealing with climate variability

For many involved in fisheries and aquaculture, climate variability is just part of the natural 'operating environment', and most operators and managers do not deal explicitly with variability beyond responding to the observed patterns, and the underlying dynamics are assumed to be constant. Within each sector, however, there are common approaches across regions, and thus the geographical approach of earlier sections is not followed here. This section is also relevant for fisheries not explicitly covered earlier in this review. In general, current practice is reactive, rather than proactive.

Fishers already adapt to changes in abundance and distribution of key species in many regions, such as the south-east (e.g. Smith and Smith 2001). The western rock lobster industry already copes with significant interannual catch fluctuations and utilises a catch prediction system that allows industry preparation for harvest fluctuations several years ahead (Caputi *et al.* 2001). Australia salmon (*Arripis georgianus*) abundance also varies along the south-west coast of Australia, and boom and bust years also occur in the scallop sector (Pearce and Caputi 1994). In such variable fisheries, one existing option is to target different species in different years.

Fishers often exhibit considerable fish-finding skills. Their target species roam widely, and are subject to considerable interannual fluctuations in distribution and relative abundance. To counter this variability, fishers use a range of technology, such as satellite information and sophisticated on-board electronic equipment (Ward and Hindmarsh 2007). These help locate suitable conditions for the species that is being targeted. Continued use and improved availability of these products and development of new predictive fish-finding tools may improve the capacity of the fishers, providing the stocks can sustain the continued or enhanced harvests.

Current approaches for managing changes in fisheries as a result of climate variability include changes in fishing ports; changes in fishery areas; changes in the quota allocated for harvest, and closures in some fisheries or fishing areas. One Australian example where environmental information is incorporated into a management response that accounts for seasonal and interannual climate variability is in the east coast pelagic longline fishery. Southern bluefin tuna (SBT) are restricted to the cooler waters south of the East Australian Current and range further north when the current contracts north to the New South Wales coast (Majkowski *et al.* 1981; Hobday and Hartmann 2006). This response to climate variability has allowed real-time spatial management to be used to restrict catches of SBT by non-quota holders in the east coast fishery by restricting access to ocean regions believed to contain SBT habitat (Hobday and Hartmann 2006; Hobday *et al.* 2009). The current distribution of the tuna habitat is derived with a near-real-time ocean model and then relayed to management during the fishing season. As the distribution of the SBT habitat changes during the season, management adjusts the location of restricted access areas throughout the season (Hobday *et al.* 2009).

Climate variability is also a fact of life for many operators in the aquaculture sector. While the environment is regulated for some of the earlier stages of a species life history (e.g. indoor hatcheries for salmon and abalone), the adults are usually exposed to a more natural environment. Even here, attempts are made to reduce the effects of climate variability via feeding (e.g. salmon, barramundi, prawns), cleaning or removal of competitors (e.g. pearl oysters), and thinning conspecifics (e.g. oysters and mussels). Responses to climate variability also occur during or after a climate-related event, such as treatment for disease (Battaglene *et al.* 2008). In the Atlantic salmon industry, the prevalence of a gill disease is increased in warm water, and bathing in fresh water is used as a treatment. In warmer summers, increased bathing is used to combat such disease outbreaks. Selective breeding is another major adaptation focus area (Battaglene *et al.* 2008). There is considerable effort in the aquaculture industry for developing strains with increased biological performance (e.g. more robust stocks with fast growth).

For many fisheries and aquaculture industries, there is considerable potential for improved use of information on climate variability, and many operators recognise that improved use of the available information will assist overall economic performance, regardless of the impact of climate change (e.g. Battaglene *et al.* 2008).

Options for dealing with projected climate change

The marine fisheries and aquaculture sectors are just beginning to develop strategies for adapting to climate change, and these efforts will gain momentum in the near future (DAFF 2008). The adaptation options are best illustrated with examples of how fisheries or aquaculture may respond to climate change. In these sectors, adaptation is being supported by central government initiatives, however mechanisms which allow and support bottom-up solutions and opportunities should also be fostered (e.g. Nelson *et al.* 2008). Additional solutions will be developed as awareness increases, and this section is designed to stimulate additional ideas for adaptation options in marine industries.

Adaptation options for aquaculture

The aquaculture industry has a range of options for adaptation to climate change including selectively breeding for tolerance to altered temperature regimes or the use of alternate species that are pre-adapted to the temperature regimes, and relocation of production facilities, including the movement of cage systems to deeper offshore waters (Preston and Poloczanska 2007; Battaglene *et al.* 2008).

The predicted temperature change in waters around Australia over the coming century is relatively slow compared to the generation times of the Australian aquaculture species that are currently considered amenable to selective breeding.

These range from a year or less for prawns (Preston *et al.* 2004), two years for oysters (Appleyard and Ward 2006) to three years for Atlantic salmon (Elliott and Riley 2003) and temperate abalone (Shepherd and Lawas 1974). Although there appears to be significant potential to adapt these aquaculture species to changes in temperature within appropriate time frames, whether this is possible will depend on the genetic diversity of the breeding population. The development of new aquaculture species to meet the growing demand for seafood products will also increase adaptability of aquaculture industries to climate change, however the choice of new species is not independent of markets, competitors, and biological limitations (Preston and Poloczanska 2007; Battaglene *et al.* 2008).

Impacts can be mitigated to some extent by foresight in planning and selection of sites and of species. Rapid response to the projected warming can be achieved in farming of caged fish, such as salmon, by moving cages offshore to deeper, cooler waters. There is increasing global interest in offshore aquaculture for a number of reasons, including the need to avoid adverse impacts on fish health and quality from pollutants discharged into coastal waters (Ryan 2004). Ameliorating the impacts of climate change, particularly for species at the limits to their thermal tolerance, may provide additional incentive to develop offshore aquaculture technology (Battaglene *et al.* 2008).

Adaptation options for marine fisheries

Climate change adaptation options for the marine fisheries sector have been seen mostly in the context of top-down management options, including building resilience through improved stock status (e.g. Jackson *et al.* 2001; Steneck *et al.* 2002). Climate change impacts on exploited species could directly affect Australian fisheries management in several ways, including setting of climate regime-specific reference points, spatial management (including closures), and predictive models for harvest levels (Hobday *et al.* 2007b). For example the Commonwealth harvest strategy policy sets benchmarks to define over-fishing, including biomass limit reference points (Rayns 2007). These benchmarks might change as fish productivity or distribution change, with subsequent impacts on catch quotas and effort levels given the direct link between the benchmarks and management decisions. The guidelines to the policy already envisage that account should be taken of changing benchmarks, but the difficulty will be in detecting the effects and determining the changes (Hobday *et al.* 2007b). Thus, it is important to develop monitoring, assessment and management strategies that are more robust in the face of this increased uncertainty. Another relevant management issue is that most fisheries are defined by jurisdictional boundaries that determine access and property rights. As species' distributions change, fishers' rights may diminish, while other fishers currently without species access rights may gain effective access to the fish.

Climate change may lead to alternative species being harvested when the primary species is less available, or to changes in the location where fishing occurs. This may lead to increased conflict between resource users. For example, perceived interaction between the commercial and recreational sector has resulted in zoning that excludes commercial fishing activities in some areas (e.g. recreational fishing zones). Management policies can differ between regions, thus movement of stocks to areas without adequate management may also lead to resource conflict (Miller 2007; Stenevik and Sundby 2007). Proactive policy development is needed to avoid such conflict.

Changes in the abundance of species that are at low levels can be positive or negative. In the case of increased abundance (population recovery), rapid industry adaptation will be likely, driven by economic opportunity. It will be a challenge to determine if the change in abundance is likely to be sustained, or is due to interannual variability. This distinction is crucial for long-term business decisions around, say, increasing fleet capacity or making technological investments. In the case of declining abundance, industry may be forced to adopt additional management measures to protect a particular species, and these may result in reduced flexibility to target the non-impacted species. Thus, adaptation options that reduce non-target capture will be useful in a changing climate, as well as improve the sustainability of fisheries in general.

Ecosystem-based fisheries management (EBFM) takes into account interrelationships between exploited fish stocks, non-target species, the environment, and human action (e.g. Link *et al.* 2002). The effect of climate on fisheries is recognised as important in this approach, and adaptation

Box 13.1: Example of the fishing industry adapting to changing spatial distribution in pelagic fish stocks

For pelagic fisheries, most of the immediate impacts of climate change will be expressed as changes in distribution of the target stocks (Hobday *et al.* 2007b, Hobday in press). As a result, the potential adaptation options considered to date are mostly around changing distributions. Fisheries already adapt to interannual variation in distribution by changing where they fish.

Adaptation strategies to cope with a permanent change in distribution include improvements in locating stocks of fish, changes in home port to minimise economic costs associated with transport of harvested catch, and zoning of fish habitats to minimise unwanted species interactions (e.g. Hobday and Hartmann 2006; Hobday et al 2009).

For example, fishers in the east coast longline fishery use a range of ports along the coast, and change location during the season as fish distribution or availability changes (Hobday *et al.* 2009). Infrastructure may vary between these ports, such as the number of berths, storage facilities, ship chandleries, and transport links. Processing facilities and airlinks are considered the main bottlenecks to when landing catch. If ports receive only occasional usage by some boats, then infrastructure is not stretched. However, if major changes in the usage resulted from changed distribution of the catch, infrastructure may prove to be inadequate in new areas (Hobday *et al.* 2007b).

Changes in species' distributions may reduce the restrictions introduced when species overlap. For example, a southward contraction in the distribution of southern bluefin tuna (SBT) on the east coast of Australia would allow longline fishers targeting other species to fish more freely in areas previously occupied by SBT (Hobday and Hartmann 2006), which may offer them an economic advantage.

As climate-driven changes in fish distribution occur, however, commercial fishers may not be able to simply follow the stocks as they may contract into different management regions. Information on the potential changes will enhance industry capability to adapt to climate change, and make sensible business and investment decisions. This information does not always exist, and where it does, it may not be accessible to those who need it.

options are being explored within ecosystem models (B. Fulton, pers. comm.). Ecosystems are extremely complex and current challenges include defining ecosystem and fisheries objectives, as well as defining indicators, reference points and implementation mechanisms. The transition to EBFM may entail short-term costs and sacrifices by the fishing industry but these will be outweighed by longer term benefits. In one example, the Northern Prawn Fishery is moving to ecosystem based management through the Northern Prawn Fishery Management Plan (1995) which implements various limits on fishery effort and catches of target and by-catch species.

Such holistic approaches to fishery management will also increase industry adaptation in the face of climate change, as increased attention to EBFM is expected to increase resilience of fisheries to a range of impacts. Other holistic approaches, including the CSIRO Ecological Risk Assessment for the Effects of Fishing (Hobday *et al.* 2007a), also support the EBFM approach, and could be adapted to provide estimates of risks due to climate change. It is recognised that adapting to climate change will be most successful when undertaken in conjunction with existing strategies (Smit and Wandel 2006), such as EBFM.

Adaptation options under the EBFM umbrella include developments in by-catch reduction, and improved targeting practises that will have the dual benefit of minimising impacts on non-target species, and provide potential alternatives to spatial closures to protect the particular species. Multi-species fisheries should continue to develop species-specific fishing equipment and targeting practices to improve future adaptability. Species-specific equipment will allow individual species to be targeted, without impacting other species that may be in decline due to climate change, and therefore protected from fishing. This move in Australia to EBFM is illustrative for climate adaptation, because it entails explicit recognition of the importance of understanding environmental relationships. Thus, an EBFM approach will also facilitate adaptation to climate change through a holistic approach (Smit and Wandel 2006).

Adaptation options for recreational and indigenous fishers

Recreational and indigenous fishers are increasingly aware that they may have a significant impact on biological populations. As a result of climate change, management of the fisheries may change. For example, recreational fishers may advocate changes to bag or size limits. They may also advocate increased support for research through licence fees, or offer targeted collection or reporting to supplement the knowledge base on particular species. Although Henry and Lyle (2003) report that the motivation of many recreational fishers is not to provide food, fishing is unlikely to take place in the absence of fish. Thus, changes in fish abundance or distribution will be of concern to fishers. Indigenous fishers, particularly those harvesting the same species as commercial operators, may seek greater involvement in integrated management, and support a reduction of fishing pressure on impacted species.

Both groups will be affected by changes in the physical environment, and must advocate that vessel and shore-based safety regulations are adhered too, support increased use of ocean forecasts, and become active in fisher education to increase awareness of environmental changes. There may be opportunities for business in new regions, and for longer seasons in the same regions. For example, game fishing targets 'warm-water species', so southward movements due to ocean warming may be an advantage with regard to longer fishing seasons, and increased availability to the south (Hobday *et al.* 2007b; Hobday in press).

Risks of maladaptation

While the potential for maladaptation exists in future strategies, the development of adaptation strategies is in the early stages for Australian fisheries and aquaculture. At this time, it is important to recognise that some strategies may have unintended consequences.

One risk of maladaptation for wild fisheries is via potential relaxation of spatial management regulations, instead of an adaptive response. This may not lead to fishers following climate-driven spatial changes in 'peak abundance' as desired; it may lead instead to serial depletion of separate stocks. This risk exists because information on fish abundance, distribution and response to climate is lacking. An adaptation response that includes reduced spatial restriction for fishers should be preceded by studies on stock responses to changing environmental conditions, fish movement responses to the environment, and fisher behaviour.

Relaxation of spatial restrictions of some fishing activities may have unforseen consequences. For example, the cost of fuel is a large proportion of the financial cost of fishing and one that is likely to increase in the future as diminishing reserves and increasing demand drive up the price of fossil fuels. In 2000, global fisheries were estimated to have burned 50 billion litres of fuel, accounting for 1.2% of global oil consumption – equivalent to the amount burned by the 18th ranked oil consuming country globally –and emitted over 130 Mt of CO_2 into the atmosphere (Tyedmers et al. 2005). Given the reliance of the modern fishing fleet on fossil fuels, future management and policy will need to consider reducing these costs, rather than subsidising them.

Aquaculture currently accounts for almost 50% of the world's food fish and has the potential to meet the estimated demand for the additional 40 Mt of aquatic food that will be required by 2030 to maintain the current per capita consumption (FAO 2006). Of concern is the potential impact of climate change on the supply of aquaculture-feed ingredients. Feeds used in the culture of carnivorous fishes (e.g. salmon and barramundi) and crustaceans (e.g. prawns) generally contain high concentrations of protein, much of which is at present obtained through the inclusion of wild harvest fish (Preston and Poloczanska 2007). Expansion of aquaculture industries is placing increasing demand on global supplies of wild harvest fishmeal to provide protein and oil ingredients for aqua-feeds. The potential for adverse impacts of climate change on global fishmeal production is well illustrated by periodic shortages associated with climate fluctuations (e.g. Klinkhardt 2006). Irrespective of the impacts of climate change, production of fishmeal and fish oil from the oceans is likely to decline as demand increases. Over the past 20 years, efforts to find cost-effective alternatives have centred on proteins from terrestrial plants, particularly soybean meal. If, as seems likely, the use of soybean meal and other terrestrial plant proteins increases, then the impacts of climate change on national and global terrestrial crop plant industries will become increasingly relevant for finfish and crustacean aquaculture (Preston and Poloczanska 2007).

Costs and benefits

Climate change will impact the biological, economic and social aspects of many fisheries, and both positive and negative impacts are expected. Fisheries will be impacted differently according to the physical changes in the regional environment. For example, south-east fisheries are most likely to be affected by changes in water temperature, northern fisheries by changes in precipitation, and western fisheries by changes in the Leeuwin Current (Voice et al. 2006; Hobday et al. 2007b). Southern fisheries will see increased opportunity where tropical species move southward, but see lost opportunity where southern species decline.

With regard to socio-economic impacts, aquaculture industries have considerable adaptation potential via selective breeding, regulating the environment, and new species opportunities (Preston and Poloczanska 2007; Battaglene et al. 2008). In common with other food production sectors, climate change is likely to have direct and indirect impacts on all Australian aquaculture production environments (Preston and Poloczanska 2007). These include freshwater ponds, brackish water and marine ponds, and industries that use rafts, lines or cages in coastal or offshore waters. Given the projected impacts of climate change on freshwater supplies in Australia (Hennessey et al. 2007), freshwater aquaculture industries may be the most vulnerable. Conversely, pond-based or open water marine aquaculture sectors could perceivably benefit from climate change and be well placed to respond to the global demand for aquatic food. Climate change is likely to favour the development of new industries such as microalgae biomass production. The emergence of global interest in the potential for mass cultivation of microalgae as a source of biofuels, feeds, chemicals, pharmaceuticals and nutraceuticals (Benemann 1992; Spolaore et al. 2006; Chisti 2007) could provide an opportunity for Australia to take advantage of increased solar radiation and elevated temperatures.

The impacts of climate change on aquaculture are likely to heighten public awareness of the need to respond to the changes. As many aquaculture ventures are in public waterways, acute impacts – such as the farmed tuna mortalities noted earlier – are likely to be highly visible. This increased awareness could lead to more concerted efforts to rigorously assess and predict the likely impacts of climate change on aquaculture. At the same time, quota reductions in the wild-capture fisheries are having major impacts on the economic viability of coastal communities outside the cities; aquaculture is one of the few alternative sources of employment generally available in coastal towns. How social and economic factors will interact in relation to climate impacts and policy setting is still emerging.

Discussion

Overall, there is a need to consider development of adaptive management structures and policies for fisheries and aquaculture that account for climate change impacts on Australia's marine resources. Coordination between agencies and jurisdictions is critical to ensure efficient consideration of the issues and informed decision-making. If these changes are implemented as soon as practical, stakeholders will have maximum flexibility in adapting to future patterns (Hulme 2005). Such coordination is just one of a number of adaptation options in several categories, ranked in order of urgency in Table 13.3.

It has been recognised for more than 100 years that physical oceanic processes affect biological processes such as recruitment to fish stocks (e.g. Hjort 1914; Hannesson 2007). Recent progress has been made in the northern hemisphere on the influence of climate change on such processes (e.g. Clark *et al.* 2003; Rose 2004; Drinkwater 2005; Perry *et al.* 2005). Considerable progress is also being made through research on the role of climate variability (such as El Niño-Southern Oscillation events) in influencing biological processes (e.g. Lehodey *et al.* 1997), which will inform how climate change may impact fish stocks in the southern hemisphere. This approach typically requires time-series of biological and physical data covering more than one cycle of the climate variability pattern (which may operate over decadal time scales). In future, observations that allow the effect of the adaptation strategy will be crucial for informed adaptive management (Table 13.3).

Predicted climate change impacts on Australian marine fisheries have been largely inferred from studies of climate variability thus far (Hobday *et al.* 2007b). This is because (1) climate change

scenarios for the coastal and pelagic environments have, until recently, lacked the spatial resolution necessary for biological studies, and (2) most fisheries and their captured species are not amenable to experimental manipulation. Climate change impacts cannot therefore be readily measured, unlike some terrestrial or benthic systems (e.g. coral reefs). Australia also suffers from a lack of long-term data for most fisheries, which further hinders research into climate impacts, despite more than a century of exploitation in some regions. Fishers were not required to complete fishery logbooks until comparatively recently; the logbooks are the most common data source used to assess resource abundance. Fishery-independent sampling is far less common. Australia's participation in several international programs focused on the climate impacts on fished species may help to overcome the temporal limitation of existing data by enabling spatial comparisons between regions. Two such programs are the GLOBEC (Global Ocean Ecosystem Dynamics) initiatives SPACC (Small Pelagics and Climate Change) and CLIOTOP (Climate Impacts on Top Ocean Predators). The Australian contribution to such international efforts should be increased in future to maximise transfer of knowledge.

In future, information on the biological relationships with climate variability must be synthesised to give insight into the impacts of climate change on fisheries and aquaculture (Table 13.3). At a policy level, development of cross-jurisdictional plans and research priorities is vital, as is recognition of the importance of increased R&D capacity for adaptation studies, particularly with regard to social and economic implications (Table 13.3). Access to climate information at finer spatial scales and coupling with ecosystem models is a high priority. Use of climate information will also be improved if decision support tools are available (Table 13.3). The high priority species and ecosystem issues include a need for integrated case studies, for example consideration of both fisheries and conservation management, and an increased understanding of how management systems cope with the uncertainty in future climate. The effort directed at adaptation will require evidence of effectiveness, and so support for observation programs providing data against which change can be evaluated is also considered a high priority (Table 13.3). With additional information,

Table 13.3: Summary of climate change adaptation options for marine fisheries and aquaculture. Priority 1 (high), 2 (medium) and 3 (low). Options relevant to fisheries or aquaculture only are indicated.

Adaptation options	Priority
Policy level	
Development of cross-jurisdictional regional management plans and research priorities appropriate for the changing distribution and abundance of the species being harvested or farmed	1
Increase R&D capacity, undertake further adaptation studies with focus on social and economic implications of biological response	1
Development of monitoring, assessment and management strategies which are robust in the face of increased uncertainty due to climate change	1
Encourage appropriate management, policy, and industry structures to enable flexibility in adapting to climate change while minimising maladaptation	2
Encourage diversification of enterprises (e.g. ecotourism, scientific charters, recreational fishing) where practicable	3
Climate information and use	
Improve resolution of climate models and couple with ecosystem models for biological predictions	1
Development of decision support tools to enable access to climate data and interpretation in relation to stakeholder goals and to analyse alternative management options	1
Ensure communication of broader climate change information	2
Species and ecosystem issues	
Undertake integrated regional case studies that inform the potential impact of climate change and the adaptation responses	1
Develop understanding of how robust current management strategies are in the face of uncertainty (fisheries)	1
Implement observational systems that can detect change or provide robust indicators for effect of adaptation approach	1
Improve predictive tools and indicators for species impacted by climate change, harvest modelling capabilities and quantitative approaches to risk management	2
Selective breeding of varieties with appropriate physical tolerances and phenological characteristics for changed environment (aquaculture)	2

assessments of future impacts can be made with greater confidence and management responses can be justified to sometimes reluctant stakeholders. Hobday *et al.* (2007b) recommended regional case studies that may provide rapid insight and generate improved understanding of climate change threats, opportunities and adaptation strategies, including (1) investigation of climate impacts on south-east fisheries, where the climate warming signal is strongest, or on western rock lobster, where a known recruitment relationship may change, and (2) investigation of socio-economic impacts of climate on fisheries and aquaculture via a regional focus study, such as on the Eastern Tuna and Billfish Fishery, where environmental variability is a key driver of fishing effort, or on an aquaculture industry such as salmon farming. Development of integrated models predicting the socio-economic impacts of climate change is also an important priority. This challenging area would require partnerships between fishers, managers, biologists, social scientists, economists and ocean modellers from a range of Australian organisations. The research partnership between climate modellers and fisheries and aquaculture scientists must be fostered to ensure that biologically relevant information at desirable spatial and temporal scales will be available from the climate models (Table 13.3).

Acknowledgements

The content and clarity of this chapter was improved by review from David Smith, Alan Butler, Philip Munday, Chris Stokes, and Mary Livingston, for which we are grateful.

References

ABARE (2007). Australian Fisheries Statistics 2006. Canberra.

Andrade HA, Garcia CAE (1999). Skipjack tuna fishery in relation to sea surface temperature off the southern Brazilian coast. *Fisheries Oceanography* **8**, 245–254.

Appleyard SA, Ward RD (2006). Genetic diversity and effective population size in mass selection lines of Pacific oyster (*Crassostrea gigas*). *Aquaculture* **254**, 148–159.

Australian Bureau of Statistics (2002). '2001 Census: Australia's Aboriginal and Torres Strait Islander population.' Report No. 4713.0. Australian Bureau of Statistics, Canberra.

Battaglene SC, Carter C, Hobday AJ, Lyne V, Nowak B (2008). 'Scoping study into adaptation of the Tasmanian salmonid aquaculture industry to potential impacts of climate change'. Report to DAFF, August 2008.

Benemann JR (1992). Microalgae aquaculture feeds. *Journal of Applied Phycology* **4**, 233–245.

Bestley S, Patterson TA, Hindell MA, Gunn JS (2008). Feeding ecology of wild migratory tunas revealed by archival tag records of visceral warming. *Journal of Animal Ecology* **77**, 1223–1233, doi: 10.1111/j.1365–2656.2008.01437.x.

Blaber SJM (2000). *Tropical Estuarine Fishes: Ecology, Exploitation and Conservation*. Blackwell Science, Oxford.

Bopp L, Monfray P, Aumont O, Dufresne J–L, Le Treut H, Madec G, Terray L, Orr JC (2001). Potential impact of climate change on marine export production. *Global Biogeochemical Cycles* **151**, 81–100.

Boyd PW, Doney SC (2002). Modeling regional responses by marine pelagic ecosystems to global climate change. *Geophysical Research Letters* **29**, 1806, doi:10.1029/2001GL014130.

Brewer DT, Rawlinson N, Eayrs S, Burridge CY (1998). An assessment of Bycatch Reduction Devices in a tropical Australian prawn trawl fishery. *Fisheries Research* **36**, 195–215.

Brodeur RD, Ralston S, Emmett RL, Trudel M, Auth TD, Phillips AJ (2006). Anomolous pelagic nekton abundance, distribution, and apparent recruitment in the northern California Current in 2004 and 2005. *Geophysical Research Letters* **33**, L22S08.

Brown A (2005). Age and growth of jack mackerel off the east coast of Tasmania. BSc (Hons) thesis, The University of Tasmania, Australia.

Bunce A (2004). Do dietary changes of Australasian gannets (*Morus serrator*) reflect variability in pelagic fish stocks? *Wildlife Research* **31**, 383–387.

Campbell RA (1999). 'Long term trends in yellowfin tuna abundance in the south-west Pacific: with an emphasis on the eastern Australian Fishing Zone.' Final report to the Australian Fisheries Management Authority, Canberra.

Caputi N, Chubb C, Pearce A (2001). Environmental effects on recruitment of the western rock lobster, *Panulirus cygnus*. *Marine and Freshwater Research* **52**, 1167–1174.

Caputi N, Chubb C, Melville-Smith R, Pearce AF, Griffin DA (2003). Review of the relationships between life history stages of the western rock lobster, *Panulirus cygnus*, in Western Australia. *Fisheries Research* **65**, 47–61.

Caputi N, Fletcher W, Pearce A, Chubb C (1996). Effect of the Leeuwin Current on the recruitment of fish and invertebrates along the Western Australian coast. *Marine and Freshwater Research* **47**, 147–155.

Catchpole A, Auliciems A (1999). Southern oscillation and the northern Australian prawn catch. *International Journal of Biometeorology* **43**, 110–112.

Chavez FP, Ryan J, Lluch-Cota SE, Niquen M (2003). From anchovies to sardines and back: multidecadal change in the Pacific Ocean. *Science* **299**, 217–221.

Chisti Y (2007). Biodiesel from microalgae. *Biotechnology Advances* **25**, 294–306.

Clark RA, Fox CJ, Viner D, Livermore M (2003). North Sea cod and climate change – modelling the effects of temperature on population dynamics. *Global Change Biology* **9**, 1669–1680.

Cowling A, Hobday AJ, Gunn J (2003). 'Development of a fishery-independent index of abundance for juvenile southern bluefin tuna and improvement of the index through integration of environmental, archival tag and aerial survey data.' FRDC Final Report 1999/118.

Cox PM, Betts RA, Jones CD, Spall SA, Totterdell IJ (2000). Acceleration of global warming due to carbon cycle feedbacks in a 3D coupled model. *Nature* **408**, 184–187.

Cury P, Bakun A, Crawford RJM, Jarre A, Quiñones RA, Shannon LJ, Verheye HM (2000). Small pelagics in upwelling systems: patterns of interaction and structural changes in 'wasp-waist' ecosystems. *ICES Journal of Marine Science* **57**, 603–618.

DAFF (2008). National Climate Change and Fisheries Action Plan. Department of Agriculture, Forestry and Fisheries.

DEH (2004). Assessment of the ecological sustainability of management arrangements for the Northern Territory Finfish Trawl Fishery. Australian Government, Department of the Environment and Heritage, Canberra.

Denman K, Hofmann E, Marchant H (1996). Marine biotic responses to environmental change and feedbacks to climate. In *Climate Change*. (Eds JT Houghton *et al.*) pp. 483–516. Cambridge University Press, New York.

Dimmlich WF, Breed WG, Geddes M, Ward TM (2004) Relative importance of gulf and shelf waters for spawning and recruitment of Australian anchovy, *Engraulis australis*, in South Australia. *Fisheries Oceanography* **13**, 310–323.

Drinkwater KF (2005). The response of Atlantic cod (*Gadus morhua*) to future climate change. *ICES Journal of Marine Science* **62**, 1327–1337.

Elliott NG, Reilly A (2003). Likelihood of bottleneck event in the history of the Australian population of Atlantic salmon (*Salmo salar* L.) *Aquaculture* **215**, 31–44.

FAO (2006). 'State of world aquaculture 2006.' FAO Fisheries Technical Paper No. 500. FAO, Rome.

Gaughan DJ (2007). Potential mechanisms of influence of the Leeuwin Current eddy system on teleost recruitment to the Western Australian continental shelf. *Deep-Sea Research II* **54**, 1129–1140.

Gordon HB, Rotstayn LD, McGregor JL, Dix MR, Kowalczyk EA, O'Farrell SP, Waterman LJ, Hirst AC, Wilson SG, Collier MA, Watterson IG, Elliott TI (2002). 'The CSIRO Mk3 Climate System Model.' CSIRO Atmospheric Research Technical Paper No. 60.

Hannesson R (2007). Geographical distribution of fish catches and temperature variations in the northeast Atlantic since 1945. *Marine Policy* **31**, 32–39.

Hart A, Skepper C, Joll L (1999). 'Growth of pearl oysters in the southern and northern areas of the pearl oyster fishery and examination of environmental influences on recruitment to pearl oyster stocks.' FRDC Final Report 95/41.

Haywood MDE, Vance DJ, Loneragan NR (1995). Seagrass and algal beds as nursery habitats for tiger prawns (*Penaeus semisulcatus* and *P. esculentus*) in a tropical Australian estuary. *Marine Biology* **122**, 213–223.

Hennessy K, Fitzharris B, Bates C, Harvey N, Howden SM, Hughes L, Salinger J, Warrick R (2007). Australia and New Zealand. Climate Change 2007: Impacts, Adaptation and Vulnerability. In *Contribution of Working Group II to the Fourth Assessment Report of the Intergovernmental Panel on Climate Change.* (Eds ML Parry, OF Canziani, JP Palutikof, PJ van der Linden, CE Hanson) pp. 507–540. Cambridge University Press, Cambridge.

Henry GW, Lyle JM (2003). National Recreational Fishing Survey. In 'The National Recreational and Indigenous Fishing Survey'. Canberra: Australian Government Department of Agriculture, Fisheries and Forestry, FRDC Project No. 99/158, pp. 27–97.

Herzfeld M, Tomczak M (1997). Numerical modelling of sea-surface temperature and circulation in the Great Australian Bight. *Progress in Oceanography* **39**, 29–78.

Hill BJ, Haywood M, Venables W, Gordon SR, Condie S, Ellis NR, Tyre A, Vance D, Dunn J, Mansbridge J, Moseneder C, Bustamante R, Pantus F (2002). 'Surrogates I – Predictors, impacts, management and conservation of the benthic biodiversity of the Northern Prawn Fishery'. Final report on FRDC Project 2000/160.

Hjort J (1914). Fluctuations in the great fisheries of northern Europe viewed in the light of biological research. *Rapports et Procès-Verbaux des Réunions du Conseil Permanent International pour l'Exploration de la Mer* **20**, 1–228.

Hobday AJ (2001). 'The influence of topography and environment on presence of juvenile southern bluefin tuna, *Thunnus maccoyii*, in the Great Australian Bight, Southern Bluefin Tuna Recruitment Monitoring Workshop Report'. CSIRO Marine Research, Hobart.

Hobday AJ (in press). Ensemble analysis of the future distribution of large pelagic fishes in Australia. *Progress in Oceanography*.

Hobday AJ, Flint N, Stone T, Gunn JS (2009). Electronic tagging data supporting flexible spatial management in an Australian longline fishery. In *Electronic Tagging and Tracking in Marine Fisheries II. Reviews: Methods and Technologies in Fish Biology and Fisheries.* Springer, The Netherlands, pp. 381–403.

Hobday AJ, Hartmann K (2006). Near real-time spatial management based on habitat predictions for a longline bycatch species. *Fisheries Management and Ecology* **13**, 365–380.

Hobday AJ, Hartmann K, Bestley S, Tsuji S, Takahashi N (2004). 'Integrated analysis project – environmental influences on the observed decline of southern bluefin tuna in the acoustic survey area. Southern Bluefin Tuna Recruitment Monitoring Workshop Report, Yokohama, Japan'. CSIRO Marine Research, Hobart.

Hobday AJ, Okey TA, Poloczanska ES, Kunz TJ, Richardson AJ (2007c). 'Impacts of climate change on Australian marine life, CSIRO Marine and Atmospheric Research'. Report to the Australian Greenhouse Office, Canberra. Available at http://www.climatechange.gov.au/impacts/publications/marinelife.html

Hobday AJ, Poloczanska ES, Matear R (2007b). 'Implications of climate change for Australian fisheries and aquaculture: a preliminary assessment, CSIRO Marine and Atmospheric Research'. Report to the Department of Climate Change, Canberra. December 2007. Available at http://www.climatechange.gov.au/impacts/publications/fisheries.html

Hobday AJ, Smith ADM, Webb H, Daley R, Wayte S, Bulman C, Dowdney J, Williams A, Sporcic M, Dambacher J, Fuller M, Walker T (2007a). 'Ecological risk assessment for the effects of fishing: methodology'. Report R04/1072 for the Australian Fisheries Management Authority, Canberra.

Hulme PE (2005) Adapting to climate change: is there scope for ecological management in the face of a global threat? *Journal of Applied Ecology* **42**, 784–794.

Hunt GL, McKinnell S (2006). Interplay between top-down, bottom-up, and wasp-waist control in marine ecosystems. *Progress in Oceanography* **68**, 115–124.

Jackson CJ, Wang Y-G (1998). Modelling growth rate of *Penaeus monodon* Fabricus in intensively managed ponds: effects of temperature, pond age and stocking density. *Aquaculture Research* **29**, 27–36.

Jackson JBC, Kirby MX, Berger WH, Bjorndal KA, Botsford LW, Bourque BJ, Bradbury RH, Cooke R, Erlandson J, Estes JA, Hughes TP, Kidwell S, Lange CB, Lenihan HS, Pandolfi JM, Peterson CH, Steneck RS, Tegner MJ, Warner RR (2001). Historical overfishing and the recent collapse of coastal ecosystems. *Science* **293**, 629–638.

Jacobson LD, De Oliveira JAA, Barange M, Cisneros-Mata MA, Félix-Uraga R, Hunter JR, Kim JY, Matsuura Y, Ñiquen M, Porteiro C, Rothschild B, Sanchez RP, Serra R, Uriarte A, Wada T (2001). Surplus production, variability, and climate change in the great sardine and anchovy fisheries. *Canadian Journal of Fisheries and Aquatic Sciences* **58**, 1891–1903.

Johnson CR, Ling S, Ross J, Shepherd SA, Miller K (2005). 'Range extension of the long-spined sea urchin (*Centrostephanous rodgersii*) in eastern Tasmania: assessment of potential threats to fisheries'. FRDC Final Report 2001/044.

Kitagawa T, Sartimbul A, Nakata H, Kimura S, Yamada H, Nitta A (2006). The effect of water temperature on habitat use of young Pacific bluefin tuna *Thunnus orientalis* in the East China Sea. *Fisheries Science* **72**, 1166–1176.

Klinkhardt M (2006). Demand for fish meal and fish oil rising. *Eurofish Magazine*, 6/2006. http://www.globefish.org/index.php?id=3531&easysitestatid=–695651663

Larcombe J, McLoughlin K (2007). 'Fishery Status Reports 2006: status of fish stocks managed by the Australian Government.' Bureau of Rural Sciences, Canberra.

Laurs RM, Fiedler PC, Montgomery DR (1984). Albacore tuna catch distributions relative to environmental features observed from satellites. *Deep-Sea Research* **31**, 1085–1099.

Lee SY (2004). Relationship between mangrove abundance and tropical prawn production: a re-evaluation. *Marine Biology* **145**, 943–949.

Lehodey P, Bertignac M, Hampton J. Lewis A, Picaut J (1997). El Niño Southern Oscillation and tuna in the western Pacific. *Nature* **389**, 715–718.

Lenanton RC, Joll LM, Penn JW, Jones K (1991). The influence of the Leeuwin Current on the coastal fisheries of Western Australia. *Journal of the Royal Society of Western Australia* **74**, 101–114.

Ling SD, Johnson CR, Frusher SD, King CK (2008). Reproductive potential of a marine ecosystem engineer at the edge of a newly expanded range. *Global Change Biology* **14**, 1–9.

Ling SD, Johnson CR, Ridgway K, Hobday AJ, Haddon M (2009). Climate driven range extension of a sea urchin: inferring future trends by analysis of recent population dynamics. *Global Change Biology* **15**, 719–731.

Link JS, Brodziak JKT, Edwards SF, Overholtz WJ, Mountain D, Jossi JW, Smith TD, Fogarty MJ (2002). Marine ecosystem assessment in a fisheries management context. *Canadian Journal of Fisheries and Aquatic Sciences* **59**, 1429–1440.

Loneragan NR, Adnan NA, Connolly RM, Manson FJ (2005). Prawn landings and their relationship with the extent of mangroves and shallow waters in western peninsular Malaysia. *Estuarine, Coastal and Shelf Science* **63**, 187–200.

Loneragan NR, Kenyon RA, Haywood MDE, Staples DJ (1994). Population dynamics of juvenile tiger prawns (*Penaeus esculentus* and *P. semisulcatus*) in seagrass habitats of the western Gulf of Carpentaria, Australia. *Marine Biology* **119**, 133–143.

Lough JM (2008). Shifting climate zones for Australia's tropical marine ecosystems. *Geophysical Research Letters* **35**, L14708.

Love G (1987). Banana prawns and the Southern Oscillation Index. *Australian Meteorological Magazine* **35**, 47–49.

Lyle J, Krusic-Golub K, Morison A (2000). 'Age and growth of jack mackerel and the age structure of the jack mackerel purse seine catch'. FRDC Final Report 1995/034.

Lyne V, Thresher R, Rintoul S (2005). 'Regional impacts of climate change and variability in south-east Australia: report of a joint review by CSIRO Marine Research and CSIRO Atmospheric Research'. CSIRO Internal Report, Hobart.

Majkowski J, Williams K, Murphy GI (1981). Research identifies changing patterns in Australian tuna fishery. *Australian Fisheries* **40**, 5–10.

Manson FJ, Loneragan NR, Harch BD, Skilleter GA, Williams L (2005). A broad-scale analysis of links between coastal fisheries production and mangrove extent: a case-study for northeastern Australia. *Fisheries Research* **74**, 69–85.

Matear RJ, Hobday AJ, Caputi N (2007). Western fisheries. In 'Climate impacts on Australian fisheries and aquaculture: implications for the effects of climate change'. (Eds AJ Hobday, ES Poloczanska, RJ Matear) Report to the Australian Greenhouse Office, Canberra.

McKinnon AD, Richardson AJ, Burford ME, Furnas MJ (2007). Vulnerability of Great Barrier Reef plankton to climate change. In *Climate Change and the Great Barrier Reef.* (Eds JE Johnson, PA Marshall) pp. 121–152. Great Barrier Reef Marine Park Authority and Australian Greenhouse Office, Townsville, Queensland.

Miller KA (2007). Climate variability and tropical tuna: management challenges for highly migratory fish stocks. *Marine Policy* **31**, 56–70.

Munday PL, Jones GP, Sheaves M, Williams AJ, Goby G (2007). Vulnerability of fishes of the Great Barrier Reef to climate change. In *Climate Change and the Great Barrier Reef.* (Eds JE Johnson, PA Marshall) pp. 357–391. Great Barrier Reef Marine Park Authority and Australian Greenhouse Office, Townsville, Queensland.

Munday PL, Jones GP, Pratchett MS, Williams AJ (2008). Climate change and the future for coral reef fishes. *Fish and Fisheries* **9**, 261–285.

Nelson R, Howden M, Stafford-Smith M (2008). Using adaptive governance to rethink the way science supports Australian drought policy. *Environmental Science and Policy* **11**, 588–601.

Pearce AF, Caputi N (1994). 'Effects of seasonal and interannual variability of the ocean environment of recruitment to the fisheries of Western Australia.' FRDC Final Report 1994/032.

Pearce A, Feng M (2007). Observations of warming on the Western Australian continental shelf. *Marine and Freshwater Research* **58**, 914–920.

Pecl GT, Jackson GD (2008). The potential impacts of climate change on inshore squid: biology, ecology and fisheries. *Reviews in Fish Biology and Fisheries*, **18**, 373–385.

Pender PJ, Willing RS, Ramm DC (1992). Northern prawn fishery bycatch study: Distribution, abundance, size and use of bycatch from the mixed species fishery. In 'Northern Territory, Department of Primary Industry and Fisheries, Fishery Report 26.' pp. 1–70.

Penn JW, Fletcher WJ, Head F (Eds) (2005). 'State of the Fisheries Report 2003/04.' Department of Fisheries, Western Australia.

Perry AL, Low PJ, Ellis JR, Reynolds JD (2005). Climate change and distribution shifts in marine fishes. *Science* **308**, 1912–1915.

Pittock B (2003). *Climate Change: An Australian Guide to the Science and Potential Impacts*. Australian Greenhouse Office, Canberra.

Poloczanska ES, Babcock RC, Butler A, Hobday AJ, Hoegh-Guldberg O, Kunz TJ, Matear R, Milton D, Okey TA, Richardson AJ (2007). Climate Change and Australian Marine Life. *Oceanography and Marine Biology Annual Review* **45**, 409–480.

Pratchett MS, Wilson SK, Graham NAJ, Cinner JE, Bellwood DR, Jones GP, Polunin NVC, McClanahan TR (2008). Effects of climate-induced coral bleaching on coral-reef fishes – ecological and economic consequences. *Oceanography and Marine Biology: An Annual Review* **46**, 251–296.

Preston NP, Crocos P, Jackson C, Duncan P, Zipf M, Koenig R (2001). Farming the Kuruma shrimp (*Marsupenaeus japonicus*) in Australia – a case history. In: *The New Wave, Proceedings of the Special Session on Sustainable Shrimp Culture, Aquaculture 2001*. (Eds CL Browdy, DE Jory) The World Aquaculture Society: Baton Rouge, USA.

Preston NP, Crocos PJ, Keys SJ, Coman GJ, Koenig R (2004). Comparative growth of selected and non-selected Kuruma shrimp *Penaeus (Marsupenaeus) japonicus* in commercial farm ponds. *Aquaculture* **231**, 73–82.

Preston NP, Macleod I, Rothlisberg PC, Long B (1997). Environmentally sustainable aquaculture production – an Australian perspective. In *Developing and Sustaining World Fisheries Resources: The State of Science and Management*. Second World Fisheries Congress, Brisbane 1996. Vol. 2: Proceedings. (Eds DA Hancock, DC Smith, A Grant, JP Beumer) pp. 471–477. CSIRO, Melbourne.

Preston NP, Poloczanska ES (2007). Aquaculture. In 'Climate impacts on Australian fisheries and aquaculture: implications for the effects of climate change.' (Eds AJ Hobday, ES Poloczanska, RJ Matear) Report to the Australian Greenhouse Office, Canberra.

Rayns N (2007). The Australian government's harvest strategy policy. *ICES Journal of Marine Science* **64**, 596–598.

Reddy R, Lyne V, Gray R, Easton A, Clarke S (1995). An application of satellite-derived sea surface temperatures to southern bluefin tuna and albacore off Tasmania, Australia. *Scientia Marina* **59**, 445–454.

Ridgway KR (2007). Long-term trend and decadal variability of the southward penetration of the East Australian Current. *Geophysical Research Letters* **34**, L13613, doi:10.1029/2007GL030393.

Ridgway KR, Condie SA (2004). The 5500-km-long boundary flow off western and southern Australia. *Journal of Geophysical Research* **109**, C04017.

Robins JB, Halliday IA, Staunton-Smith J, Mayer DG, Sellin MJ (2005). Freshwater-flow requirements of estuarine fisheries in tropical Australia: a review of the state of knowledge and application of a suggested approach. *Marine and Freshwater Research* **56**, 343–360.

Rose GA (2004). Reconciling overfishing and climate change with stock dynamics of Atlantic cod (*Gadus morhua*) over 500 years. *Canadian Journal of Fisheries and Aquatic Sciences* **61**, 1553–1557.

Rose GA (2005). On distributional responses of North Atlantic fish to climate change. *ICES Journal of Marine Science* **62**, 1360–1374.

Ryan J (2004). 'Farming the deep blue.' A report prepared by Bord Iascaigh Mhara-Irish Sea Fisheries Board and Irish Marine Institute. Presented at 'Farming the Deep Blue,' Limerick, Ireland, October 6–7, 2004.

Sarmiento JL, Gruber N, Brzezinski MA, Dunne JP (2004a). High-latitude controls of thermocline nutrients and low latitude biological productivity. *Nature* **427**, 56–60.

Sarmiento JL, Slater R, Barber R, Bopp L, Doney SC, Hirst AC, Kleypas J, Matear R, Mikolajewicz U, Monfray P, Soldatov V, Spall SA, Stouffer R (2004b). Response of ocean ecosystems to climate warming. *Global Biogeochemical Cycles* **18**, doi:10.1029/2003GB002134.

Schick RS, Goldstein J, Lutcavage ME (2004). Bluefin tuna (*Thunnus thynnus*) distribution in relation to sea surface temperature fronts in the Gulf of Maine (1994–96). *Fisheries Oceanography* **13**, 225–238.

Sheaves M (1998). Spatial patterns in estuarine fish faunas in tropical Queensland: a reflection of interaction between long-term physical and biological processes? *Marine and Freshwater Research* **49**, 31–41.

Shepherd SA, Lawas HM (1974). Studies on southern Australian abalone (genus Haliotis). 2. reproduction of five species. *Australian Journal of Marine and Freshwater Research* **25**, 47–62.

Smit B, Wandel J (2006). Adaptation, adaptive capacity and vulnerability. *Global Environmental Change* **16**, 282–292.

Smith ADM, Smith DC (2001). A complex quota-managed fishery: science and management in Australia's South-east Fishery. Introduction and overview. *Marine and Freshwater Research* **52**, 353–360.

Spolaore P, Joannis-Cassan C, Duran E, Isambert A (2006). Commercial applications of microalgae. *Journal of Bioscience and Bioengineering* **101**, 87–96.

Steneck RS, Graham MH, Bourque BJ, Corbett D, Erlandson JM, Estes JA, Tegner MJ (2002). Kelp forest ecosystems: biodiversity, stability, resilience and future. *Environmental Conservation* **29**, 436–459.

Stenevik EK, Sundby S (2007). Impacts of climate change on commercial fish stocks in Norwegian waters. *Marine Policy* **31**, 19–31.

Stobutzki I, McLoughlin K (2007). Northern prawn fishery. In 'Fishery Status Reports 2006: Status of Fish Stocks Managed by the Australian Government.' (Eds J Larcombe, K McLoughlin) pp. 35–48. Bureau of Rural Sciences, Canberra.

Stobutzki I, Blaber S, Brewer D, Fry G, Heales D, Miller M, Milton D, Salini J, van der Velde T, Wassenberg T (2000). 'Ecological sustainability of bycatch and biodiversity in prawn trawl fisheries.' FRDC Final Report 96/257.

Suppiah R, Hennessy KJ, Whetton PH, McInnes K, Macadam I, Bathols J, Ricketts J, Page CM (2007). Australian climate change projections derived from simulations performed for the IPCC 4th Assessment Report. *Australian Meteorological Magazine* **56**, 131–152.

Thresher R, Koslow JA, Morison AK, Smith DC (2007a). Depth-mediated reversal of the effects of climate change on long-term growth rates of exploited marine fish. *Proceedings of the National Academy of Sciences* **104**, 7461–7465.

Thresher R, Klaer N, Lyne V, Bax N, Rintoul S (2007b). South-east Demersal Fisheries. In 'Climate impacts on Australian fisheries and aquaculture: implications for the effects of climate change.' (Eds AJ Hobday, ES Poloczanska, RJ Matear) Report to the Australian Greenhouse Office, Canberra, Australia.

Tyedmers PH, Watson R, Pauly D (2005). Fueling global fishing fleets. *Ambio* **34**, 635–638.

Vance DJ, Bishop J, Dichmont C, Hall N, McInnes K, Taylor B (2003) 'Management of common banana prawn stocks of the Gulf of Carpentaria: separating the effects of fishing from those of the environment.' AFMA Final Report, Project No: 98/0716.

Vance DJ, Haywood MDE, Staples DJ (1990). Use of a mangrove estuary as a nursery area by postlarval and juvenile banana prawns, *Penaeus merguiensis* Deman, in Northern Australia. *Estuarine Coastal and Shelf Science* **31**, 689–701.

Voice M, Harvey N, Walsh K (Eds) (2006). 'Vulnerability to climate change of Australia's coastal zone: analysis of gaps in methods, data and system thresholds.' Report to the Australian Greenhouse Office, Canberra. June 2006.

Ward P, Hindmarsh S (2007). An overview of historical changes in the fishing gear and practices of pelagic longliners, with particular reference to Japan's Pacific fleet. *Reviews in Fish Biology and Fisheries* **17**, 501–516.

Young JW, Bradford RW, Lamb TD, Lyne VD (1996). Biomass of zooplankton and micronekton in the southern bluefin tuna fishing grounds off eastern Tasmania, Australia. *Marine Ecology Progress Series* **138**, 1–14.

Young JW, Jordan AR, Bobbi C, Johannes RE, Haskard K, Pullen G (1993). Seasonal and interannual variations in krill (*Nyctiphanes australis*) stocks and their relationship to the fishery for jack mackerel (*Trachurus declivis*) off eastern Tasmania, Australia. *Marine Biology* **116**, 9–18.

Young JW, Lamb TD, Bradford R, Clementson L, Kloser R, Galea H (2001). Yellowfin tuna (*Thunnus albacares*) aggregations along the shelf break of southeastern Australia: links between inshore and offshore processes. *Marine and Freshwater Research* **52**, 463–474.

Young JW, Lansdell MJ, Hobday AJ, Dambacher JM, Griffiths SPECIES, Cooper S, Kloser R, Nichols PD, Revill A (2009). 'Determining ecological effects of longline fishing in the Eastern Tuna and Billfish Fishery.' FRDC Final Report 2004/063.

14

AGRICULTURAL GREENHOUSE GASES AND MITIGATION OPTIONS

JC Carlyle, E Charmley, JA Baldock, PJ Polglase and B Keating

KEY MESSAGES:

■ Agricultural practices release considerable quantities of greenhouse gases (GHGs) into the atmosphere and many practices involve more than one GHG. Australian agriculture accounts for 87.9 million tonnes CO_2-equivalent (Mt CO_2-e), 16.8% of total national GHG emissions (which includes all gases from all sectors): CH_4 and N_2O account for 76% and 24% of the agricultural total respectively, with emissions of CH_4 from livestock being the dominant agricultural GHG source (67% of agricultural emissions).

■ Mitigation options fall into three broad categories: (1) emissions reduction, consisting of technologies, practices or a combination of both that directly reduce GHG emissions; (2) enhancing GHG sinks: this option largely relates to increasing the capture and storage of photosynthetically derived C in plant biomass, soil, or harvested products, and (3) fossil fuel substitution: use of crops and residues for electricity generation or liquid fuel production.

■ Short-term options for mitigation of CH_4 emissions from extensive livestock systems are mainly limited to approaches based around livestock and pasture management (including plant species selection), and offsetting emissions through increasing C stocks. Solutions such as alternate hydrogen sinks in the rumen, vaccines and livestock breeding are long-term goals with significant hurdles to overcome.

■ Many of the actions required to reduce N_2O emissions are identical to best-practice strategies to maximise the efficiency of N use and minimise undesirable environmental impacts. There are two main options: (1) managing the availability of mineral N relative to crop requirements to minimise nitrification and accumulation of nitrate, and (2) managing soil water by improved drainage or better scheduling irrigation to minimise water logging and the potential for denitrification.

■ As a stop-gap measure (over about 20 years), significant opportunities exist to offset Australian GHG emissions by increasing the C stored in Australian landscapes, but the full system implications of the necessary land use changes require much deeper investigation. Climate change, particularly fire and drought, poses a significant risk to stored C and there is a need to assess this impact for the most probable climate change scenarios.

■ Mitigation strategies need to be based on whole-systems emission budgets and to consider trade-offs with other factors and enterprise-scale economics. The distinctive need around this issue is for integration of scientific effort in key aspects of the soil, plant, animal system, forestry and agricultural products, and new work in the interaction of current and potential management systems. This will lead to 'best management practice' systems operating at an integrated landscape scale with C sequestration and GHG mitigation as an integrated outcome considered alongside other economic and environmental outcomes.

Introduction

Reducing the rate of increase in greenhouse gas (GHG) emissions is imperative if the worst impacts of climate change are to be avoided. Nationally, this has been underscored by Australia's ratification of the Kyoto Protocol, and the Garnaut (Garnaut 2008a and 2008b) and McKinsey (McKinsey & Company 2008) reports highlighting the need for immediate integrated action and options for mitigation. There is no doubt that Australia's primary focus must address direct reductions of emissions from the Stationary energy and Transport sectors, and in rates of land clearing. However, the magnitude of emission reduction required means that a wide range of mitigation options will need to be considered and these need to include agriculture and land use.

Agricultural practices release considerable quantities of GHGs to the atmosphere, with many practices involving more than one GHG (Johnson *et al.* 2007; Smith *et al.* 2007). Fluxes of GHGs are often diffused and highly variable in space and time across varying scales. This is a challenge for precise measurement, and considerable uncertainty exists around many flux estimates, and the ability to quantify the impact of mitigation options (Dalal *et al.* 2003; Harvey *et al.* 2008; Lassey 2008). A range of mitigation options exists now and others are under development. Some of these use current technology and can be implemented immediately and in many cases generate a range of environmental, economic, and social benefits beyond those relating to mitigation (e.g. Howden and Reyenga 1999; Lal 2004).

The dominant agricultural GHGs are CO_2, CH_4, and N_2O (Johnson *et al.* 2007; Smith *et al.* 2007). CO_2 is released primarily from microbial decomposition of plant residues and soil organic matter, burning of crop residues, and fossil fuel use. CH_4 is released when organic matter is decomposed by microbes under anoxic conditions, primarily ruminant digestion, rice production, and manure storage. N_2O is produced by microbial transformation of N in soils and manures, most commonly when N availability exceeds plant requirement and in association with wet (anoxic) conditions. Clearing land of native vegetation for agricultural use has released, and continues to release, enormous quantities of CO_2 and other GHGs to the atmosphere. For example, tropical deforestation continues to account for

around 20% of global anthropogenic GHG emissions (Gullison *et al.* 2007). The GHGs released by clearing come from (1) the cleared vegetation as it is decomposed, burnt, or both, and (2) from the accelerated decomposition of soil organic matter as a consequence of burning and soil cultivation.

Agricultural CO_2 emissions will not be considered further in this review because, unlike CH_4 and N_2O, they are not the focus of mitigation strategies and because they are reported on separately by national GHG inventories. This is because (1) while the total release of CO_2 from agricultural soils is huge, it is assumed to be balanced by crop uptake so that the net release is zero, and (2) CO_2 emissions associated with agricultural fuel and energy use are accounted for under the energy sector (Stationary energy, Transport, and Fugitive emissions) (Smith *et al.* 2007). This chapter will focus on the amounts and principal emission sources of the dominant agricultural GHGs CH_4 and N_2O, options for their mitigation, and opportunities for managing agricultural systems for enhanced removal of CO_2 from the atmosphere.

Sizes and sources of agricultural non-CO_2 GHG emissions

The impact of a GHG on global warming is a function of three factors: (1) radiative forcing (i.e. the effect of a unit of gas on the earth's radiation balance); (2) longevity in the atmosphere (i.e. how long the radiative forcing of a unit of gas is expected to continue), and (3) total quantity of gas emitted (Johnson *et al.* 2007). The first two define a gas's Global Warming Potential (GWP); GWPs for CO_2, CH_4, and N_2O are 1, 25 and 298 respectively. On a global scale the relative contribution of CO_2, CH_4 and N_2O to global warming is around 57%, 27% and 15% respectively (Johnson *et al.* 2007).

Global agricultural emissions

Recent estimates, summarised by Smith *et al.* (2007), place annual global emissions of non-CO_2 GHGs from agriculture in the range of 5120 Mt CO_2-e to 6116 Mt CO_2-e; CH_4 contributed 3300 Mt CO_2-e and N_2O 2800 Mt CO_2-e, representing about 50% and 60% respectively of global anthropogenic emissions of these gases, equivalent to around 10–12% total global anthropogenic emissions of all greenhouse gasses (including CO_2). Agricultural emissions of non-CO_2 GHGs are dominated by N_2O released from soils (38%), and CH_4 produced by enteric fermentation in ruminants (32%). The remainder is made up of biomass burning (12%), rice management (11%), and manure management (7%).

Australian agricultural emissions

Australian emissions are derived from the National GHG Inventory Report for 2005 (AGO 2007). In 2005 agriculture accounted for 87.9 Mt CO_2-e, 16.8% of Australia's total GHG emissions. CH_4 is by far the dominant agricultural GHG and represents 76% of total agricultural GHG emissions, or about 13% of total Australian emissions. The dominant sources of CH_4 emissions from agriculture are (1) enteric fermentation from ruminants, 58.7 Mt CO_2-e (87% of agricultural CH_4 emissions) and, (2) savanna burning, 6.1 Mt CO_2-e, (9% of agricultural CH_4 emissions). N_2O accounts for 24% of total agricultural GHG emissions, or about 4% of total Australian emissions. There are three major N_2O emissions categories: (1) agricultural soils, 16.6 Mt CO_2-e (80% of agricultural N_2O emissions); (2) savanna burning, 2.6 Mt CO_2-e (13% of agricultural N_2O emissions); (3) manure management, 1.5 Mt CO_2-e (7% of agricultural N_2O emissions). The dominant emission of N_2O from agricultural soils is derived from a diverse range of cropping and production systems subject to varying management intensity.

Australian forestry and GHGs

Unlike agriculture, production forestry activities represent a significant net sink for GHGs and have played an important role in offsetting the equivalent of around 9% of total Australian GHG emissions (DAFF BRS 2008). Under UNFCCC guidelines, which are followed by Australia's national GHG inventory for reporting purposes, forestry is not reported as a separate activity but is included in the Land Use, Land Use Change and Forestry sector (LULUCF). The 'Forest Lands' component of this sector comprises emissions and removals from managed native forests and plantations, emissions from fuel wood consumption and non-CO_2 GHG emissions from prescribed burning and wildfire in forests (AGO 2007). In 2005 'Forest Lands' constituted a net sink of 51.5 Mt CO_2-e with a further net sink of

5.0 Mt CO_2-e in harvested wood products (accounted as 'other' under the LULUCF sector). Production forestry activities therefore constituted a net sink of 56.5 Mt CO_2-e. This almost wholly balances the increase in emissions (1990–2010) from the Industrial + Fugitive energy sectors (AGO 2007).

Options for reducing agricultural GHG emissions

Options for mitigating agricultural GHG emissions can be considered under three main categories (Smith *et al.* 2007). (1) Emissions reduction, consisting of technologies, practices or a combination of both that directly reduce GHG emissions at source. (2) Enhancing GHG sinks: this largely relates to increasing the capture and storage of photosynthetically derived C. The two main options are (a) increased C storage in (woody) plant biomass, such as tree planting or management to allow increased coverage of native woody vegetation, and (b) changes in management or land use that lead to increased soil C content. (3) Fossil fuel substitution: this consists of two options: (a) the direct use of crops and residues for electricity generation or liquid fuel production, and (b) the substitution of agricultural or forestry products for other products from more greenhouse-intensive production processes resulting in an indirect reduction in fossil fuel emissions (e.g. using wood instead of steel for house construction). The net benefit is a function of the amount of fossil fuel-derived emissions replaced less the emissions resulting from growing, transporting and converting crops or residues into fuel or products.

In this section we summarise the principal options available for direct reduction of CH_4 and N_2O emissions, we then consider the potential for enhancing agricultural C sinks in the following section. While we will deal with the major GHGs individually, it is crucial to remember that many mitigation practices will simultaneously influence more than one GHG and not always in the same direction. Thus, in assessing the net impact of mitigation strategies, it is vital that all GHGs are considered and that these are aggregated up to an appropriate scale. In most cases, this is probably at the enterprise/farm level, but in others it may require consideration on a larger scale.

CH_4 emission reduction technologies and practices

Enteric CH_4 is Australia's largest agricultural GHG source accounting for around 67% of total agricultural GHG emissions. CH_4 is a natural by-product of the ruminant digestive process and is carried out by a group of rumen microbes called methanogens that utilise surplus hydrogen, produced during microbial metabolism, to reduce CO_2 to CH_4 which is either breathed or eructed by the animal (McAllister and Newbold 2008). This process is essential to ruminant digestion as even small concentrations of hydrogen inhibit rumen function. Thus, simply inhibiting methanogen activity is only a partial solution, as an alternate mechanism is then required to prevent hydrogen build up. CH_4 represents a significant loss of dietary energy so that reducing its production would not only benefit the environment but could also increase feed efficiency. The extent of loss varies, but for grazing livestock that typically digest 60–80% (50–70% for Australian grazing conditions) of their diet, enteric CH_4 accounts for at least 8% of digested energy (Lassey 2008).

Estimates of global and national enteric CH_4 emissions are based on a 'bottom-up' methodology that combines estimates of emissions from an average individual animal in a given category (e.g. dairy cows or beef cattle) with estimates of animal numbers in that category (Lassey 2008). Per-animal emissions are derived by a procedure that (1) estimates the energy required for maintenance of animal condition and production of product, e.g. milk or beef, and converts this to feed intake based on the metabolic conversion efficiency for a given animal/diet combination. (2) It then relates CH_4 production empirically to feed intake based on available experimental information (Lassey 2008). Global estimates of CH_4 yield from this approach have an uncertainty of ±15% which, combined with a similar uncertainty around livestock numbers, gives an uncertainty in global CH_4 emissions of ±21% (Lassey 2008). This, coupled with the absence of accurate cost-effective methods for directly estimating CH_4 emissions from individual animals or herds, presents a challenge for emissions trading and quantifying the impact of mitigation options that may be applied at the farm scale.

CH_4 abatement options can be classified as: (1) improved management and feeding practices; (2) dietary additives; (3) manipulation of rumen microbial populations; (4) inhibition of methanogens, e.g. by bacteriophages or vaccines, and (5) livestock breeding for reduced CH_4 emissions. CH_4 emissions from grazing ruminants are best expressed per unit of dry matter intake or unit product (meat, milk) as certain dietary additives or management options influence dry matter intake, animal productivity, or both.

Improved feeding practices can achieve emission reductions, either directly through reduced emissions per unit of feed intake, or indirectly by reducing emissions per unit of product (meat, milk or wool). The direct effect is related to the general relationship whereby CH_4 emissions per unit of the diet consumed decrease as diet quality (digestibility) increases (Beauchemin et al. 2008). For example, replacing roughage with concentrate reduces the proportion of dietary energy converted to CH_4 and improves animal performance. This occurs mainly through starch rather than fibre being the substrate for microbial fermentation, and through a reduction in rumen pH. CH_4 production per animal may actually be increased, but emissions per unit feed intake and per unit product are invariably decreased (Beauchemin et al. 2008; Smith et al. 2007). Replacing grass silage with maize or cereal silage may give similar results due to the latter's higher starch content. Depending on the production system for concentrate, the amount fed, and other husbandry practices, the use of concentrate can reduce whole-farm emissions and increase profitability (Beauchemin et al. 2008). Improved pasture quality as a consequence of better management or the introduction of superior pasture species and cultivars has the potential to reduce emissions in much the same manner as concentrates. However the scale of reduction is less certain for systems that are already well managed (Beauchemin et al. 2008) but may have greater impact in rangelands and tropical savannas (Charmley et al. 2008).

Indirectly, improved feeding practices can reduce CH_4 emissions per unit of product, simply by making the production system more efficient. These aspects improve animal productivity per unit of CH_4 emissions by reducing the maintenance overhead associated with unproductive or low production livestock (Beauchemin et al. 2008; Charmley et al. 2008; Eckard et al. 2000; Smith et al. 2007) resulting in, for example, earlier finishing of the slaughter generation (i.e. younger age at slaughter), or higher productivity per animal (e.g. milk production). In fact any management practice that improves stock condition and health has the potential to reduce emissions per unit of product through earlier finishing and a reduction in unproductive animal numbers (Howden and Reyenga 1999). However, if increases in production efficiency are offset by enhanced stocking rates, total CH_4 emissions may remain high or even increase.

There is a range of dietary additives that can reduce CH_4 emissions mainly through suppressing methanogenesis. These have been reviewed by Beauchemin *et al.* (2008) and Smith *et al.* (2007). Among the best known are ionophores, such as monensin, that act as antibiotics and have been fed to milk and dairy cattle to improve efficiency of milk and beef production. However, CH_4 reduction requires high doses and the effects are transient. This, plus the fact that monensin use is banned in Europe, means it is unlikely to be a viable means of reducing CH_4 emissions in Australia. Diet supplementation with lipids (dietary oils) can reduce enteric CH_4 emissions by over 40% per unit of dry matter intake, but a range of 10–25% is more likely operationally. A review by Beauchemin *et al.* (2008) indicates a 6% reduction in CH_4 emissions for every 1% increase in lipids up to a maximum of 6–7% of diet, beyond which further increases lead to a depression in dry matter intake which negates the benefit of further increasing the energy density of the diet. Other additives include (1) halogenated compounds that inhibit methanogenic bacteria, but again their effect is transient and there are side effects. (2) Plant secondary compounds such as condensed tannins, saponins or essential oils can be fed as the extracted compound, supplementing with the whole plant containing these compounds (e.g. forage legumes), or by having the plant present in the grazing system. At high concentrations these compounds can reduce digestibility of the diet. (3) Yeast products have been added to diets with improved dry matter intake and animal performance. Low cost makes these attractive as a CH_4 mitigating agent but this effect is largely untested. (4) Propionate precursors such as malate and fumarate can replace CO_2 as hydrogen acceptors, but high doses are required making this option expensive. (5) Enzymes such as cellulase and hemicellulase have been shown to substantially increase

digestibility and reduce CH_4 production and may be viable for intensive livestock. Many of the additives described above are already in use by the livestock industry (e.g. ionophores and dietary lipids), and therefore unlikely to be included as a mitigation technique that can be accounted for under a trading scheme. Many of the novel additives are poorly tested and practical means of adoption, particularly under extensive systems, will be a major hurdle.

New molecular techniques are increasing our understanding of the genetic basis for hydrogen transfer in rumen (Attwood and McSweeney 2008). This understanding will allow for targeted development of anti-methanogenic interventions and elucidation of alternate hydrogen transfer pathways that do not terminate with CH_4. Currently, vaccines against methanogenic bacteria are being developed but their efficacy under operational conditions remains to be proven and none are commercially available. Trial vaccines developed to date have proved unsuccessful under farm conditions (Leslie *et al.* 2008). Future vaccine development may be assisted by recent advances in methanogenomics with the aim of identifying a common gene or genes that occur solely in methanogens. However, if methanogens are controlled an alternate sink for hydrogen needs to be established to maintain rumen function. Options include encouraging microbes that carry out reductive acetogenesis (CO_2 + hydrogen to acetate) or promotion of organisms that consume reducing equivalents during conversion of metabolic intermediates (Attwood and McSweeney 2008).

Within a population there is strong evidence that individuals on the same diet show differences in CH_4 emissions. This offers the opportunity to lower emissions via selective conventional or marker-assisted breeding. A significant impediment to this approach has been the difficulty of cost-effectively screening large numbers of animals for CH_4 emissions as input into a breeding program. Recently in New Zealand, quantitative trait loci have been tentatively identified that could allow rapid screening (Leslie *et al.* 2008).

Considerable uncertainty exists around the scale of reduction in CH_4 emissions that might be achieved through the mitigation options described above. There is also the issue of what may be technically feasible (i.e. the upper

Table 14.1: Technical reduction (%) potential for enteric CH_4 emissions in Oceania (Smith *et al.* 2007).

Livestock type	Improved feeding practices	Specific agents and dietary additives	Long-term structural/ management change and breeding
Dairy cows	22	14	6
Beef cattle	8	8	4
Sheep	5	3	0.4

potential) as opposed to economically feasible. The latter will depend on the nature of the farm enterprise and the C price that results from emissions trading. Based on a summary of available literature, Smith *et al.* (2007) estimated that technical reduction potentials for livestock CH_4 emissions are in the range 0.4–22% (Table 14.1).

Clearly the majority of dietary and feed additive options are best suited to intensive livestock production but are unlikely to be operationally feasible or cost effective in the extensive livestock systems that characterise much of northern Australia unless novel delivery mechanisms, such as in drinking water or introduction of CH_4-reducing forage species into pastures, prove feasible. In the short term, an approach based on livestock and pasture management is probably most practical for these systems, including offsetting of emissions through increasing C stocks (Howden and Reyenga 1999). A progressive shift to feed lotting for finishing might also lower overall system CH_4 emissions (Beauchemin *et al.* 2008), but a complete life cycle assessment of GHG emissions from this approach would be required (Howden and Reyenga 1999). Solutions such as vaccines and breeding are long-term goals with significant hurdles to overcome.

Leslie *et al.* (2008) provide a succinct and realistic assessment of the current state of play in New Zealand for implementation of on-farm CH_4 mitigation, elements of which are applicable to agricultural GHG mitigation more generally. 'Finally, we want to emphasise that reducing emissions at the business-unit level – the farm – will take time. Solutions developed at the research level have to be proved at the farm scale, appropriate delivery methods have to be developed, putative products have to be commercialised and adopted by farmers, and the resulting emission reduction incorporated into national emissions inventories.'

N_2O emission reduction technologies and practices

Agricultural N_2O represents only 4% of total Australian GHG emissions and in comparison to CH_4 is significantly less important nationally. However, N_2O emissions assume much greater significance for certain production systems and, under emissions trading, could represent a considerable liability for some enterprises. Significant uncertainties are associated with estimates of N_2O emissions; at least -30 to +50% for agricultural soils, -50 to +115% for savanna burning, and ≈10% for manure management (AGO 2007). These uncertainties reflect the difficulty of measuring emissions that vary enormously over small distances (<1 metre) and time scales (hours), so that deriving robust estimates over large areas through time is challenging (Dalal *et al.* 2003; Harvey *et al.* 2008). To simplify these difficulties emissions are estimated using best-bet information and assumptions (AGO 2007).

In many instances, the actions required to reduce N_2O emissions are identical to best practice strategies to maximise the efficiency of N use and minimise undesirable environmental impacts such as nitrate contamination of waterways (Dalal *et al.* 2003; Galbally *et al.* 2005). Consideration of mitigation options for N_2O requires an appreciation of the key processes responsible for N_2O production. Emissions from agricultural soils depend primarily on the microbially mediated processes of nitrification (oxidation of ammonium to nitrate) and denitrification (reduction of nitrate through a sequence of gaseous N forms, one of which is N_2O, to N gas) (de Klein and Eckard 2008). Nitrification is favoured by a high availability of ammonium (e.g. excessive fertiliser use) and well-oxygenated conditions. Denitrification is favoured by high nitrate availability and low oxygen conditions (most commonly waterlogging). In the case of burning, the amount of N_2O produced depends on the nature of the residue burnt and fire intensity, both of which are influenced by season (Russell-Smith *et al.* 2004).

There are three main options for reducing N_2O emissions from soil. (1) The availability of mineral N relative to crop requirements can be managed to minimise nitrification and accumulation of nitrate (Smith *et al.* 2007). This can be achieved in a number of ways the applicability of which will vary with production system. For example by (a) controlling N inputs from fertiliser (i.e. avoid-

Table 14.2: Annual N_2O mitigation potentials for major agricultural activities and practices (Smith *et al.* 2007).

Activity	Practice	Mean N_2O reduction[1] $(t\ CO_2–e\ ha^{-1}\ yr^{-1})$
Croplands	Agronomy	0.10
Croplands	Nutrient management	0.07
Croplands	Tillage & residue management	0.02
Croplands	Water management	0
Croplands	Land use change	2.3
Croplands	Agro-forestry	0.02
Grasslands	Grazing, fertilisation, fire	0
Manure/ biosolids	Application	0

[1]The reduction in N_2O expressed as tonnes of CO_2 equivalent per hectare land area per year.

ing excess fertiliser use), residue (including incorporation through tillage), and biological N fixation; (b) reducing N inputs in urine under intensive grazing; directly, by balancing the carbohydrate: protein ratio in feed, or indirectly, by strategies that reduce the concentration of protein in grazed pasture (de Klein and Eckard 2008), and (c) applying nitrification inhibitors (e.g. dicyandiamide) to reduce the conversion of ammonium to nitrate. Recent New Zealand research has shown that dicyandiamide may be effective in controlling emissions from livestock urine and simultaneously improve pasture production (de Klein and Eckard 2008; Smith *et al.* 2008), but a similar study in Victoria, Australia was less conclusive (Kelly *et al.* 2008). A problem with nitrification inhibitors is that their effect is transitory. In the future, it may prove feasible to utilise plant-based production of nitrification inhibitors within a sward via new species, gene manipulation, or both (Fillery, pers.comm.). (2) Soil water can be managed by improving drainage or better scheduling of irrigation to minimise water logging and the potential for denitrification. A related strategy would be avoiding applying N fertiliser to wet or waterlogged soils.

Achieving significant reductions in N_2O emissions is challenging and there is no single, high-impact strategy. This reflects the complexity of factors influencing emissions from a given land use system, the wide range of contributing systems,

and the limited opportunities for cost-effective management intervention in some of these systems. Considerable uncertainty exists around the scale of reduction in N_2O emissions that might be achieved through the mitigation options described above. A major driver is likely to be the sharp rise in global oil and gas prices that is substantially increasing the cost of N fertiliser and thus the need to improve on-farm N use efficiency. Based on a summary of available literature Smith *et al.* (2007) estimate that technical reduction potentials for N_2O emissions are in the range 0–2.3 t CO_2-e/ha/yr (Table 14.2).

Enhancing agricultural carbon sinks

Increasing C storage in agricultural landscapes depends on the capture of atmospheric CO_2 by plants and subsequent storage of that C in plants, soil, and products. Carbon stored in plants may be retained as a permanent sink or managed in a sustainable cycle of harvest and replanting with ongoing mitigation achieved through (1) storage of C in long-lived biological products (e.g. wood) and (2) substitution of hydrocarbon-based or GHG-intensive products with natural materials (e.g. fossil fuel replacement, plastics, bulk chemicals, concrete or steel) (Smith *et al.* 2007). Any carbon sequestered will only be retained if the enhanced stores of C in plants, soils and products are maintained in perpetuity.

Strategies to increase C storage in agricultural landscapes do not address the primary cause of increased atmospheric GHG concentrations. Nonetheless, they are useful in terms of 'buying time' for the development and deployment of low emissions technologies (e.g. Lal 2004). They could be deployed relatively rapidly, do not require major technological breakthroughs, and in some cases have the potential to generate a range of environmental, economic, and social benefits beyond those relating to mitigation (Dalal and Chan 2001; Lal 2004; Smith *et al.* 2007). Such options are receiving widespread public and political exposure.

Carbon sequestration in forests

The major biologically-based management interventions that can increase C storage, and therefore decrease net C emissions are: (1) the planting of new forests and woody perennials including trees

Table 14.3: The potential for different types of agriculture to store additional carbon after changed practices under broad climatic conditions (from Table 8.4, Smith *et al.* 2007).

Activity	Practice	Mean C storage potential (t CO_2-e/ha/yr)			
		Cool-dry	**Cool-moist**	**Warm-dry**	**Warm-moist**
Croplands	Agronomy	0.29	0.88	0.29	0.88
Croplands	Nutrient management	0.26	0.55	0.26	0.55
Croplands	Tillage and residue management	0.15	0.51	0.33	0.70
Croplands	Water management	1.14	1.14	1.14	1.14
Croplands	Set-aside and LUC	1.61	3.04	1.61	3.04
Croplands	Agro-forestry[1]	0.15	0.51	0.33	0.70
Grasslands	Grazing, fertilisation, fire	0.11	0.81	0.11	0.81
Degraded lands	Restoration	3.45	3.45	3.45	3.45
Manure/biosolids	Application	1.54	2.79	1.54	2.79

[1]Trees planted as narrow belts or scattered groups. Much higher rates occur with broadscale tree plantings.

on farms, environmental plantings, biodiversity corridors joining native forests, and salinity management plantings. These could be managed as permanent C sinks or used as feedstock for biofuels, electricity generation, or long-lived products. (2) Changes in agricultural management practices that increase soil C storage, such as reducing the rate of loss of soil carbon and/or enhancing the inputs of carbon in the form of plant residues. (3) Protection and management of existing rangeland and forest (fire and grazing) to increase native tree cover, 'C density', and soil C. Estimates of the C storage potentials for major agricultural activities and practices have been compiled by Smith *et al.* (2007) and are summarised in Table 14.3. These rates are much lower than those achievable with broadscale tree planting (e.g. the example from Polglase *et al.* 2008 described below).

Globally forests account for the majority of C stored in live vegetation, equivalent to the amount of C in the atmosphere, or 2.7 times atmospheric C if forest soils are included (Dixon *et al.* 1994; Falkowski *et al.* 2000). They are the dominant terrestrial sink for anthropogenic emissions, sequestering ≈2.23 Gigatonnes (Gt) C per year, or 23% of anthropogenic emissions (Canadell and Raupach 2008). This sequestrated carbon is mostly in new plantations and a result of the 'CO_2-fertilisation effect'. Forests are comparable to the oceans as a C sink (Dixon *et al.* 1994; Falkowski *et al.* 2000). The Australian forest estate contained 6.56 Gt C in 2004 in its live components (above-ground biomass and roots), and managed forests sequestered a net 43.5 Mt CO_2-e in 2005 (DAFF BRS 2008). Plantations offset about 3.5% and managed

native forests about 5.5% of total GHG emissions in 2005 (DAFF BRS 2008).

Tree planting can potentially sequester large quantities of C (Canadell and Raupach 2008) and, if properly integrated into the landscape, achieve biodiversity and environmental benefits, with minimal impact on water resources. Biosequestration in managed forests is relatively straightforward in the policy sense since tree establishment and growth can be readily estimated, monitored and audited (Garnaut 2008a). This is better understood for plantations, but effort is now being directed to improving Australia's capacity to report for heterogeneous native forests, a far more challenging prospect. Plantation biosequestration activities are generally flagged for early inclusion in national emission trading schemes (Garnaut 2008a). The potential for tree planting to offset GHG emissions is set by the area of suitable land, while the reality will be determined by economics and the need to balance against land use for other purposes such as food production and water yield. The most comprehensive global estimate to date (Nabuurs *et al.* 2007) suggests that by 2030 afforestation could sequester 1618 Mt CO_2-e per year at a C price of ≤US$20 per tonne CO_2, increasing to 4015 at a C price of US$100 per tonne CO_2. A number of sources suggest that large-scale tree planting in Australia could offset greenhouse gas emissions by 20–172 Mt CO_2-e per year (McKinsey and Company, 2008; Chief Scientist of Australia, 2007), equivalent to around 4%–33% of Australia's total 2005 GHG emissions. The wide range reflects the differences in assumptions made concerning land suitability and economic drivers.

Achieving significant carbon benefits through large-scale tree plantings that deliver associated biodiversity and economic benefits, have minimal impact on water resources, and are well-integrated with other land uses, will require science-based scenario analysis and planning tools. Currently the first generation of such tools are allowing 'prospecting' of potentially suitable areas based on a range of economic and environmental criteria. They also give a first approximation of the amounts of carbon that could potentially be sequestered under defined scenarios that objectively consider multiple criteria. A preliminary country-wide modelling exercise has been undertaken (Polglase *et al.* 2008) to identify areas where environmental tree planting can simultaneously deliver (1) a profitability of $150 ha/yr above the current land use, and (2) reductions in water yield of less than 150 mm per year. The exercise was restricted to privately owned, cleared, agricultural land (excluding public land, national parks, or extant tree plantings) and assumed a carbon value of $20 per tonne of CO_2 equivalent. Some 9 Mha of land meet the criteria under this scenario. Modelling estimates that tree planting over this area would sequester an average 143.4 Mt CO_2-e per year over at least the first 20 years, annually equivalent to $\approx$25% of Australia's total 2005 greenhouse gas emissions. Carbon accumulation rates in these trees would eventually plateau, meaning they would not represent a significant ongoing net sink for carbon after attaining maximum biomass. However, if the trees were harvested and replanted or coppiced and the harvested material (e.g. wood) was converted into a permanent C storage product (e.g. charcoal), a perpetual C sink could result. Additionally, if even a portion of the harvested biomass were used for bioenergy or biofuel, it would deliver a net C benefit equivalent to the amount of fossil fuel-derived emissions replaced less the emissions resulting from the growing, transporting and converting of harvested material. The sequestration rates per unit of land area in the preceding example equate to around 16 t CO_2-e/ha/yr, which is substantially higher than estimates for the potential C storage that might be achieved by a range of agricultural options (Table 14.3).

A major consideration for all biosequestration, including trees, is the potential risk to sequestered C. Fire and drought are key risks, while in much of tropical Australia cyclone damage would be important. Currently fire has relatively minimal impact on Australian planted forests, partly reflecting the geography of planting, and intensive management over a relatively small (1.82 Mha) area. The situation for much of Australia's 147 Mha of native forest is very different, e.g. bushfires in 2003 burnt $\approx$4.5 Mha of native forest (DEWHA 2006), and similar areas were burnt in 2006 and 2007. Rough calculations would indicate that collectively these fires may have released an amount of CO_2 equivalent to that sequestered by 6 Mha of plantations (about three times the current plantation estate). It is probable that the fire risk faced by very large areas of planted forest managed at low intensity would be higher than currently experienced by most Australian plantations.

Climate change itself poses significant uncertainty and risk to tree-based biosequestration (see Booth *et al.* Chapter 9 this volume). (1) All scenarios of potential tree growth and C sequestration in Australia to date are based on past/current climate information. They need to be repeated using future climate scenarios. (2) Risks relating to fire, drought, cyclones, and pests are likely to increase under climate change. (3) Atmospheric CO_2 concentrations will continue to rise. The impact of this on growth of Australian tree species, and therefore potential C sequestration, is poorly known.

While this preliminary assessment suggests tree-based C sequestration could be potentially very significant in terms of reducing Australia's net GHG emissions, it is important to note that the full system implications of such large-scale land use changes need much deeper investigation. Key impacts that need to be further investigated include food production and security, water flows in landscapes and river basins, environmental conservation and biodiversity, and regional communities and economies.

Carbon sequestration in agricultural soils

The global soil C pool of 2500 Gt (1550 Gt of organic C and 950 Gt of inorganic C) is 3.3 times the size of the atmospheric pool (760 Gt) and 4.5 times the size of the biotic pool (560 Gt) (Falkowski *et al.* 2000). The soil organic C pool to 1 m depth ranges from 30 t/ha in arid climates to 800 t/ha in organic soils in cold regions, but the majority of soils lie in the range 50 to 150 t/ha. Australian soils are estimated to contain 34 Gt of C to 1 m depth (Grace *et al.* 2006), spanning the range 18–447 t C/ha.

In the absence of significant erosion, soil C content is the net result of a dynamic equilibrium of gains (primarily plant residues) and losses (primarily microbial decomposition) and is sensitive to anything that influences these two variables (e.g. Lal 2004; Hutchinson *et al.* 2007). Gains can be increased by any practice that increases the input of plant residues, such as residue retention. Losses can be minimised by avoiding practices that accelerate decomposition, such as cultivation. Climate and rising atmospheric CO_2 influence both variables. For non-irrigated soils in Australia, water availability defines an upper limit for plant productivity and residue inputs and therefore soil C content. Soil C content is also influenced by clay content (generally increases with), soil depth, and bulk density (Dalal and Chang 2001; Lal 2004). It may not be biologically possible, nor certainly economically feasible, to increase C storage in soils that already have a high C content (Stewart *et al.* 2007), although this is unlikely to be an issue for the vast majority of Australian agricultural soils.

Conversion of natural ecosystems to agriculture is generally associated with loss of soil C and depletion of the soil organic C pool; globally by as much as 60% in topsoils of temperate regions and 75% or more in cultivated topsoils of the tropics (Lal 2004; Hutchinson *et al.* 2007). In Australia, reductions in soil C content on conversion to agriculture and subsequent long-term cropping may be in the range of 18 to 72% over a 20–70 year period (Dalal and Chang 2001). The technical potential to increase the C content of the world's agricultural and degraded soils has been estimated at 50 to 66% of their historic C loss (Lal 2004). The increase in soil C content that can be achieved economically is likely to be considerably lower (Bangsund and Leistritz 2008) and will depend on soil texture and structure, rainfall, temperature, farming system, and soil management.

Strategies to increase the soil C pool include no-till farming, increased residue retention, cover crops, ley-pasture rotation, cereal-legume and other crop rotation, increasing plant productivity (e.g. nutrient management, water conservation and harvesting, efficient irrigation), improved grazing practices, manuring and sludge application, agro-forestry, and planting or regeneration of perennial (predominantly woody) vegetation (Dalal and Chang 2001; Hutchinson *et al.* 2007). Assumptions around some of these options require more complete testing and evaluation. For example, Six *et al.* (2004) compiled all available data on soil-derived GHG emission comparisons between conventional tilled and no-tillage systems for humid and dry temperate climates and concluded that (1) the effects on GHG mitigation by adopting no-tillage practices were much more variable and complex than had been previously realised, and (2) policy plans to reduce global warming through this land management practice needed further scrutiny to ensure success. Since zero tillage/stubble retention is already practised in much of Australian agriculture, other options for increasing C inputs would need to be considered.

There is no doubt that a significant proportion of Australian agricultural soils could potentially store more C. The key issues are how much, how rapidly, and the extent to which an increase can be achieved while maintaining a viable farm enterprise. A key constraint to increasing soil C content under current cropping systems is how to achieve the large permanent increase required in organic matter inputs. Large and sustained increases in the C content of Australian agricultural soils will require similar large and sustained increases in C inputs, in addition to current inputs. The significance of this apparently simple statement is best appreciated by considering an example.

- A 10 cm deep topsoil with 1% C contains 12 t C/ha (assumes a bulk density of 1.2 g/cm^3). Doubling this to 2% C (24 t C/ha) requires an additional 12 t C/ha to be added to the soil.

- Since plant residues contain ≈ 45% C (Skjemstad *et al.* 2004) this would equate roughly to 26 t/ha of dry matter (above and below ground). But, since at least 50% of the added plant residues will decompose within one year (Amato *et al.* 1987; Cogle *et al.* 1987), this actually requires 52 t/ha dry matter.

- Thus, achieving an increase in soil C from 1% to 2% in five years would require annual residue additions of about 10 t/ha additional to that currently being added. This increase is very large in comparison to the potential to increase residue inputs, e.g. the average above-ground shoot dry mass (i.e. potential residue) for most major Australian cereals is around 10 t/ha or less. However, at locations where water use efficiency (expressed in terms of kg CO_2-C captured per mm available water) is low, designing plant systems with a higher water use efficiency may provide options for enhancing soil carbon content.

- Soil carbon is typically associated with other nutrients (e.g. N and P). The C/N and C/P ratios of soil organic matter remain constant at values of approximately 10 and 120, respectively, across agricultural soils irrespective of soil carbon content. The implication of this is that in order to build carbon, nutrients must also be supplied. The scenario depicted above (increasing soil C from 1% to 2% in five years at a bulk density of 1.2 g/cm^3) would require the addition of 1200 kg N/ha and 100 kg P/ha.

There are about 20 Mha of rain-fed cropping land in Australia (Dalal and Chang 2001). The C content of these soils is often below 1%, and in lighter textured soils of WA or the Mallee area of the eastern cropping belt closer to 0.5% (Dalal and Chang 2001; Grace *et al.* 2006). Prior to clearing for agriculture, these soils would have supported higher C contents. For example, most remnant 'virgin' soils in these cropping areas have C contents of ≈1.5% (AGO 2002; Dalal and Chang 2001). Increases in the C content of these soils could potentially be achieved through the range of practices described above. However, the increased C inputs required to drive an increase in soil C are large (see example above) and changes will only occur at a slow rate of increase, plateauing at a new equilibrium content. Importantly, maintenance of a higher equilibrium C content will require continued higher C inputs that could become economically unsustainable, and in many cases would be hard to achieve without major shifts in management and cropping systems. Estimates of the potential to increase the C content of these soils vary. Grace *et al.* (2007) modelled typical crop rotations and C inputs (residues) for these systems and estimated potential increases in soil C of 100–200 kg C/ha per year (or 0.01% increase in soil C). Over 20 Mha of cropping land this equates to 2-4 Mt C per year. Dalal and Chang (2001) estimated that if soil C losses from the Australian cereal belt could be fully reversed this would represent a sink of 50 Mt CO_2-e per year for

a period of 20 years: annually equivalent to 10% of the total 2005 Australian greenhouse gas emissions. However, the same study concludes that only 23% of this potential C sink could be realised through improved management practices under cereal cropping; 11.8 Mt CO_2-e per year, annually equivalent to 2.3% of total 2005 Australian greenhouse gas emissions. The remaining 77% is only achievable by increased cropping intensity, higher plant and animal biomass inputs, and conversion to perennial crops, agroforestry, and forestry crops. Higher rates of sequestration are possible but these are likely to be for high clay content soils in high rainfall areas under perennial pasture, i.e. unrepresentative of the vast majority of Australian cropping land.

The major risk to such sequestered soil C would be a change in management practices that either reduced C inputs or increased rates of decomposition. An increase in soil C will only perpetuate as long as the management practices that generated it remain in place. Climate change itself poses a potential risk to soil C (Smith *et al.* 2007) through its probable impact on plant growth (inputs) and decomposition (losses). There have been few attempts to predict the impact of likely Australian climate change scenarios on soil C stocks. More pessimistic scenarios predict a net reduction in the continental store of soil C under climate change (Grace *et al.* 2006).

A significant challenge for soil C storage as a mitigation option relates to the issues of monitoring and verification. These have been succinctly summarised by the Soil Science Society of America's position statement on soil C sequestration (Soil Science Society of America 2001): 'Improvement in monitoring and verification protocols for C sequestration in soil-plant ecosystems is needed for quantitative economic and policy analyses. These protocols need to be quantitatively defensible and readily applicable to fields and watersheds with differing land uses and weather conditions. Such protocols must be acceptable, both domestically and internationally, to scientists, policymakers, landowners, and business groups. These protocols must be suitable for use by employees of government agencies and licensed professionals'.

GHG mitigation potential in rangelands

Rangelands cover around 70% of the Australian continent and encompass a range of climatic zones and vegetation types (Ash *et al.* 1995; Fisher *et al.* 2004) of which the savannas of northern Australia are particularly significant in terms of the magnitude and frequency of fire and associated GHG emissions. Some 300 000 km^2 of savanna can burn annually, predominantly late in the March to December dry season, and largely as a result of human-induced fire as opposed to natural causes such as lightning (Russell-Smith *et al.* 2004; Russell-Smith *et al.* 2007). Such fires release very large quantities of GHGs. The combined annual release of CH_4 and N_2O from savanna burning is estimated to be 8.7 Mt CO_2-e, 9.9% and 1.7% respectively of Australia's 2005 agricultural and total GHG emissions (AGO 2007).

Significant GHG abatement can be achieved through fire management by reintroducing the indigenous community practice of early season prescribed burning to limit the extent and spread of late-season wildfires that lead to high emission rates of GHGs. For each hectare, emissions from fires early in the dry season can be up to 50% less than late season fires due to lower intensity and greater fire patchiness (Russell-Smith *et al.* 2004). Since such abatement effectively reduces emissions at source, it is ongoing, and not subject to the issue of permanence and risk associated with C storage in trees and soil.

The role of, and potential for, savannas in C sequestration is less certain. The principle management interventions by which C storage could be manipulated are fire and grazing. These would act principally through influencing the C stored in trees, grasses, and soil (Henry *et al.* 2002; Hill *et al.* 2003; Chen *et al.* 2005; Cook *et al.* 2005; Beringer *et al.* 2007). Fire and its season of occurrence can significantly influence vegetation and soil C in savannas. For example, Ansley *et al.* (2006) found that soil C at 0–20 cm depth increased from 2044 g C/m^2 in controls to 2393–2534 g C/m^2 in the three treatments that included summer fire. In contrast, winter fires had no effect on soil C. Williams *et al.* (2004) have estimated that reductions in fire frequency from a probability of 0.5 per year to a probability of 0.25 per year, could increase net C sequestration as measured by net biome productivity by about 0.25 t C/ha per year. Ash *et al.* (1995) have suggested that improved grazing management practices that increase the perennial grass component of rangeland in northern Australia could sequester 320 Mt C in the top 10 cm of soil. Averaged over 30 years, this equates to 39 Mt CO_2-e per year which is around 7.5% of total Australian 2005 GHG emissions. Both fire and grazing influence the biomass and growth of woody perennials in rangeland and savanna systems (Gifford and Howden 2001; Henry *et al.* 2002; Hill *et al.* 2003). Both can be manipulated to increase above ground and soil C storage, but, in systems where livestock production is the main economic activity, the trade-offs with pasture productivity and livestock carrying capacity need to be quantified (Howden *et al.* 2001).

Climate change has the potential to impact very significantly on GHG emissions from, and C storage in, savanna. There is a need to assess this impact for the most probable climate change scenarios.

Discussion

Non-CO_2 GHG emissions from Australian agriculture account for 87.9 Mt CO_2-e, 16.8%, of total national GHG emissions (which include emissions of all gases from all sectors). In terms of direct mitigation, livestock CH_4 offers the greatest scope for reduction but also the biggest challenge in terms of a lack of options proven to operate at farm scale, particularly for more extensive rangeland grazing systems. In the short to medium term, a mixture of approaches tailored to specific sectors of the industry is likely to yield most benefit. This is likely to include dietary manipulation and additives for more intensive systems while mitigation approaches in extensive systems are likely to be limited to aspects of livestock and pasture management, including offsetting emissions through increasing C stocks. Such approaches are consistent with many of the best-practice objectives of sustainable grazing management. Solutions such as vaccines and breeding are long-term goals with significant hurdles to overcome.

Many of the actions required to reduce N_2O emissions are identical to best-practice strategies to maximise the efficiency of N use and minimise undesirable environmental impacts such as nitrate contamination of waterways. Significant reductions in N_2O can almost certainly be achieved in some intensively managed cropping and production systems (e.g. sugar cane and dairy) and it makes sense to concentrate efforts on these.

There is undoubtedly significant potential to offset agricultural and other sector GHG emissions through increasing the C stored in agricultural landscapes, sequestering C in long-lived products, or substituting agricultural products for fossil fuels. For example, the C sequestered by production forestry activities in Australia has almost wholly balanced the increase in emissions (1990–2010) from the Industrial + Fugitive energy sectors. Almost any land-based C storage option can appear very significant in the Australian context, even where storage increases per unit land of area are very small, simply because of the huge 'multiplier' effect of land area. While this should not discount the value of small changes in C stocks over big areas, it is certainly the case that these will be much harder to monitor and verify under any form of emissions trading. Of the options available, tree planting has the highest C sequestration rates per unit of land area. Climate change poses significant uncertainty and risk to vegetation and soil-based biosequestration. Virtually all scenarios of C sequestration in Australia to date are based on past/current climate information. They need to be repeated using future climate scenarios that consider changed risk profiles (fire, drought, cyclones, and pests) and elevated atmospheric CO_2 concentrations.

Mitigation strategies need to be based on whole-systems emission budgets and consider trade-offs with other factors and enterprise-scale economics. The distinctive need around this issue is for integration of scientific effort in key aspects of the soil, plant, animal system, forestry and agricultural products, and new work in the interaction between current and potential management systems. This will lead to 'best management practice' systems operating at an integrated landscape scale with C sequestration and GHG mitigation as an integrated outcome considered alongside other economic and environmental outcomes.

References

Amato M, Ladd JN, Ellington A, Ford G, Mahoney JE, Taylor AC, Walsgott D (1987). Decomposition of Plant Material in Australian Soils. IV Decomposition in situ of ^{14}C-and ^{15}N-Labelled Legume and Wheat Materials in a Range of Southern Australian Soils. *Australian Journal of Soil Research* **25**, 95–105.

Ansley RJ, Boutton TW, Skjemstad JO (2006). Soil organic carbon and black carbon storage and dynamics under different fire regimes in temperate mixed-grass savanna. *Global Biogeochemical Cycles* **20**, GB3006, doi:10.1029/2005GB002670.

Ash, AJ, Howden SM, McIvor JG (1995). Improved rangeland management and its implications for carbon sequestration. In *Rangelands in a sustainable biosphere. Proceedings of the Fifth International Rangeland Congress, Salt Lake City, Utah, USA, 23–28 July, 1995. Volume 1.*

Attwood G, McSweeney C (2008). Methanogen genomics to discover targets for methane mitigation technologies and options for alternative H2 utilisation in the rumen. *Australian Journal of Experimental Agriculture* **48**, 28–37.

Australian Greenhouse Office (2002). 'Pre-clearing soil carbon levels in Australia'. NCAS Report No 12.

Australian Greenhouse Office (2007). National Inventory Report 2005 – Volume 1. The Australian Government Submission to the UN Framework Convention on Climate Change April 2007.

Bangsund DA, Leistritz FL (2008). Review of literature on economics and policy of carbon sequestration in agricultural soils. *Management of Environmental Quality* **19**, 85–99.

Beauchemin KA, Kreuzer M, O'Mara F, McAllister TA (2008). Nutritional management for enteric methane abatement: a review. *Australian Journal of Experimental Agriculture* **48**, 21–27.

Beringer J, Hutley LB, Tapper NJ, Cernusak LA (2007). Savanna fires and their impact on net ecosystem productivity in North Australia. *Global Change Biology* **13**, 990–1004.

Canadell J, Raupach MR (2008). Managing forests for climate change mitigation. *Science* **320**, 1456–1457.

Charmley E, Stephens ML, Kennedy PM (2008). Predicting livestock productivity and methane emissions in northern Australia: development of a bio-economic modelling approach. *Australian Journal of Experimental Agriculture* **48**, 109–113.

Chen X, Hutley LB, Eamus D (2005). Soil organic carbon content at a range of north Australian tropical savannas with contrasting site histories. *Plant and Soil* **268**, 161–171.

Chief Scientist of Australia (2007) Fact Sheet. Carbon storage: biosequestration.

Cogle AL, Saffigna PJ, Strong WM, Ladd JN, Amato M (1987). Wheat straw decomposition in subtropical Australia. I. A Comparison of ^{14}C labelling and two weight-loss methods for measuring decomposition. *Australian Journal of Soil Research* **25**, 473–479.

Cook GD, Liedloff AC, Eager RW, Chen X, Williams RJ, O'Grady AP, Hutley LB (2005). The estimation of carbon budgets of frequently burnt tree stands in savannas of northern Australia, using allometric and isotopic discrimination. *Australian Journal of Botany* **53**, 621–630.

Dalal RC, Chan KY (2001). Soil organic matter in rainfed cropping systems of the Australian cereal belt. *Australian Journal of Soil Research* **39**, 435–464.

Dalal RC, Wang W, Robertson GP, Parton WJ (2003). Nitrous oxide emissions from Australian agricultural lands and mitigation options: a review. *Australian Journal of Soil Research* **41**, 165–195.

De Klein CAM, Eckard RJ (2008). Targeted technologies for nitrous oxide abatement form animal agriculture. *Australian Journal of Experimental Agriculture* **48**, 14–20.

Department of Agriculture, Fisheries and Forestry, Bureau of Rural Sciences (2008). 'Australia's state of the forests report, five-yearly report 2008'.

Department of the Environment, Water, Heritage and the Arts (2006). State of the Environment 2006.

Dixon RK, Brown S, Houghton RA, Solomon AM, Trexler MC, Wiesniewski J (1994). Carbon pools and flux of global forest ecosystems. *Science* **263**, 185–190.

Eckard RJ, Dalley D, Crawford M (2000). Impacts of potential management changes on greenhouse gas emissions and sequestration from dairy production systems in Australia. In *Management Options for Carbon Sequestration in Forest, Agricultural and Rangeland Ecosystems*. 25 May 2000. (Eds R Keenan, AL Bugg, H Ainslie) pp. 58–72. Workshop Proceedings CRC for Greenhouse Accounting.

Falkowski P, Scholes RJ, Boyle E, Canadell J, Canfield D, Elser J, Gruber N, Hibbard K, Hogberg P, Linder S, Mackenzie FT, Moore B, Pedersen Y, Rosenthal Y, Seitzinger S, Smetacek V, Steffen W (2000). The global carbon cycle: a test of our knowledge of Earth as a system. *Science* **290**, 291–296.

Fisher A, Hunt L, James C, Landsberg J, Phelps D, Smyth A, Watson I (2004). Review of total grazing pressure management issues and priorities for biodiversity conservation in rangelands: a resource to aid NRM planning. Desert Knowledge CRC and Tropical Savannas Management CRC: Alice Springs.

Galbally I, Meyer C, Bentley S, Weeks I, Leuning R, Kelly K, Phillips F, Barker-Reid F, Gates W, Baigent R, Eckard R, Grace P (2005). A study of environmental management drivers of non-CO2 greenhouse gas emissions in Australian agro-ecosystems. *Environmental Sciences* **2**, 133–142.

Garnaut R (2008a). Garnaut Climate Change Review. Issues Paper 1. Climate Change: Land Use – Agriculture and Forestry. February 2008.

Garnaut R (2008b). Garnaut Climate Change Review Draft Report June 2008.

Gifford RM, Howden M (2001). Vegetation thickening in an ecological perspective: significance to national greenhouse gas inventories. *Environmental Science & Policy* **4**, 59–72.

Grace P (2007). Carbon Farming – Facts and Fiction. Proceedings of the Healthy Soils *Symposium, July 2007*, Sunshine Coast, Queensland. pp. 92–99.

Grace PR, Post WM, Hennessy K (2006). The potential impact of climate change on Australia's soil organic carbon resources. *Carbon Balance and Management* **1**, 14.

Gullison RE, Frumhoff PC, Canadell JG, Field CB, Nepstad DC, Hayhoe K, Avissar R, Curran LM, Friedlingstein P, Jones CD, Nobre C (2007). Tropical forests and climate policy. *Science* **316**, 985–986.

Harvey M, Pattey E, Saggar S, Bromley T, Dow D (2008). Verification techniques for N_2O emission at the paddock scale in New Zealand: FarmGas2006. *Australian Journal of Experimental Agriculture* **48**, 138–141.

Henry BK, Danaher T, McKeon GM, Burrows WH (2002). A review of the potential role of greenhouse gas abatement in native vegetation management in Queensland's rangelands. *Rangelands Journal* **24**, 112–132.

Hill MJ, Braaten R, McKeon GM (2003). A scenario calculator for effects of grazing land management on carbon stocks in Australian rangelands. *Environmental Modelling & Software* **18**, 627–644.

Howden SM, Reyenga PJ (1999). Methane emissions from Australian livestock' implications of the Kyoto Protocol. *Australian Journal of Agricultural Research* **50**, 1285–1291.

Howden SM, Moore JL, McKeon GM, Carter JO (2001). Global change and the mulga woodlands of southwest Queensland: greenhouse gas emissions, impacts, and adaptation. *Environment International* **27**, 161–166.

Hutchinson JJ, Campbell CA, Desjardins RL (2007). Some perspectives on carbon sequestration in agriculture. *Agricultural and Forest Meteorology* **142**, 288–302.

Johnson M-F, Franzluebbers AJ, Lachnicht Weyers S, Reicosky DC (2007). Agricultural opportunities to mitigate greenhouse gas emissions. *Environmental Pollution* **150**, 107–124.

Kelly KB, Phillips FA, Baigent R (2008). Impact of dicyandiamide application on nitrous oxide emissions from urine patches in northern Victoria, Australia. *Australian Journal of Experimental Agriculture* **48**, 156–159.

Lal, R (2004). Soil carbon sequestration impacts on global climate change and food security. *Science* **304**, 1623–1627.

Lassey KR (2008). Livestock methane emission and its perspective in the global methane cycle. *Australian Journal of Experimental Agriculture* **48**, 114–118.

Leslie M, Aspin M, Clark H (2008). Greenhouse gas emissions from New Zealand agriculture: issues, perspectives and industry response. *Australian Journal of Experimental Agriculture* **48**, 1–5.

McAllister TA, Newbold CJ (2008). Redirecting rumen fermentation to reduce methanogenesis. *Australian Journal of Experimental Agriculture* **48**, 7–13.

McKinsey and Company (2008). An Australian Cost Curve for Greenhouse Gas Reduction. McKinsey Climate Change Initiative, McKinsey Pacific Rim, Inc. http://www.mckinsey.com/clientservice/ccsi/pdf/Australian_Cost_Curve_for_GHG_Reduction.pdf

Nabuurs GJ, Masera O, Andrasko K, Benitez-Ponce P, Boer R, Dutschke M, Elsiddig E, Ford-Robertson J, Frumhoff P, Karjalainen T, Krankina O, Kurz WA, Matsumoto M, Oyhantcabal W, Ravindranath NH, Sanz Sanchez MJ, Zhang X (2007). Forestry. In *Climate Change 2007: Mitigation. Contribution of Working Group III to the Fourth Assessment Report of the Intergovernmental Panel on Climate Change*. (Eds B Metz, OR Davidson, PR Bosch, R Dave, LA Meyer). pp. 541–584. Cambridge University Press, Cambridge.

Polglase P, Paul K, Hawkins, C, Siggins A, Turner J, Booth T, Crawford D, Jovanovic T, Hobbs T, Opie K, Almeida A, Carter J (2008). 'Regional opportunities for agroforestry systems in Australia (2008)'. Report to the Joint Venture Agroforestry Program. Rural Industries Research and Development Corporation, Canberra (in review).

Russell-Smith J, Edwards A, Cook GD, Brocklehurst P, Schatz J (2004). 'Improving greenhouse emission estimates associated with savanna burning in northern Australia: Phase 1. Final Report to the Australian Greenhouse Office'. Tropical Savannas CRC, Darwin.

Russell-Smith J, Yates CP, Whitehead PJ, Smith R, Craig R, Allan GE, Thackway R, Frakes I, Cridland S, Meyer MCP, Gill AM (2007). Bushfires 'down under': patterns and implications of contemporary Australian landscape burning. *International Journal of Wildland Fire* **16**, 361–377.

Six J, Ogle SM, Breidt FJ, Conant R, Mosier AR, Paustian K (2004). The potential to mitigate global warming with no-tillage management is only realized when practised in the long term. *Global Change Biology* **10**, 155–160.

Skjemstad JO, Spouncer LR, Cowie B, Swift RS (2004). Calibration of the Rothamsted organic carbon turnover model (RothC ver. 26.3), using measurable soil organic carbon pools. *Australian Journal of Soil Research* **42**, 79–88.

Smith LC, de Klein CAM, Monaghan RM, Catto WD (2008). The effectiveness of dicyandiamide in reducing nitrous oxide emissions from a cattle-grazed, winter forage crop in Southland, New Zealand. *Australian Journal of Experimental Agriculture* **48**, 160–164.

Smith P, Martino D, Cai Z, Gwary D, Janzen H , Kumar P, McCarl B, Ogle S, O'Mara F, Rice C, Scholes B, Sirotenko O, Howden M, McAllister T, Pan G, Romanenkov V, Schneider U, Towprayoon S, Wattenbach M (2007). Agriculture. In *Climate Change 2007: Mitigation. Contribution of Working Group III to the Fourth Assessment Report of the Intergovernmental Panel on Climate Change*. (Eds B Metz, OR Davidson, PR Bosch, R Dave, LA Meyer) pp. 497–540. Cambridge University Press, Cambridge.

Soil Science Society of America (2001). Position statement on soil carbon sequestration. https://www.soils.org/science-policy/testimony-statements-resolutions

Stewart CE, Paustian K, Conant RT, Plante AF, Six J (2007). Soil carbon saturation: concept, evidence and evaluation. *Biogeochemistry* **86**, 19–31.

Williams RJ, Hutley LB, Cook GD, Russell-Smith J, Edwards A, Chen X (2004). Assessing the carbon sequestration potential of mesic savannas in the Northern Territory, Australia: approaches, uncertainties and potential impacts of fire. *Functional Plant Biology* **31**, 415–422.

15

ENHANCING ADAPTIVE CAPACITY

NA Marshall, CJ Stokes, SM Howden and RN Nelson

KEY MESSAGES:

- Adaptation processes occur across all scales from individual to community, industry, regional, national and global. Understanding the adaptation process and supporting or enhancing it at various scales will be important for maintaining effective functioning of social and economic systems in the face of climate-driven changes.

- The effectiveness of adaptive responses will be influenced by (1) the operating context within which responses occur (e.g. the policy and governance setting); (2) the availability of effective adaptation options, and (3) the capacity of individuals to access support and implement adaptation options. For each of these areas we suggest actions that can be taken respectively by policy-makers, research and development providers, and enterprise managers.

- Strategies to enhance adaptive capacity at broad policy scales include investing in (1) climate-related natural, physical, financial, human and social assets; (2) financial incentives and other incentives that provide social status and/or reward early adopters; (3) climate research; (4) adaptive governance approaches that encourage climate-learning and adaptive management through developing effective environmental, economic and social feedback mechanisms; (5) 'mainstreaming' climate change into all relevant government policies, particularly drought and natural resource management policies, and (6) addressing other key stressors that can interact with those of climate change.

- Research and extension agencies can help identify effective adaptation options by (1) applying existing knowledge in more effective and innovative ways including greater collaboration with decision-makers; (2) broadening the array of research approaches used to identify practical solutions, and (3) continuing basic research that fills fundamental knowledge gaps, tests the validity of key assumptions, and evaluates the effectiveness of adaptation options.

- At the enterprise level, adaptive capacity can be enhanced by (1) managing climate risk and uncertainty; (2) planning, learning and reorganising for climate-driven change; (3) building flexibility to absorb the costs of change and experiment with options for the future; (4) developing the capacity for individual evaluation of climate change adaptations through access to climate information, expertise and technology, and (5) reducing dependency on the climate-sensitive resource (such as through increasing strategic skills).

Introduction

As the preceding chapters have discussed, climate change has the potential to dramatically alter agricultural production and the natural resource base on which it depends. In the face of such large potential change, agricultural communities and industries will explore a range of pathways and means to adapt. This book is an attempt to make that exploration more efficient and effective by identifying a range of adaptation options, their costs, benefits and other consequences that agricultural, forestry, fishery and water managers may consider in the adaptation process. This requirement for effective adaptation to support sustainable production will be emphasised by anticipated increasing demand from a growing population and the rising aspirations from emerging economies. More than ever, agricultural managers will need to anticipate, and prepare for climate-driven change.

Systemic changes such as those possible from climate change require responses not just from primary producers, but will also require strong support from government and industry institutions if agricultural resources and the extended social systems dependent on them are to be sustained (Nelson *et al.* 2008). The challenge faced by people working on the land is to build the productivity and profitability of their agricultural enterprises in the face of climate uncertainty without depleting the soils and water upon which they depend (Stafford Smith and Foran 1992). Agricultural communities around the world are highly dynamic systems that have evolved to cope with a wide range of disturbances from market fluctuations, changing cultural norms and natural disasters. However, experience has shown that some individuals and communities will cope more easily than others (Salinger *et al.* 2005; Ziervogel and Calder 2003). Important human factors, both in terms of the capacity of managers and the context within which they operate (e.g. policy and institutional support), determine the extent to which adaptation occurs. Ultimately, reducing the impacts of climate change is dependent on managers effectively applying appropriate adaptation measures on the ground. Adaptation is therefore more than just a technical task of defining the impacts of climate change and indentifying technical fixes (Howden *et al.* 2007). An effective response to climate change therefore needs to not only address the technical challenges of adaptation (as discussed in preceding chapters), but also needs to ensure that land managers and supporting institutions have the capacity to implement adaptation options and that these options are appropriately designed to be policy- and management-relevant from the outset.

Strategies aimed at building and supporting capacity to cope with climate change will play a crucial role in ensuring that primary producers can prevail, and continue to provide the goods and services upon which society depends. However, effective strategies can only be designed and delivered with some knowledge of adaptive capacity and the factors that influence it (Brooks and Adger 2004; Lebel *et al.* 2006). This chapter examines the factors that confer adaptive capacity and explores ways of enhancing the capacity of agricultural communities, industries, regions, resource users and supporting institutions, to adapt to climate change. However, in order to appreciate the need for adaptive capacity, we first take a brief look at the linked flow of biophysical and social factors that determines the ultimate extent to which agricultural communities will be disrupted by climate change, and the related concepts of vulnerability and adaptive capacity.

Vulnerability and adaptive capacity

Vulnerability has been described as the 'capacity to be wounded' (Kates *et al.* 2000) or as the opposite of resilience (Gallopin 2006) and is defined by the IPCC Third Assessment Report (IPCC 2001) as 'the degree to which a system is susceptible to, or unable to cope with, adverse effects of climate change, including climate variability and extremes. Vulnerability is a function of the character, magnitude, and rate of climate variation to which a system is exposed, its sensitivity, and its adaptive capacity' (Figure 15.1). The potential impact of climate change on a current agricultural activity depends on both the magnitude of climate change in the particular location (exposure: e.g. the percentage change in rainfall) and how strongly each unit change in climate impacts that activity (sensitivity: e.g. the change in enterprise income for each 1% change in rainfall) (Figure 15.1). However, since land managers are likely to take actions in response to these changes (adaptation), the potential impacts are likely to be modified. The ultimate vulnerability of communities and enterprises to climate change is therefore strongly dependent not just on potential biophysical impacts but also on the way people respond and their potential to moderate those impacts. The effectiveness of adaptive responses will be influenced by factors such as (Figure 15.1): (1) the operating context within which responses occur (e.g. existing non-climate stressors, the policy and governance setting, and access to resources that enable effective responses); (2) the availability of effective adaptation options, as evaluated and provided by

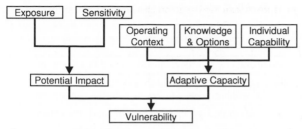

Figure 15.1: Linked human and biophysical factors that determine the ultimate vulnerability of agriculture systems to climate change. After Allen Consulting Group (2005)

research and other sources of knowledge and technology, and (3) the capacity of individuals to access support and implement adaption options.

The concept of adaptive capacity has its origins in hazard and disaster research (King 2000; Flint and Luloff 2005), complex systems research (Folke 2003; Janssen and Ostrom 2006), and the social sciences (Buttel 1987; Abel 1998; Scoones 1999; McCay 2000). Adaptive capacity has been described as the ability to adjust to, or live with, climate change (Folke *et al.* 2002; IPCC 2007) but there are numerous other definitions or approaches (Turton 1999; Adger 2003; Adger *et al.* 2003). For the purposes of this chapter, we define adaptive capacity as the preconditions and ability to mobilise effective responses without losing options for the future (Folke *et al.* 2002; Olsson *et al.* 2004; Brooks *et al.* 2005). These 'preconditions' generally reflect characteristics such as learning, the flexibility to experiment and adopt novel solutions, and the overall ability to respond effectively to a broad range of challenges (Levin *et al.* 1998; Gunderson 2000). As well as technical, economic and policy aspects, adaptive capacity depends on the ability of individuals, communities and institutions to adapt to adversity and stressful events through networks or institutions that learn, store knowledge and experience, and that are creative, flexible and novel in their approaches to problem solving (Vayda and McCay 1975; McCay 1981; Sonn and Fisher 1998). These are crucial human and social assets that contribute significantly to adaptive capacity to climate risk in Australia (Nelson *et al.* 2008).

In determining the ultimate outcomes of climate change for agricultural communities there are two main factors that humans can influence. The first is mitigation: by reducing global emissions

of greenhouse gases (Chapter 16) we can deal with the root cause of the problem and limit the magnitude of human-induced global climate change risk to which the planet is exposed. The second, and the focus of this book, is adaptation: by building the capacity to adjust climate-sensitive activities to plausible future climate scenarios, we can limit our vulnerability to the climate change that does occur. The two are linked in that the more effort that is applied up-front to mitigation efforts, the less effort will be required for adapting to climate changes (Howden *et al.* 2007). However, while successful mitigation requires global collaboration to be effective, adaptation measures can be taken by individuals and communities to achieve local outcomes (Nelson *et al.* 2007a). Since past emissions of greenhouse gases have already committed the planet to some level of climate change, a corresponding level of adaptation will have to occur to best reposition agriculture for the prevailing climate. The main choice for society to make is whether to take pre-emptive adaptive action to reduce the impacts of climate change or to respond reactively after the impacts have already been realised.

In the following sections we consider three components to building adaptive capacity in Australian primary industries (Figure 15.1): (1) policy and governance, including approaches for dealing with uncertainty; (2) building and applying the necessary knowledge base, and (3) adaptation at the manager/enterprise level.

Building adaptive capacity

Policy, governance and supporting institutions

Governments and other institutions that support primary industries will have a vital role to play in assisting rural communities to adapt to climate change. In order for them to do so, one step is understanding what makes some communities better able to adapt to climate change than others, and to be able to identify where the most vulnerable communities are and why they are vulnerable so as to provide targeted assistance. Creating supportive policy environments that enable more effective or lower-risk change pathways and minimise barriers and impediments to adoption is likely to be important. In addition, any efforts to improve adaptive capacity will require

Table 15.1: An integrated approach for understanding and building adaptive capacity (based on Ellis (2000) and Cinner et al. (2009)).

Dimension	Description
Flexibility	The level of social, cultural, political, economic and environmental flexibility within a community
Capacity to organise	The formal and informal capacity to develop adaptation plans, establish and maintain community organisations, make decisions and manage large-scale responses following extreme climate events
Capacity to learn	The extent to which the capacity to organise is moderated by adaptive learning and subsequent modification of adaptation plans
Assets (The five types of capital from Ellis' (2000) Rural Livelihoods Approach)	
Human	The skills, health and education of individuals that contribute to the productivity of labour and capacity to manage land
Social	Reciprocal claims on others by virtue of social relationships, the close social bonds that facilitate cooperative action and the social bridging, and linking via which ideas and resources are accessed
Natural	The productivity of land, and actions to sustain productivity, as well as the water and biological resources from which rural livelihoods are derived
Physical	Capital items produced by economic activity from other types of capital that can include infrastructure, equipment and improvements in genetic resources (crops, livestock)
Financial	The level, variability and diversity of income sources, and access to other financial resources (credit and savings) that together contribute to wealth

approaches for dealing with uncertain climate change information in decision-making.

The preconditions that enable adaptation at the community scale can be considered as comprising four essential characteristics: flexibility, the capacity to re-organise, the capacity to learn, and an adequate asset base (Cinner et al. 2009) (Table 15.1). The 'assets' that enable a community to adapt effectively can be usefully thought of in terms of the five types of capital defined by Ellis (2000) in the Rural Livelihoods Approach (Table 15.2). The basis of Ellis' (2000) approach is that rural livelihood strategies comprise sets of activities that are continuously invented, adapted and adopted in response to changing access to human, social, natural, physical and financial capitals.

For example, social capital and community empowerment reflect the level of social interaction, social networks and social relations that exist within a community (Putnam 1993; Worthington and Dollery 2000; Adger et al. 2002), providing some indication of the capacity for a community to cope with change and adapt (King 2005; Korpi 2001; Meffe 2001; Tindall and Wellman 2001). Communities with increased social capital typically have reciprocal networks of community interactions and increased social trust that are directed towards mutual benefit (Cernea 1993; Hofferth and Iceland 1998; Dasgupta and Maler 2001; Brunckhorst 2002). Social capital

contributes to the effective operation of networks and channels through which government programs can effectively support rural communities in their efforts to improve land management practices. It includes generating and maintaining social norms and behaviours that contribute positively to natural resource management outcomes, while sanctioning behaviours that may be detrimental. In its review of the role of social capital in policy and program development, the Productivity Commission noted that benefits generated include the reduction of transaction costs, promoting cooperative behaviour, and the diffusion of knowledge and innovations (Productivity Commission 2003). Policy can develop social capital through, for example, the support of institutions such as Landcare.

Based on the Rural Livelihoods Approach, Nelson and colleagues (Nelson et al. 2009a; Nelson et al. 2009b) have proposed a method for mapping the adaptive capacity of Australian rural communities in terms of Ellis' (2000) five capitals. There is an expectation that such information is useful to policy-makers because it can highlight communities that are ill-equipped to adapt and may be vulnerable to even moderate impacts of climate change. Furthermore, knowing which types of capital are lacking for effective adaptive capacity can assist in deciding which forms of support and intervention are most appropriate.

Aside from targeted local interventions, it will also be important to create a broader policy environment that supports adaptation. The context within which property-level adaptation occurs will, in large part, be driven by decisions made at higher institutional levels such as intergovernmental agreements; national policies, planning and strategies, and supporting industry and government institutions. Where these broader decision-making processes include considerations of climate change, there will be opportunities to enable and promote on-farm adaptation, but there will also most likely be conflicting priorities that could provide impediments and barriers to adoption. For example, increased development of farm dams to improve on-farm climate risk management might cumulatively have significant impacts on river health downstream due in particular to major flow reductions in dry periods. In addition, many institutions can generate 'perverse incentives' and create maladaptations (Gunderson et al. 2006). The preceding chapters have highlighted particular areas where maladaptations could occur in each primary industry as a forewarning of the issues that might require particular policy attention.

One of the major barriers to implementing adaptation strategies is the desire to wait until more reliable information is available (Sarewitz and Pielke 2007). If timely, proactive action is to be taken, it will be crucial to find ways of making decisions using imperfect knowledge. We suggest the following approach to start dealing with this uncertainty, building adaptive capacity and changing the decision environment to promote adaptation actions (Howden et al. 2007).

(1) To change their management practices, enterprise managers should be convinced that projected climate changes are real and are likely to continue. This is more likely to occur if existing trends in climate are consistent with projected changes in climate and where the underlying processes are well understood and communicated. A corollary of this approach is that assessment of climate risk and opportunity should no longer be based on the past 100 years of climate records but rather those of the recent past *and* projections of climate in the medium term (i.e. 2030) which will need to account for decadal variability as well

(McKeon et al. 2004). This adaptation element will be facilitated by policies that maintain climate monitoring and communicate this information effectively.

(2) Managers need to be confident that the projected changes will significantly impact on their enterprise and that they understand the range of consequences of different adaptation options singly and in combination. Policies that support the research, systems analysis, extension capacity, industry and regional networks that provide this information could thus be strengthened. This includes modelling capabilities such as APSIM (Keating et al. 2003) that allow scaling up knowledge, for example from gene to cell to organisms and eventually to management systems and national policy scales. Managers and policy-makers also need to understand the limitations of adaptations so that they do not underestimate potential vulnerability.

(3) The technical and other options necessary to respond to the projected changes need to be available. In many cases these will be extensions or intensifications of existing climate risk management or production enhancement activities in response to a potential change in the climate risk profile. Where the existing technical options are inadequate, investment in new technical or management strategies may be required such as developing improved crop germplasm that is more suited to expected atmospheric and climatic conditions. Another example is upgrading climate forecasts in terms of reliability, lead times and utility so as to allow year-by-year 'tracking' adaptation to a changing climate as first proposed by McKeon et al. (1993).

(4) Where climate impacts may lead to major transformations such as land use change, there may be demands to support transitions such as industry adjustment and enterprise relocation. This may be achieved through direct financial and material support, creating alternative livelihood options with reduced dependence on agriculture, supporting community partnerships such as Landcare, enhancing capacity to develop

social capital and share information and retraining. Effective planning for and management of such transitions may result in less habitat loss, less risk of carbon loss and also lower environmental costs compared to unmanaged, reactive transitions.

(5) New infrastructure, policies and institutions could be developed to support changes in management and land and water use arrangements. Options include addressing climate change in terms of sustainable development and natural resource management; enhancing investment in irrigation infrastructure and efficient water use technologies; encouraging appropriate transport and storage infrastructure and establishing more efficient markets for products and inputs. Adaptation options that may be more difficult to implement will probably include those that (1) deal with lag effects; (2) entail significant trade-offs involving other parties; (3) impact on social welfare issues and policies, and (4) deal with climate extremes that by their nature are infrequent but which may require substantial up-front investment.

(6) Importantly, farmers and policy-makers in industry and government must maintain the capacity to make continuing adjustments and improvements in adaptation by 'learning by doing' via targeted monitoring of adaptations to climate change and their costs, benefits and effects. A participatory approach that cycles systematically between the biophysical and the socio-economic aspects is likely to most effectively harness the substantial scientific knowledge of many agricultural systems, while retaining a focus on the values important to stakeholders, thereby achieving relevance, credibility and legitimacy (Howden et al. 2007). The inclusion of an adaptive loop in such frameworks is critical to develop flexible, dynamic policy and management that can accommodate climate surprises or changes in the underlying knowledge base. Participatory engagement with decision makers can, by bringing their practical knowledge into the assessment, also identify a more comprehensive range of adaptations than are typically explored by scientists, as well as being able to assess the practicality

of options and contribute to more realistic assessment of the costs and benefits involved in management or policy change (Crimp et al. 2008).

Knowledge generation and application

As raised above, a consistent theme that has emerged throughout this book is that adaptation actions will be more effective if they are taken pre-emptively (rather than reacting to climate change after it has occurred) and this will require taking action while considerable uncertainty remains over the impacts of climate change and the effectiveness of adaptation options. A recent review of the most pressing environmental issues in Australia echoed this view, suggesting that it will become increasingly important to 'use the present state of scientific knowledge to supply reasonable inferences for action based on imperfect knowledge' in developing policy and management strategies to address environmental problems (Morton et al. 2009). Indeed, of the 22 most pressing environmental issues these authors identified, half were related to climate change. This was not only because of the inherent importance of climate change, but also because the science in this area is currently so poorly developed. The Fourth Assessment Report (IPCC 2007) highlighted the particular paucity of impact and adaptation research for the Australian continent. This lack of knowledge contributes to the vulnerability of Australia to climate change, particularly for the tropics (Rosenzweig and Liverman 1992). Tackling climate change issues within these constraints will require both (1) applying existing knowledge in more effective and innovative ways and (2) answering critical unresolved issues where targeted research can reduce uncertainty (Morton et al. 2009). There is also a need to shift the emphasis of research from simply understanding the problem of climate change to finding practical solutions, which will require a broadening in the array of research approaches that are used.

In order to apply climate change knowledge more effectively, there is an urgent need to reconcile the supply and demand of integrated (socio-economic and biophysical) adaptive capacity assessments for Australian agriculture (Nelson et al. 2008). The supply response of integrated, policy-relevant vulnerability assessments from the science community is only just beginning to emerge. Historically, the course of climate science has closely followed the linked processes in Figure 15.1 as, for example, reflected in the Intergovernmental Panel on Climate Change series of climate assessment reports (e.g. IPCC 2001; 2007). The earliest research concentrated on understanding the causes of climate change. This led to concerns over the sensitivity of agricultural systems to climate change and, when climate projections became available, combining the two research areas allowed risk assessments of potential impacts. Growing awareness of the threats posed by climate change has given rise to a strong policy demand for integrated measures of the potential impacts of climate change on agricultural industries and rural communities. However, most of the types of science available to support adaptation to climate change in Australia are built on a heritage of hazard and impact modelling (Nelson *et al.* 2008). Much of this research continues to pursue incrementally greater predictive skill and to extend the timeframes over which projections can be made. By itself, however, such research does not address the multi-dimensional uncertainty surrounding climate change faced by decision-makers in government and agricultural industries.

We have referred to a similar disconnect between science and drought policy in Australian agriculture as a policy-relevance gap (Meinke *et al.* 2007), and demonstrated that methodological (Kokic *et al.* 2007; Nelson *et al.* 2007b) and institutional options (Nelson *et al.* 2008) for closing it do exist. While ongoing improvements in the precision of climate change impact assessments will continue to be useful, some uncertainty will always remain. The growing need is now to produce highly relevant, demand-driven science to develop flexible approaches that build the individual and institutional capacity to cope with uncertainty, with the expectation that plans will have to be adjusted as improved information becomes available through subsequent monitoring, learning, research and collaboration (Howden *et al.* 2007). This will require greater engagement with policy-makers and managers, greater reliance on collaborative and participatory research approaches and inclusion of a more diverse set of disciplines (Morton *et al.* 2009; Nelson *et al.* 2009b).

While there is a growing demand to make policy-relevant inferences from current, imperfect knowledge, it will remain important to maintain the supply of rigorous evidence and evaluation to inform decision making. This will require research that tests the validity of key inferences and assumptions, evaluates the effectiveness of policy and management recommendations, and identifies and fills fundamental knowledge gaps. For example in climate science, key research areas could include providing projections of management- and policy-relevant weather metrics (e.g. cold indices for stone fruit); downscaling regional projections to management scales, and combining information on both climate variability and trend in seasonal and medium-term (decadal) forecasts. Each of the preceding primary industry chapters provided a set of adaptation priorities, which included important research areas such as adaptation options requiring evaluation and key knowledge gaps that need to be addressed. The following chapter summarises the adaptation options that are broadly relevant across primary industries (Table 16.1). Some of these options include research components such as broad-based cost-benefit analyses of adaptation options; improving agricultural systems models to better simulate climate change impacts and evaluate adaptation options, and monitoring and evaluating climate change and the effectiveness of adaptation measures.

To provide the necessary knowledge to support adaptation, the balance in emphasis of climate change research will need to shift from providing the basic understanding of climate change to providing policy-relevant assessments of impacts and evaluations of adaptation options. This will require a greater recognition of the socio-economic dimensions of adaptation, collaboration across a wider range of disciplines, selecting appropriate research approaches from a wider range of options, and closer engagement with policy-makers and managers. Such approaches are more likely to generate practical adaptation options that are effective, accepted by end users and implemented in decision making (Howden *et al.* 2007).

Adaptive capacity at the enterprise/individual scale

Ultimately the impact of climate change on agriculture will only be moderated if adaptation options are effectively implemented on the ground by individual managers. We therefore consider what actions can be taken by managers at the enterprise scale to improve their capacity to adapt to climate change. At an individual level adaptive capacity can be considered in terms of five key characteristics and influences: (1) perceptions of risk associated with change and the capacity to manage for uncertainty; (2) the ability to plan, learn, and reorganise; (3) the extent of financial and emotional flexibility; (4) the level of interest in adapting to climate change, and (5) the level of resource dependency (Marshall *et al.* 2007; Marshall and Marshall 2007).

Based on this approach, some examples of actions that individuals can take to enhance their adaptive capacity include (1) taking a risk-based approach in preparing management options for a range of foreseeable future scenarios (Chapter 16); (2) participating in collaborative learning opportunities to identify novel adaptation options and management strategies or to reinforce or enhance existing strategies; (3) seeking alternative and supplementary opportunities to generate income and increasing the level of transferrable skills (Harris *et al.* 1998; Colding *et al.* 2004; Olsson *et al.* 2004); (4) developing strategic business skills to increase profitability over the longer term; (5) planning for financial flexibility; (6) increasing climate awareness and accessing climate information; (7) continually monitoring, adjusting management and implementing adaptation options in response to changes in local climatic, environmental, economic and political conditions, and (8) increasing involvement in information sharing and learning networks within the community.

The actions listed in the earlier 'Policy, governance and supporting institutions' section of this chapter will assist individuals in managing climate threats by directly addressing key characteristics of an individual's adaptive capacity and their level of dependency on the climate sensitive resource. For example, in order to develop a risk-based approach to preparing for climate change, individuals are likely to be able to respond positively if appropriate training programs are implemented. Similarly, individuals will develop an interest in climate change and adapting through accessing readily available climate expertise, information and technology.

In some cases, the risks associated with climate change may mean that rural enterprises become unviable. In these cases, it will be important to help landholders to transform within their industry or beyond by adjusting and diversifying their management practices or occupations or the location of their agricultural activities. However, there will be significant costs to transforming. For example, specialised people with specialised skills tend to perform their task very efficiently and are in a sound position to understand and manage risk (for a well-understood activity). With diversification there is perhaps a tendency to become a 'jack-of-all-trades' and potentially 'master-of-none'. Realistic strategies that could be implemented at a policy level that avoid the issue of broad diversification outside of the resource system include 'value adding' by identifying and helping create additional and profitable markets.

Conclusions

The vulnerability of Australian primary industries to climate change can be reduced by enhancing the capacity of enterprise managers to moderate any negative impacts and take advantage of new opportunities. This capacity is determined by the institutional or operating context, the availability of (and ability to generate) effective adaptation options, and the capabilities of individual land managers. This chapter has provided some practical approaches for building adaptive capacity in the face of uncertain climate change through actions that can be taken by government institutions, research and extension providers, and enterprise managers. The next step will be to develop, prioritise and implement comprehensive adaptation strategies for Australia's primary industries. Through policies, information sharing and engagement, industry leaders, policy-makers, climate researchers and other stakeholders can have a major influence in ensuring that Australian primary industries are well-equipped to cope with a variable and changing climate.

References

Abel N (1998). Complex adaptive systems, evolutionism, and ecology within anthropology: interdisciplinary research for understanding cultural and ecological dynamics. *Georgia Journal of Ecological Anthropology* **2**, 6–29.

Adger WN (2003). Building resilience to promote sustainability: an agenda for coping with globalisation and promoting justice. *Newsletter of the International Human Dimensions Programme on Global Environmental Change* **02/1003**.

Adger WN, Kelly PM, Winkels A, Huy LQ, Locke C (2002). Migration, remittances, livelihood trajectories, and social resilience. *Ambio* **31**, 358–366.

Adger WN, Saleemul H, Brown K, Conway D, Hulme M (2003). Adaptation to climate change in the developing world. *Progress in Development Studies* **3**, 179–195.

Allen Consulting Group (2005). 'Climate change risk and vulnerability'. Department of the Environment and Heritage, Canberra.

Brooks N, Adger WN (2004). Assessing and enhancing adaptive capacity. In *Adaptation Policy Frameworks for Climate Change: Developing Strategies, Policies and Measures*. (Eds B Lim, E Spanger-Siegfried) pp. 165–181. Cambridge University Press, Cambridge.

Brooks N, Adger WN, Kelly PM (2005). The determinants of vulnerability and adaptive capacity at the national level and the implications for adaptation. *Global Environmental Change* **15**, 151–163.

Brunckhorst DJ (2002). Institutions to sustain ecological and social systems. *Ecological Management and Restoration* **3**, 108–116.

Buttel FH (1987). New directions in environmental sociology. *Annual Review of Sociology* **13**, 465–488.

Cernea MM (1993). The sociologist's approach to sustainable development. *Finance and Development* **30**, 1–7.

Cinner JM, Fuentes BP, Randriamahazo H (2009). Exploring social resilience in Madagascar's marine protected areas. *Ecology and Society* **in press**.

Colding J, Elmqvist T, Olsson P (2004). Living with disturbance: building resilience in social-ecological systems. In *Navigating Social-Ecological Systems. Building Resilience for Complexity and Change*. (Eds F Berkes, J Colding, C Folke) pp. 163–173. Cambridge University Press, Cambridge.

Crimp S, Nidumulo U, Gaydon D, Howden SM, Hayman P (2008). Examining the value of dynamic seasonal forecasts in managing farm-level production and environmental outcomes in a variable climate. In *Proceedings of the 14th Australian Society of Agronomy Conference, Adelaide*.

Dasgupta P, Maler K (2001). Wealth as a criterion for sustainable development. *World Economics* **2**, 199–244.

Ellis F (2000). The determinants of rural livelihood diversification in developing countries. *Journal of Agricultural Economics* **51**, 289–302.

Flint CG, Luloff AE (2005). Natural resource-based communities, risk, and disaster: an intersection of theories. *Society and Natural Resources* **18**, 399–412.

Folke C (2003). Conservation, driving forces, and institutions. *Ecological Applications* **6**, 370–372.

Folke C, Carpenter S, Elmqvist T, Gunderson L, Holling CS, Walker B (2002). Resilience and sustainable development: building adaptive capacity in a world of transformations. *Ambio* **31**, 437–440.

Gallopin G (2006). Linkages between vulnerability, resilience and adaptive capacity. *Global Environment Change* **16**, 293–303.

Gunderson L (2000). Ecological resilience – in theory and application. *Annual Review of Ecological Systems* **31**, 425–439.

Gunderson L, Carpenter S, Folke C, Olsson P, Peterson G (2006). Water RATs (resilience, adaptability, and transformability) in lake and wetland social-ecological systems *Ecology and Society* **11**, 16.

Harris CC, McLaughlin WJ, Brown G (1998). Rural communities in the interior Columbia Basin. How resilient are they? *Journal of Forestry* **96**, 11–15.

Hofferth SL, Iceland J (1998). Social capital in rural and urban communities. *Rural Sociology* **63**, 574–598.

Howden SM, Soussana J, Tubiello FN, Chhetri N, Dunlop M, Meinke H (2007). Adapting agriculture to climate change. *Proceedings of the National Academy of Sciences* **104**, 19691–19696.

IPCC (2001). *Climate Change 2001: Impacts, Adaptation and Vulnerability*. Cambridge University Press, Cambridge.

IPCC (2007). *Climate Change 2007 – Impacts, Adaptation and Vulnerability – Contributions of Working Group II to the Fourth Assessment Report of the International Panel on Climate Change*. Cambridge University Press, Cambridge.

Janssen MA, Ostrom E (2006). Resilience, vulnerability, and adaptation: a cross-cutting theme of the international human dimensions programme on global environmental change. *Global Environmental Change* **16**, 237–239.

Kates RW, Clark WC, Corell R, Hall M, Jaeger CC, Lowe I, McCarthy JJ, Schnellnhuber HJ, Bolin B, Dickson NM, Faucheux S, Gallopin GC, Greubler A, Huntley B, Jager J, Jodha NS, Kasperson RE, Mabogunje A, Matson P, Mooney HA, Moore III B, Riordan T, Svedin U (2000). '"Sustainability science" research and assessment systems for sustainability program discussion paper 2000–33.' Environment and Natural Resources Program, Belfer Centre for Science and International Affairs, Harvard University.

Keating BA, Carberry PS, Hammer GL, Probert ME, Robertson MJ, Holzworth D, Huth NI, Hargreaves JNG, Meinke H, Hochman Z, McLean G, Verburg K, Snow V, Dimes JP, Silburn M, Wang E, Brown S, Bristow KL, Asseng S, Chapman S, McCown RL, Freebairn DM, Smith CJ (2003). An overview of the APSIM, a model designed for farming systems simulation. *European Journal of Agronomy* **18**, 267–288.

King D (2000). Uses and limitations of socioeconomic indicators of community vulnerability to natural hazards: data and disasters in northern Australia. *Natural Hazards* **24**, 147–156.

King D (2005). Planning for hazard resilient communities. In *Disaster Resilience: An Integrated Approach*. (Eds D Paton, D Johnston). Charles C. Thomas, Springfield, Illinois.

Kokic P, Nelson R, Meinke H, Potgieter A, Carter J (2007). From rainfall to farm incomes-transforming advice for Australian drought policy. I. Development and testing of a bioeconomic modelling system. *Australian Journal of Agricultural Research* **58**, 993–1003.

Korpi T (2001). Good friends in bad times? Social networks and job search among the unemployed in Sweden. *Acta Sociologica* **44**, 157–170.

Lebel L, Anderies JM, Campbell B, Folke C, Hatfield-Dodds S, Hughes TP, Wilson J (2006). Governance and the capacity to manage resilience in regional social-ecological systems. *Ecology and Society* **11**, 19.

Levin S, Barrett S, Baumol W, Bliss C, Bolin B, Dasgupta P, Ehrich PR, Folke C, Gren I, Holling CS, Jansson A, Jansson B, Maler K, Martin D, Perrings C, Sheshinski E (1998). Resilience in natural and socioeconomic systems. *Environment and Development Economics* **3**, 222–235.

Marshall NA, Fenton DM, Marshall PA, Sutton S (2007). How resource-dependency can influence social resilience within a primary resource industry. *Rural Sociology* **72**, 359–390.

Marshall NA, Marshall PA (2007). Conceptualising and operationalising social resilience within commercial fisheries in northern Australia. *Ecology and Society* **12**, http://www.ecologyandsociety.org/vol12/iss1/art1/

McCay BJ (1981). Optimal foragers or political actors? Ecological analyses of a New Jersey fishery. *American Ethnologist* **11**, 356–382.

McCay BJ (2000). Environmental anthropology at sea. In *New Directions in Anthropology and Environment*. (Eds CL Crumley, AE van Deventer, JJ Fletcher) pp. 254–271. Altamira Press, Walnut Creek, California.

McKeon GM, Hall WB, Henry BK, Stone GS, Watson IW (2004). Pasture degradation and recovery in Australia's rangelands: learning from history. Queensland Department of Natural Resources, Mines and Energy.

McKeon GM, Howden SM, Abel N, King JM (1993). Climate change: adapting tropical and sub-tropical grasslands. In *Proceedings of 17th International Grassland Congress, Palmerston North, New Zealand*. pp. 1181–1190.

Meffe GK (2001). Crisis in a crisis discipline. *The Journal of the Society for Conservation Biology* **15**, 303.

Meinke H, Sivakumar MVK, Motha RP, Nelson R (2007). Preface: climate predictions for better agricultural risk management. *Australian Journal of Agricultural Research* **58**, 935–938.

Morton SR, Hoegh-Guldberg O, Lindenmayer DB, Harriss Olson M, Hughes L, McCulloch MT, McIntyre S, Nix HA, Prober SM, Saunders DA, Andersen AN, Burgman MA, Lefroy EC, Lonsdale WM, Lowe I, McMichael AJ, Parslow JS, Steffen W, Williams JE, Woinarski JCZ (2009). The big ecological questions inhibiting effective environmental management in Australia. *Austral Ecology* **34**, 1–9.

Nelson DR, Adger WN, Brown K (2007a). Adaptation to environmental change: contributions of a resilience framework. *Annual Review of Environment and Resources* **32**, 395–419.

Nelson R, Howden SM, Stafford Smith M (2008). Using adaptive governance to rethink the way science supports Australian drought policy *Environmental Science & Policy* **11**, 588–601.

Nelson R, Kokic P, Crimp S, Martin P, Meinke H, Howden M, Devoil P, Nidumolu U (2009a). The vulnerability of Australian agriculture to climate variability & change: Part II – Integrating impacts with adaptive capacity. *Environmental Science and Policy*. Submitted.

Nelson R, Kokic P, Crimp S, Meinke H, Howden SM (2009b). The vulnerability of Australian agriculture to climate variability & change: Part I – Conceptualising and measuring vulnerability. *Environmental Science and Policy*, Submitted.

Nelson R, Kokic P, Meinke H (2007b). From rainfall to farm incomes-transforming advice for Australian drought policy. II. Forecasting farm incomes. *Australian Journal of Agricultural Research* **58**, 1004–1012.

Olsson P, Folke C, Hahn T (2004). Social-ecological transformation for ecosystem management: the development of adaptive co-management of a wetland in southern Sweden. *Ecology and Society* **9**, online at: www.ecologyandsociety.org/vol9/iss4/art2

Productivity Commission (2003). Social capital: reviewing the concept and its policy implications. Productivity Commission, Canberra.

Putnam RD (1993). The prosperous community: social capital and public life. *American Prospect* **13**, 35–42.

Rosenzweig C, Liverman D (1992). Predicted effects of climate change on agriculture: a comparison of temperate and tropical regions. In *Global Change: Implications, Challenges and Mitigation Measures*. (Ed. SK Majundar) Pennsylvania Academy of Sciences, Pennsylvania.

Salinger MJ, Sivakumar MVK, Motha R (2005). Reducing vulnerability of agriculture and forestry to climate variability and change: workshop summary and recommendations. *Climatic Change* **70**, 341–362.

Sarewitz D, Pielke RA (2007). The neglected heart of science policy: reconciling supply of and demand for science. *Environmental Science & Policy* **10**, 5–16.

Scoones I (1999). New ecology and the social sciences: what prospects for a fruitful engagement? *Annual Review of Anthropology* **28**, 479–507.

Sonn CC, Fisher AT (1998). Sense of community: community resilient responses to oppression and change. *Journal of Community Psychology* **26**, 457–472.

Stafford Smith DM, Foran B (1992). An approach to assessing the economic risk of different drought management tactics on a south Australian pastoral sheep station. *Agricultural Systems* **39**, 83–105.

Tindall DB, Wellman B (2001). Canada as social structure: social network analysis and Canadian sociology. *Canadian Journal of Sociology* **26**, 265–305.

Turton AR (1999). 'Water scarcity and social adaptive capacity: towards an understanding of the social dynamics of water demand management in developing countries.' MEWREW Occasional Paper No. 9. University of London, School of African Studies, London.

Vayda AP, McCay BJ (1975). New directions in ecology and ecological anthropology. *Annual Review of Anthropology* **4**, 293–306.

Worthington AC, Dollery BE (2000). Can Australian local government play a meaningful role in the development of social capital in disadvantaged rural communities? *Australian Journal of Social Issues* **35**, 349–361.

Ziervogel G, Calder R (2003). Climate variability and rural livelihoods: assessing the impact of seasonal climate forecasts in Lesotho. *The Geographical Journal* **170**, 6–21.

16
SUMMARY

CJ Stokes and SM Howden

KEY MESSAGES:

- There is a clear imperative for action to prepare agriculture to adapt to climate change. The agricultural sector in Australia, which is already constrained by a harsh environment, is particularly vulnerable to climate change with projected negative impacts on the amount, quality and reliability of production. Past greenhouse gas emissions have already committed the earth to at least several decades of continued warming, while emissions and climate indicators are currently tracking the most pessimistic IPCC scenarios.

- The benefits and positive opportunities presented by climate change may start to peak during the initial stages (possibly as late as mid-century), but the negative impacts are likely to lag behind, becoming progressively stronger over time and with greater build up of greenhouse gases in the atmosphere. Caution is therefore needed not to underestimate the long-term challenge of climate change based on initial, more moderate experiences.

- Given the inherent uncertainty in the effects of climate change, there is a clear need to develop enhanced adaptive capacity in agricultural systems (including socio-economic and cultural/ institutional structures) that is sufficiently robust to cope with a broad range of plausible scenarios. Synergies with existing government policies such as self-reliance in drought and related supporting programs as well as with institutions such as Landcare are needed to develop this capacity.

- This book has identified a number of potential options for primary industries to adapt to climate change. Few of these have been fully evaluated, but those that have suggest the benefits of adaptation are so significant that further systems analyses are warranted.

- Several adaptation priorities with broad applicability across the agricultural sector have been identified. In particular, a common adaptation option, at least in the short term, will be to improve and promote existing management strategies for dealing with climate variability. This will enhance capacity to deal with extreme events and incrementally track early stages of climate change until longer-term trends become clearer.

- Marginal production areas are among the most vulnerable and will most likely be among the first areas in which the impacts of climate change will exceed adaptive capacity. It will be important to identify areas where climate change risks and opportunities require strong policy intervention (beyond simple incremental adjustments to existing agricultural practices) so that affected communities can be appropriately supported through any disruptive transition periods.

Introduction

Australia's climate is changing and further change seems unavoidable, even if global actions are taken to reduce greenhouse gas emissions. The continent's hot, dry, variable climate and fragile, weathered soils already present significant challenges for primary industries. The agricultural sector will therefore be particularly vulnerable to projected warming, drying and increased variability of the climate (Garnaut 2008; Hennessy *et al.* 2007; IPCC 2007b). Climate change could negatively impact the amount and quality of produce, reliability of production and the natural resource base on which agriculture depends. This vulnerability requires high levels of adaptive responses. For primary industries to continue to thrive in the future the sector needs to anticipate these changes, be prepared for uncertainty, and develop adaptation strategies now.

There is a clear imperative for a greater, coordinated effort to tackle the challenges a changing climate will pose. For agriculture in particular, there are several reasons why it is becoming increasingly urgent to develop options and capacity to adapt to climate change (Howden *et al.* 2007):

(1) Past emissions of greenhouse gases have committed the globe to further warming of approximately 0.2°C per decade for several decades, making some level of impact, and necessary adaptation responses, already unavoidable.

(2) The emissions of the major greenhouse gases are continuing to increase, with the resultant changes in atmospheric CO_2 concentration, global temperature, and sea level observed today already at the high end of those implied by the scenarios considered by the Intergovernmental Panel on Climate Change (IPCC 2007c). Furthermore, some climate change impacts are happening faster than previously considered likely. If these trends continue, then more proactive and rapid adaptation will be needed.

(3) There is currently slow progress in developing global emission-reduction agreements beyond the Kyoto Protocol, leading to concerns about the level of future global emissions and hence climate changes and associated impacts.

(4) The high end of the scenario range for climate change has increased over time, and these potentially higher global temperatures may have nonlinear and increasingly negative impacts on existing agricultural activities.

(5) Climate changes may also provide opportunities for agricultural investment, rewarding early action taken to capitalise on these options.

Furthermore, in preparing adaptation responses to climate change, it is also important to note that impacts are unlikely to increase linearly over time (Chapter 1). Current experiences of the initial stages of climate change may therefore be a poor indicator of the challenges that lie ahead. The largest benefits and opportunities presented by climate change are likely to occur early on. For example, the beneficial effects of rising atmospheric CO_2 are expressed almost immediately (within months to decades) and level out as CO_2 continues to rise. At current CO_2 levels we may already be experiencing between a third and a half of the maximum potential benefit that higher CO_2 will ever provide (Chapter 1). Similarly, in the early stages of global warming, some locations with cooler climates may initially experience some benefits, for example through increased growth periods and reduced frost risk. In contrast, the negative impacts of climate change are likely to lag behind the benefits (with some of these lagged negative effects already 'locked in' for the future, even if greenhouse gas emissions are reduced). Furthermore, with continued accumulation of greenhouse gases in the atmosphere, there will be an increase in magnitude and likelihood of negative impacts (such as greater warming, changes in weather-generating circulation patterns, and the possibility of crossing 'tipping points' or other unknown non-linearities in climate change). The initial impacts of climate change, as we are currently experiencing may on balance be relatively benign to slightly positive, but are likely to become progressively negative and more severe through the latter part of this century. Caution is therefore needed not to underestimate the long-term challenge of climate change based on initial, more moderate experiences.

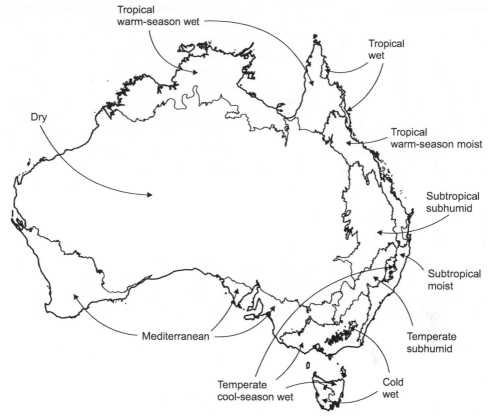

Figure 16.1: Regional variation in impacts of climate variation is discussed using the agro-climate zones of Hobbs and McIntyre (2005).

Regional impacts of climate change

As a means of synthesising the industry-specific impacts of climate change from the previous chapters we provide a geographic overview of how the groups of current agricultural activities may be affected in different regions across Australia. Given the uncertainties in regional-scale climate change projections (CSIRO and Australian Bureau of Meteorology 2007) and their subsequent impacts, these summaries should be taken as indicative scenarios (certainly not as predictions) illustrating the variety of ways climate change could affect regional mixes of agricultural activities. Nonetheless, they are still useful in demonstrating the diversity of solutions that will be required, and in emphasising the need to respond to regionally unique collections of challenges, sometimes with competing demands between industries (particularly over allocation of declining water resources). For this summary, we use the agro-climate zones defined by Hobbs and McIntyre (2005) (Fig 16.1).

Cold wet

Runoff from catchments in the 'cold wet' region is projected to decrease, and declining snowpack may make stream flows less reliable. The risk scores for catchments (Chapter 12) are very low to low in Tasmania, but moderate to high on the mainland (including the Kosciusko Plateau, an important irrigation source for the Murray-Darling Basin). Much of this region is included in National Parks, so native tree species may be at risk from climate change, but few if any commercial forest areas would be affected. There is very little cereal cropping but, for those crops that are grown, rising temperatures may increase crop growth and expand the growing season. Dairy is also likely to benefit from rising temperatures. For the intensive livestock industry more broadly, energy demand for heating production sheds may decrease.

Temperate cool-season wet

Catchment runoff models project a slight decrease (coastal) to a moderate decrease (inland) in water resources of the 'temperate cool-season wet' region. The northern part of this region is an important water source for the Murray–Darling Basin. Flood risk is likely to increase in the east. The catchment risk score is moderate to very high. Declining supplies of irrigation water could significantly reduce winter crop yields and may push existing irrigated systems into a more opportunistic mode. Adaptations include a range of changes in crops and crop management and increased opportunistic cropping. Some areas that were previously too cool for grapes may become suitable for viticulture. Existing grape varieties will ripen in a warmer part of the season, affecting quality, but alternate varieties that would not ripen in the present climate and are suitable for longer growth seasons may become more viable. With lower spring rainfall, the incidence of pests and disease may decline for both grapevines and horticulture. Some marginally cool regions may also become more suitable for horticulture, but extreme hot temperatures may cause more damage in others. This region contains many important plantation areas, including all or part of the Green Triangle (southern South Australia and south-west Victoria), Tasmania, Central Victoria, Murray Valley, Central Gippsland, East Gippsland, Southern Tablelands, Central Tablelands and Northern Tablelands National Plantation Inventory regions. The major plantation species are *Pinus radiata* (radiata pine) and *Eucalyptus globulus* (blue gum). Likely impacts of climate change on these plantations are currently being analysed. Pasture-based dairy is likely to benefit from slight warming, particularly in Tasmania, but irrigated dairy may be exposed to risks of lower water availability.

Temperate subhumid

A moderate decrease is projected for water resources in the 'temperate subhumid' region. Most of the water for this region is sourced from

further east where a moderate decrease in runoff is also projected. The water storages in this region are not as large as on the Murray system and water stress will be greater, so there will be less carry-over water. The catchment risk scores for this region are high to very high. Yields of winter crops such as wheat could decline significantly due to likely reductions in rainfall. Irrigated systems may have to become more opportunistic to cope with reduced availability of water. Adaptations include a range of changes in crops and crop management, increased opportunistic cropping, increased water use efficiency, increased attention to managing stored soil moisture and the use of seasonal climate forecasts. Warmer temperatures may improve growing conditions (longer seasons) for cotton leading to possible industry expansion, if irrigation water is available. Rice growers may have to consider aerobic and alternate-wet-and-dry production options, as reduced water supplies threaten current ponded culture. The likelihood of cold damage during rice flowering may decline, but the risk of crop heat-damage may increase. A wide range of potential farming system changes need to be considered, including greater utilisation/understanding of seasonal climate forecasts. Forestry is not a major activity in the region at present, but there may be future oil mallee plantings for carbon sequestration. Care should be taken in establishing and managing these plantations, since being in already relatively low rainfall areas they would be potentially vulnerable to any further rainfall reductions. Irrigated dairy is likely to be impacted by reduced water allocation, and increased temperatures. There will be a demand for more energy efficient designs of production sheds as the need for cooling in intensive livestock production systems increases.

Mediterranean

Moderate to large decreases in runoff are projected for catchments in the south-west part of the 'Mediterranean' region. In the south-east, where most water is sourced from upstream in the Murray-Darling Basin, substantial decreases in runoff are also projected. The catchment risk scores are moderate to very high (west) and low to very high (east). All cropping in this region is likely to be negatively affected by reduced rainfall, declining water supplies and increasing plant water demand. However, the risks of dryland salinisation may decrease. Grain cropping will become more challenging at the current dry margins but may expand into areas currently too wet for regular cropping. For rice, cold damage during flowering may decline, but risks of crop heat damage may increase. Alternate-wet-and-dry rice production systems could be considered for the future. Winegrape vines may ripen in a warmer part of the season, affecting quality. Reduced chilling over winter may affect suitability for growing of some perennial fruit crops, while extreme hot temperatures will have to be managed. There are major areas of commercial forestry, particularly in Western Australia and South Australia. Bioclimatic analysis should be used to identify the most vulnerable *Eucalyptus globulus* (blue gum), *Pinus radiata* (radiata pine), *P. pinaster* (maritime pine) and oil mallee plantings, so these can be monitored to provide early warning of any problems and to guide future plantings. Priorities for the dairy industry will be cooling solutions to reduce animal heat stress.

Dry

Runoff is projected to increase slightly in the central south and decrease slightly on the fringes of the 'dry' region, especially in the east. The catchment risk scores are very low (western half) and low to moderate (eastern half). This vast arid area is likely to experience the greatest warming and drying trends within the rangelands. This will further stress many grazing enterprises that are already only marginally viable and where few opportunities for adaptation exist. Most of this region is too arid for commercial forestry though there may be some potential for oil mallee and carbon sequestration plantings in the higher-rainfall edges of the region. Care should be taken in establishing and managing these plantations as, being in already relatively low rainfall areas, they would be potentially vulnerable to any further rainfall reductions. Small irrigated cropping areas may be challenged by increasing water demand and decreasing supplies.

Subtropical moist

Runoff from catchments in the 'subtropical moist' region is projected to decrease slightly, but flood risk is likely to increase. Most water storages are not as substantial as those inland. The catchment risk score is moderate to very high. Sugarcane producers in this region will likely have to improve efficiency of water use to deal with declining water supplies while sea level rise may threaten very low-lying production areas. Projected warming will increase the duration of the growing season, so planting sugarcane earlier in the season could be considered. Suitability for other crops may increase, particularly from short-duration annual crops, providing opportunities for diversification. Horticulture may benefit from reduced frosts, making the climate suitable for possible expansion. There are significant areas of hardwoods particularly *Eucalyptus pilularis* and *E. grandis* in this region. Preliminary analyses suggest neither species is likely to be at high risk. Irrigated dairy will have to deal with reduced water allocations and greater need for cooling of production sheds (requiring more thermally efficient designs).

Subtropical subhumid

Runoff from 'subtropical subhumid' catchments is projected to greatly decrease and warmer conditions will increase water stress. Self-extracted water sources will be less well buffered than distributive systems. The catchment risk score is high to very high. There could be potentially significant reductions in yield and quality of winter crops such as wheat due to likely reductions in rainfall and existing exposure to high temperatures in the northern part of this region. Potential reductions in frost may increase crop options. Summer rainfall seems equally likely to increase as to decrease and this also may provide some options to alter the balance of cropping. Cotton will be sensitive to reduced irrigation water and higher temperatures with negative effects on yield and fibre quality. Improved water use efficiency (irrigation practice and variety choice) and tolerance to heat stress (variety choice) and modification of crop management (planting date, row configurations, irrigation scheduling) will need to be reconsidered to help offset these issues. Plantation forestry is not a major activity in the region. The dominant forestry species *Acacia harpophylla* (Brigalow) is widely distributed and is unlikely to be at high risk from climate change. In grazed savanna areas, the negative effects of declines in rainfall and increasing incidence of drought on pasture productivity may initially be offset by the benefits of higher CO_2 and a prolonged growing season from warming (but would be negatively affected later if rainfall declines significantly). More intense rainfall may increase the risks of soil erosion, rising CO_2 may favour trees at the expense of pasture production, and pasture quality may decline. Feedlots will have to offset the potential for greater animal heat stress and production sheds for intensive livestock industries will require more cooling.

Tropical warm-season moist

A slight to moderate decrease is projected for runoff from catchments in the 'tropical warm-season moist' region. Localised areas of irrigated tropical agriculture are at risk of drought. The catchment risk scores in this region are moderate to very high. Sugarcane, which is the dominant crop in this area, falls into three production regions (Chapter 8). In all regions, crop damage from wind and cyclones may increase. Soil erosion may increase, with greater nutrient and sediment runoff into the Great Barrier Reef lagoon. In the Northern region, increased waterlogging may limit paddock access, particularly during the growing season. Reduced spring rain would negatively impact crop establishment. In the Herbert and Burdekin region, the security of water supply from the Burdekin Dam may be reduced. Improved irrigation practices will be required to deal with rising water tables and salinity issues, problems that will be exacerbated by rising sea levels. Declines in winter and spring rain may increase trafficability, improving harvesting efficiency. In the Central region, limited water supplies may be further strained by

projected drying. Warming will extend growing seasons and improve crop growth in the frost-prone western districts. Poor drainage and tidal intrusion in the lower floodplains are likely to be exacerbated by projected sea level rise.

Horticulture might expand in this region to take advantage of decreased frost risk. Heat stress, flooding, erosion and cyclones can all have devastating impacts and risk assessments will need to be undertaken with regard to these. Cropping cycles will change with increasing temperatures, requiring corresponding adjustments in intake scheduling and marketing. South-east Queensland is the major forestry centre in this region, with pine species accounting for about 83% of the plantation area. The main species, *Pinus elliottii* (slash pine) *P. caribaea* (Caribbean pine) and their hybrid, as well as the native *Araucaria cunninghamii* (hoop pine) are also grown further north, so selected sites should be monitored to provide some early warning of any problems associated with climate change. The productivity of rangelands in this region may be negatively affected by a slight decline in rainfall and increasing incidence of drought. More intense rainfall may increase the risks of soil erosion, rising CO_2 may favour trees at the expense of pasture production, and pasture quality may decline. Irrigated dairy is likely to be impacted by reduced water allocations and increased temperatures. Cloud cover during the wet season will continue to be an issue for pasture and sugarcane growth in the north. Production sheds for the intensive livestock industry are likely to require more cooling to cope with animal heat stress.

Tropical warm-season wet

Runoff projections for catchments in the 'tropical warm-season wet' region show a slight increase on Cape York to slight decrease further west. There is only one substantial water storage, being Lake Argyle. Catchment risk scores for this region are very low to low in the west and low to moderate on Cape York. Cropping is limited in this region, but there is some dryland cropping, with summer crops such as sorghum, and some irrigated cropping. All crops are likely to be increasingly negatively affected by high temperatures. But if rainfall increases (about 30% is possible) this may alleviate restricted growing seasons for the dryland crops in particular. Options for the establishment of a sizeable cotton industry are being considered within the Ord Irrigation Area, and climate change impacts should be taken into account in planning and assessment. Horticulture may expand as frost risk decreases. However, heat stress, flooding, erosion and cyclones can all have devastating impacts and risk assessments will need to be undertaken with regard to these. Growing cycles will change with increasing temperatures, so corresponding adjustments will be required for intake scheduling and marketing. The area of plantations in the Northern Territory is relatively small, but expanding rapidly. The total plantation area in 2005 was about 16 000 ha, of which 85% was *Acacia mangium*. Similar areas of *Khaya senegalensis* (African mahogany) are likely to be planted in coming years. Of the rangelands, the northern savannas are among the least likely to experience productivity declines under climate change. But more intense rainfall may increase the risks of soil erosion, rising CO_2 may favour trees at the expense of pasture production, pasture quality may decline and animal heat stress will become more of a challenge. Production sheds in the intensive livestock industry will require more cooling, and more energy efficient designs could be considered for this purpose.

Tropical wet

Runoff projections for the 'Tropical wet' region catchments are for a slight increase and catchment risk scores are low to moderate. There is no notable irrigated agriculture in the region and sugarcane is the dominant crop. Prospects of lower rainfall may increase sugarcane yields through increased sunshine, but there could be increased risks of soil erosion and pests and diseases. Adaptations include altered planting regimes and changing harvest chains. The plantation area of the Ingham-Cairns region of north Queensland is not large, but it may provide some useful early warning of possible climate warming problems for important pine plantations further south in south-east Queensland. Conserving highly species-diverse tropical rainforests in their current condition may be difficult under climate change, but the diversity of species may provide some capacity to adapt to climate change.

Fisheries and aquaculture

For the northern fisheries region (refer to Fig. 13.1 for fisheries regions), changes in patterns of rainfall and freshwater flows may impact important nursery areas, such as estuaries and mangroves, indirectly impacting fisheries catch. Biological responses to increased temperature include enhanced recruitment and growth of some species such as prawns. Sea level rise and more intense extreme events (storms and cyclones) could increase risks to infrastructure for aquaculture, such as prawn ponds, and nursery areas for wild fisheries.

In the western region, a number of commercially fished species are closely linked to Leeuwin Current dynamics. Changes to this current are at too fine a scale to be captured by existing climate models, but some weakening is expected and this would affect fish populations. However, the overall pattern in ocean temperature change is uncertain. An increase in extreme events may impact aquaculture operations (e.g. oyster farming) and changes in sea level need to be considered by coastal aquaculture operations.

In the south-eastern region, a strengthening of the East Australia Current will exacerbate increases in water temperatures in south-east Australia (projected to be the greatest anywhere in the southern hemisphere). Southward shifts in species' ranges have already been recorded in conjunction with ocean changes over the past 30 years. Possible increases in wind-driven upwelling in the Great Australian Bight, could result in increased productivity at the base of the food chain. Wind influences recruitment cycles of some wild-fished species, however how these may change in the future has not been resolved. Warming waters also have implications for aquaculture operations, such as salmon farming.

In the sub-Antarctic region, it is expected that sea ice will become more seasonal and decline in northern extent. Increased temperatures will increase growth rates of some species (up to a point). Ocean acidification is expected to reduce productivity at the base of the food chain.

For pelagic fisheries in offshore parts of all regions, fish distributions are expected to follow changing ocean temperatures, particularly the strong warming in south-east Australia. Tropical species may increase in availability, while temperate species that move southwards are expected to become less available.

Coping with uncertain change

Risk-based approaches

An essential element in adapting to climate change will be to accept the inherent uncertainty in future climate change scenarios, including uncertainties in the geographic variation in projected changes. There are many sources of uncertainties that form a cascade, each flowing and compounding, one after another. For example, we do not know for certain what actions humans will take globally to increase or reduce future greenhouse gas emissions (IPCC 2007a). For any given emissions scenario, there are uncertainties in the global carbon cycle and hence how much of the greenhouse gas remains in the atmosphere. For a given change in atmospheric composition, the resultant climate changes are only generally known and these become less reliable when moving from global to local scales (Chapters 1 and 2). For any given climate change projection, there are further uncertainties about what the impact will be on current land (and ocean) use practices and the underpinning natural resources (Chapters 3 to 13). It is then uncertain how affected communities will respond, or how their responses will either reduce or exacerbate the impacts of climate change (Chapters 14 and 15). For most of these sources of uncertainty there is greatest confidence in broad generalisations at a global scale, and confidence declines progressively (due to spatial variation) in moving down to national, regional, local and enterprise scales.

Given this uncertainty, it will initially be much more important to focus on preparing a suite of adaptation options to encompass the range of projected impacts that may occur, rather than trying to tailor location-specific adaptation strategies to precise local projections of climate change. What is needed is a flexible, risk-based approach that prepares and equips primary producers to be able to cope with a range of foreseeable scenarios, and develops the capacity to implement, monitor and adjust these as needed. Such changes will need to be supported by synergies with government policies such as self-reliance during drought and the community-based activities such as Landcare. Likewise, similar flexibility and adaptability will be required from supporting institutions, in that they will need to continuously monitor and evaluate climate data and adaptation measures to make appropriate adjustments to supporting strategies and policies.

It may be tempting to delay action until there is certainty about the impacts of climate change and the effectiveness of adaptive responses. But it is unlikely that we will have that luxury. While there will be continuing technical advances in the science of projecting climate change and its impacts, some uncertainty will always remain. Uncertainty can only be eliminated by waiting until climate change impacts are actually realised (or are unavoidable), which undermines the basis and effectiveness of pre-emptive adaptation. This would leave primary industries on the back foot having to reactively deal with climate change impacts after the fact, while still having to prepare to deal with future climate change (which would remain uncertain). An essential aspect of adapting in the face of uncertainty will be to accept that decisions made on the best available, imperfect knowledge may, in retrospect, turn out to be suboptimal. It will therefore be equally important to collectively learn from such experiences and incorporate this and other new knowledge in iteratively improving adaptive responses over time.

Means to extremes

Throughout this book there has been a clear theme demonstrating the strong links between adaptations to climate change and strategies for coping with climate variability. There are several reasons why building capacity to cope with climate variability will serve as a strong starting foundation in preparing to deal with climate change. (1) In the short term, land managers are more likely to be concerned about the immediate challenge of year-to-year variation, because climate change effects will largely be within the bounds of existing climate variability; (2) adjusting management for seasonal differences from year to year builds in automatic tracking of longer-term climate trends; (3) these approaches encourage a shift from less flexible management styles to ones where managers adaptively monitor and respond to changes over time, and (4) managing for extreme events is already important in agriculture and will become more so as extreme events become more common under climate change.

As the agricultural sector has become more adept at coping with the risks of climate variability, there has been growing awareness of the importance of managing for extreme events relative to inflexible strategies suited to 'normal' (or 'mean') years. Differences in the long-term viability of enterprises are often distinguished more by a manager's ability to respond to extreme events than their ability in 'mean' years. For example, in pastoral enterprises, severe, rapid degradation can occur from inappropriate management responses to drought events, undermining future enterprise viability (McKeon *et al.* 2004). While on the positive side, most profits are made in the minority of highly favourable years (McKeon *et al.* 2004). Management based on past experience of 'mean' years can lead to negative impacts (environmental and financial) in extremely bad years and missed opportunities in extremely favourable years. With climate change, as conditions deviate more and more from mean conditions over time, past experience of mean climate conditions will become increasingly less relevant. The enterprises that are most likely to remain viable are those that are responsive to the altered occurrence of extreme events and are willing to forgo preconceptions based on past experiences to make the necessary, possibly novel, management adjustments. For primary industries, where managers are often characterised by a strong sense of self-reliance and trust in empirical knowledge derived from past personal experience, it will be essential to broaden existing networks of knowledge-sharing and learning beyond traditional rural community groups to collaborate more closely with those institutions that can assist in providing information and developing strategies to adapt to climate change.

Enhanced strategies for coping with climate variability will provide a way for encouraging gradual, incremental adjustments to climate change primarily using existing management options. However, by itself, this approach may be insufficient in preparing primary industries to face emerging novel climatic conditions. It will also be necessary to build in considerations of longer-term climate change trends, including preparing novel management options, particularly for those

communities where climate change poses the greatest threat. Marginal production areas are likely to be among the most vulnerable to climate change. In locations where primary industries are already under financial stress, even slight increases in the severity or frequency of extreme weather events could exceed their capacity to cope. It is these enterprises that are most likely to provide the early warning signs of climate change. Preferably, the most vulnerable industries and communities will be identified and assisted in advance. Such situations are likely to require more intensive policy intervention than simply supporting adaptation of existing land uses and could involve transformation to quite different types of activities. Such transformative changes in land use should not be viewed only in a negative, reactive way, since climate change may also produce new opportunities where alternative land uses are more productive or desirable than existing ones. In both cases it will be important to recognise where the greatest changes in land use and management practices may be required so that the communities can be supported in dealing with the social upheaval and risks that these transitions will create.

Common adaptation priorities

The preceding chapters have identified a number of potential options for individual primary industries to adapt to climate change. We have summarised those that have emerged as having broad potential applicability across the agricultural sector (Table 16.1). Many adaptation options are extensions or enhancements of existing activities that are aimed at managing the impacts of climate variability and improving the sustainability and efficiency in the use of natural resources. However, less than a dozen potential adaptation options have been fully evaluated (e.g. technical feasibility, end user acceptability, costs and benefits) to determine their effectiveness in responding to climate change. The few analyses that have considered the broader costs and benefits of adaption have shown that practicable and financially viable adaptations will have very significant benefits in ameliorating risks of negative climate changes and enhancing opportunities where they occur (Chapter 3). The benefit-to-cost ratio of undertaking research and development into these adaptations appears to be very large

Table 16.1: Summary of climate change adaptation issues that are shared across primary industries.

Policy
Develop links to existing government policies and initiatives (e.g. Greenhouse Gas Abatement Program, Greenhouse Challenge Plus, Landcare, salinity, water quality, rural restructuring) and into integrated catchment management so as to enhance the capacity to adapt to climate change
Managing transitions
Develop policies and mechanisms to provide technical and financial support as adaptation measures are adopted. Managers' past personal experience will become decreasingly relevant as ecosystems, land uses, and management practices change, requiring greater support by and collaboration with institutions that can facilitate transitions to new systems that are better adapted to the changing climate
Accepting uncertainty and change
Enhance capacity for land and marine managers and supporting institutions to deal with uncertainty. Current and future actions will have to be taken based on uncertain enterprise- / regional-scale predictions and observations of climate changes. Adaptation strategies will need to enhance adaptive capacity by ensuring that rural communities are equipped to cope with a range of possible, but uncertain changes in local climatic conditions
Communication
Ensure communication of broader climate change information as well as industry-specific and region-specific information as it becomes available
Monitoring climate change and adaptation responses
Maintain effective climate data collection, distribution and analysis systems to link into ongoing evaluation and adaptation. Monitor climate conditions and relate these to yield and quality aspects to support/facilitate adaptive management. Develop climate projections that can be downscaled so as to be relevant to farm, catchment and coastal scales. Consideration could be given to the introduction of climate change adaptation into Environmental Management Systems
Research and development and training
Undertake further adaptation studies that include broad-based costs and benefits to inform policy decisions. Maintain the research and development base (people, skills, institutions) to enable ongoing evaluation of climate/CO_2/(cultivar, species or land use)/management relationships, and to streamline rapid R&D responses (for example, to evaluate new adaptations or new climate change scenarios). This R&D needs to be developed in a participatory way so that it can contribute to training that improves self-reliance in the agricultural sector and provides the knowledge base for farm-scale adaptation
Breeding and selection
Maintain public sector support for agricultural biotechnology and conventional breeding with access to global gene pools so as to have suitable varieties and species for higher CO_2 and temperature regimes and changed moisture availability
Model development and application
Develop further systems modelling capabilities such as APSIM for crops and AussieGrass and GrazFeed for grazing that link with meteorological data distribution services, and can use projections of climate and CO_2 levels, natural resource status and management options to provide quantitative approaches to risk management for use in several of these cross-industry adaptation issues. These models have been the basis for successful development of participatory research approaches that enable access to climate data and interpretation of the data in relation to farmers' records and alternative management options. Such models can assist proactive decision-making on-farm, inform policy and extend findings from individual sites to large areas
Coping with climate variability
Facilitate the adoption of seasonal climate forecasts (e.g. those based on El Niño and La Niña, sea-surface temperatures, etc.) to help farmers, industry and policy incrementally adapt to climate change while managing for climate variability. Maximise the usefulness of forecasts by combining them with on-ground/water measurements (e.g. soil moisture, nitrogen, ocean temperature), market information and systems modelling
Pests, diseases and weeds
Maintain or improve quarantine capabilities, sentinel monitoring programs and commitment to identification and management of pests, diseases and weed threats. Improve the effectiveness of pest, disease and weed management practices through predictive tools such as quantitative models, integrated pest management, area-wide pest management, routine record keeping of climate and pest/disease/weed threats, and through development of resistant species and improved management practices
Plant and animal nutrition
Adjust nutrient supply to maintain grain, fruit, fibre and pasture quality through application of fertiliser, enhanced legume-sourced nitrogen inputs or through varietal selection or management action. Note however, that this may have implications for greenhouse emissions (via field-based emissions of nitrous oxide or emissions of CO_2 during manufacture). Any increases in nutrient supply will have to be carefully managed to minimise soil acidification, waterway eutrophication or runoff into estuaries and marine systems

Table 16.1: Summary of climate change adaptation issues that are shared across primary industries. (Continued)

Water
Increase water use efficiency by (1) a combination of policy settings to encourage development of effective water-trading systems that allow for climate variability and climate change and that support development of related information networks; (2) improving water distribution systems to reduce leakage and evaporation; (3) developing farmer expertise in water management tools (crop models, decision support tools), and (4) enhancing adoption of appropriate water-saving technologies
Land use/location change and diversification
Undertake risk assessments to evaluate needs and opportunities for changing varieties, species, management or land use/location in response to climate trends or climate projections. Support assessments of the benefits (and costs) of diversifying farm enterprises
Environmental interactions
Determine the impact of climate change (interacting with land management) on environmental problems such as salinity risk (both dryland and irrigated), water quality, land degradation and river sediment loads. Inform policies, such as those on salinity and water quality, accordingly

(indicative ratios greatly exceed 100:1). There is thus a clear need to evaluate the system-wide costs and benefits of implementing adaptations (including socio-economic aspects and potential feedbacks through greenhouse emissions) for a wider range of cases. Such evaluations should be undertaken in a way that engages with industry groups so as to deal effectively with their key concerns, draw on their valuable expertise and contribute to enhanced knowledge in the agricultural community.

Successful adaptation to climate change will need both strategic preparation and tactical response strategies. Adaptation measures will have to reflect and enhance current 'best-practices' designed to cope with adverse conditions such as drought. Adoption of these new practices will require, among other things (Howden *et al.* 2007) (1) confidence that the climate really is changing; (2) the motivation to change in order to avoid risks or seize opportunities; (3) demonstration of the benefits of new adaptation options; (4) support during transitions to new management or new land use; (5) building capacity for communities to develop and implement adaptation strategies; (6) altered transport and market infrastructure, and (7) an effective system to monitor climate change impacts and human adaptive responses to iteratively evaluate and adjust policy and management decisions.

This book has highlighted some positive options and opportunities for taking immediate action to prepare agriculture for the challenges of climate change. By resolving to face these challenges now we can lessen the future impact of climate change, and avoid the more hasty and disruptive policy interventions and management adjustments that would be required if responses were delayed.

Acknowledgements

Maps of the agro-climate zones were prepared by Brett Abbott. Each chapter author provided contributions for their primary industry on the projected impacts of climate change across the agro-climate zones. We are also very grateful to Pauline Simonetti and Janet Walker for their assistance in proofreading and editing all the chapters in this book, including this one.

References

CSIRO and Australian Bureau of Meteorology (2007). 'Climate change in Australia: technical report 2007'. CSIRO and Australian Bureau of Meteorology, Melbourne.

Garnaut S (2008). 'The Garnaut climate change review: final report'. Cambridge University Press, Cambridge.

Hennessy K, Fitzharris B, Bates BC, Harvey N, Howden SM, Hughes LSJ, Warrick R (2007). Australia and New Zealand. In *Climate Change 2007 – Impacts, Adaptation and Vulnerability – Contribution of Working Group II to the Fourth Assessment Report of the International Panel on Climate Change*. (Eds ML Parry, OF Canziani, JP Palutikof, PJ van der Linden, CE Hanson) pp. 507–540. Cambridge University Press, Cambridge.

Hobbs RJ, McIntyre S (2005). Categorizing Australian landscapes as an aid to assessing the generality of landscape management guidelines. *Global Ecology and Biogeography* **14**, 1–15.

Howden SM, Soussana JF, Tubiello FN, Chhetri N, Dunlop M, Meinke HM (2007). Adapting agriculture to climate change. *Proceedings of the National Academy of Sciences* **104**, 19691–19696.

IPCC (2007a). *Climate Change 2007 – Mitigation of Climate Change – Contribution of Working Group III to the Fourth Assessment Report of the International Panel on Climate Change.* Cambridge University Press, Cambridge.

IPCC (2007b). *Climate Change 2007 – Synthesis Report – Fourth Assessment Report of the International Panel on Climate Change.* Cambridge University Press, Cambridge.

IPCC (2007c). *Climate Change 2007 – The Physical Science Basis – Contribution of Working Group I to the Fourth Assessment Report of the International Panel on Climate Change.* Cambridge University Press, Cambridge.

McKeon G, Hall W, Henry B, Stone G, Watson I (2004). 'Pasture degradation and recovery in Australia's rangelands: learning from history'. Queensland Department of Natural Resources, Mines and Energy, Brisbane.

17
LOOKING FORWARD

AJ Ash, CJ Stokes and SM Howden

KEY MESSAGES:

- We highlight and discuss some priority challenges to be faced in adapting to climate change: managing risk and uncertainty; balancing incremental adaptation and transformational change; considering mitigation-adaptation interactions; conserving biodiversity and natural resources, and building adaptive capacity.

- Given the imperative for immediate action, decisions will have to be made while there is still some uncertainty about climate change and its impacts on primary industries. While climate change projections and assessments of climate change impacts will improve over time, these are unlikely to ever be entirely accurate and some uncertainty will always remain. Policy and management strategies will therefore have to take broad risk-based approaches that accommodate uncertainty (rather than focussing too narrowly on technical solutions to location-specific climate change projections).

- In the early stages of climate change, incremental changes in agricultural practices may be sufficient. But with greater levels of climate change, or for areas that are particularly vulnerable to climate change, transformative adaptations will be required. Where major transitions in agricultural practices are necessary, communities may require additional assistance.

- Climate change adaptation strategies will have to strike a balance that addresses both the technical and capacity issues. There will be a need to develop technical management options that are appropriate for altered future climate scenarios. However, it will also be essential to (1) support these technical developments by building capacity for communities to plan and implement adaptation strategies, and (2) address institutional and policy constraints to adaptation.

- While much of this book has focussed on initial actions that can be taken with what we currently know, significant knowledge gaps still need to be addressed before more comprehensive and effective adaptation strategies can be implemented.

Introduction

This book has clearly demonstrated how no primary industry or rural region of Australia is immune from the effects of climate change. The need for adaptation is clear and while there are huge challenges, Australia's primary industries, which already have to manage extreme climatic variability, are in a position to lead Australian industry in innovative and proactive responses to climate change.

There is a relatively small window of opportunity over the next two to three decades where the negative impacts of climate change on primary industries will be modest, and for some regions and sectors there will be opportunities to take advantage of rising temperatures and carbon dioxide concentrations. This period of opportunity should be used to develop new technologies and new practices that will allow agriculture to cope with the more severe consequences of climate change that appear likely to play out over the next 50 years.

Priority challenges

In developing these new technologies and options, five priority challenges will need to be addressed.

Managing risk and uncertainty

Primary industries have always had to manage with uncertainty and make decisions with imperfect information, whether that be uncertainty about future markets and commodity prices or the climate for the next season. Flexible, risk-based approaches to management have been developed in response to these uncertainties and the emergence of climate change as a major production and resource management risk strengthens the need to ensure that appropriate approaches to risk management are in place.

In this context a common theme throughout this book is the need to better adapt to climate variability as a means to assist with managing the effects of climate change. There is much more that primary industries could be doing to adapt to climate variability, from increasing the flexibility of management strategies to more explicitly incorporating seasonal climate forecasts into management decisions. In many regions of Australia, the lack of seasonal forecast skill is limiting its use (Ash *et al.* 2007) and there is a clear need to improve forecast accuracy. More sophistication in the delivery and use of climate information at the seasonal and annual scale will build capacity in the primary industries and pave the way for more informed use of longer term climate change projections.

However, there is a risk in being over-reliant on climate predictions and projections. A mindset can develop that action cannot be taken until climate science can deliver climate change projections that have more precision and accuracy. This can have the unintended consequence of delaying adaptation in the face of climate change.

The accuracy of climate change projections is limited by fundamental, irreducible uncertainties (Dessai *et al.* 2009) brought about by an incomplete understanding of the climate system, the inability to model chaotic processes and uncertainty about future greenhouse gas emissions. So rather than pursuing as a sole strategy of ever-increasing precision and accuracy, more effort needs to be put into developing approaches to delivering climate scenarios for risk management, which deal less with precision and more with encompassing a realistic diversity of possible futures at finer resolution (Dessai *et al.* 2009) and a more comprehensive understanding of the possible responses and their utility. This also is akin to the experience that seasonal forecasting science has found in Australia in distinguishing model skill from the net economic benefit of decisions supported: the two are not necessarily directly related. This would argue for two developments: (1) a very small number of credible scenarios for generalised risk management (e.g. 3–4 scenario storylines, which are regionalised and which help decision thinking in less formal analyses), and (2) mechanisms to support comprehensive exploration of decision-making spaces with many model runs (of low precision) in a robust decision-making approach.

Balancing incremental adaptation and transformational change

There are many potential incremental adaptation options available to offset projected impacts (Howden *et al.* 2007). These incremental adaptation options often involve building on existing approaches to manage existing climate variability. They range from new plant varieties, improved water use efficiency and other paddock-scale options to much more significant change in enterprise mix and land use change. Implementation of these options is likely to have substantial benefits under moderate climate change for some cropping systems (Howden and Crimp 2005; Howden and Jones 2004).

However, it is highly unlikely that even well planned incremental innovation will suffice in response to some of the future projected climate changes. Recent syntheses of the value of adaptation options to agriculture suggest that there are significant limits to the benefits of adaptations within existing agricultural systems (Easterling *et al.* 2007; Howden *et al.* 2007). These studies concluded that more transformational change will be required to adapt to significant climate change. So a major challenge for agriculture is to plan ahead and determine where and when incremental adaptation will not be sufficient to cope with climate change and to develop approaches that will be required to address transformational change in land management and land use. This level of forward planning is unprecedented in Australian agriculture and needs to be supported by science, industry groups and government policy.

This is just not a simple case of acquiring more certain climate change projections and determining which areas may no longer be climatically suitable for current agricultural, forestry or fishing activities. Other factors that need to be considered include social and economic vulnerability, adaptive capacity, new technologies, governance systems, and policies such as structural adjustment. For example, a possible response to changing distributions of pelagic fish stocks is for commercial fishermen to change their home port. This brings into consideration whether there is adequate infrastructure to support such a locational change and the impacts on the social and economic wellbeing of the communities from which the industry is departing and those it will be entering.

To illustrate this point further there have been over the last year or so calls for some farming regions in marginal environments to be eased out of farming ahead of the projected climate changes. However, many factors other than climate determine vulnerability (Nelson *et al.* 2005) and within marginal environments there are some farmers who are struggling and are not viable while others are at the forefront of innovation and risk management and are both economically viable and environmentally sustainable. This highlights the complexity of vulnerability and adaptive capacity and the potential dangers in a 'one size fits all' policy. Clearly climate change will put some regions of Australia under immense economic, social and environmental stress with implications for both individual farmer and rural community viability.

Transformational change to address these climate changes needs to be developed with a full systems understanding of the climatic, economic and social forces at work. This is partly because the combined economic and environmental drivers of global change may interact in surprising ways. For example, the significant ongoing changes in climate predicted by the IPCC will impact an Australian agricultural sector that is under constant change from international market forces. This is dramatically illustrated by a continuous consolidation of farms since the mid 1960s associated with long-term declines in farmers' terms of trade. Farm numbers in Australia have fallen from around 200 000 in 1966–67 to around 130 000 in 2004–05, resulting in significant economic and social change in rural communities. In this transformational change process, larger and more efficient farms have been able to increase their profitability through technical and management innovations that increase their productivity. However, these changes have also led to social impacts in some regions, such as the depopulation and economic decline of some rural communities.

The development of new on-farm technologies and management practices and exploration of the economic, environmental and social costs and benefits of transformation options will require a significant investment in research and development. Currently there is considerable emphasis on mitigation and the challenges and opportunities for agriculture and forestry in reducing greenhouse gas concentrations in the atmosphere. However, given the inevitability of some climate change and the vulnerability of Australian agriculture to climate change our efforts in adaptation must be accelerated.

Adaptation-mitigation interactions

Australia is set to introduce in 2010 an emissions trading scheme with an emissions reduction target of 5–15% by 2020. While forestry can participate in the Carbon Pollution Reduction Scheme (CPRS) on an opt-in basis it is still not clear whether agriculture will be included in the CPRS, and even if it is, the suggested start date for agriculture is 2015. According to the National Greenhouse Gas Inventory, agriculture is responsible for 16% of the National emissions directly, another 11% in net emissions from land clearing and an additional unknown amount from the use of energy, transport and other processes. Thus, agriculture is associated with at least a quarter, and up to a third, of the National greenhouse gas emissions. This significant emissions profile provides a strong rationale for focussing on emission reductions through a variety of pathways.

Regardless of whether agriculture is included in the CPRS, there is inevitably going to be significant investment in carbon offsets in rural landscapes, most commonly through afforestation but also possibly through soil carbon and altered fire regimes. These offsets will be driven in part by participants in the CPRS endeavouring to achieve their emission caps but also by non-participants of the CPRS looking to reduce their net greenhouse gas emissions or as a business proposition.

Adaptation to climate change in agriculture and forestry needs to be undertaken with a sound understanding of its interactions with mitigation. These adaptation-mitigation interactions will most likely be played out in four main areas: (1) increased farm input costs as a result of the CPRS; (2) the need to reduce emissions from agriculture whether formally as part of a CPRS or through another incentive and/or regulatory approach; (3) changes in land use through widespread use of carbon sequestration in agricultural lands, and (4) carbon sequestration initiatives and the security of stored carbon in response to climate change.

Farm costs will increase with the introduction of the CPRS through the flow-on effects of increased fuel and energy costs. Agriculture is on the whole a price taker and is often unable to even partly pass on increased costs. Estimates of the impact of the CPRS indicate a reduction in farm gross margins of between 3% and 9% (Keogh and Thompson 2008). There is a risk that reduced farm profitability will make it more difficult to implement innovative adaptation practices and technologies.

Whether agriculture is part of the CPRS or not, it will need to reduce emissions to help achieve National greenhouse gas targets. Strategies to reduce emissions may interact with climate change impacts and adaptation in a number of ways. For example, reductions in rainfall in a cropping zone may make livestock a more attractive enterprise option but ruminant livestock are a major contributor to agricultural emissions through methane production; i.e. a negative interaction emerges between adaptation and mitigation. However, there are also examples of synergistic adaptation-mitigation interactions; e.g. use of zero tillage as a strategy to conserve water has the positive effect of storing slightly more carbon in the soil surface layers than more conventional tillage practices.

There are likely to be significant changes in land use in response to the CPRS and other initiatives to reduce net greenhouse gas emissions through offsets. These changes may possibly see a significant amount of land that is currently being used for agriculture converting to forests or other forms of permanent vegetation. At the enterprise level, decisions on investment in adaptation (e.g. infrastructure, new technologies, type of enterprise) need to be made in the context of this significant 'offset' driver as in some situations it may be more profitable to focus on emission offsets rather than continue with an adapted form of agriculture and

vice versa. Non-business social factors such as farming tradition and lifestyle will likely play a significant role in these decisions.

An important adaptation-mitigation interaction will be how well adapted some of the carbon sequestration measures are to the direct and indirect impacts of climate change that will play out over the next few decades. Considerable effort has gone into the establishment of plantation forests in recent years and it is not clear how much thought has gone into selecting species suitable for a future climate. Forests, especially in southern Australia, are likely to be exposed to increased risk of bushfire in the future and again it is unclear how well adapted these plantings are to this changing risk. In addition, most studies suggest an increased risk of pests and disease under a changed climate and this will pose some particular threats for the sustained growth and health of forest offsets. Similar stresses may occur through reduced water availability. Evidence of these sorts of threats is already emerging in Canada with warming contributing to increased outbreaks of the mountain pine beetle that are damaging large areas of high-elevation forest and turning them from carbon sinks to carbon sources (Kurz *et al.* 2008).

Given the need to ensure the long-term security of these carbon stocks, both from a carbon trading and a national greenhouse inventory perspective, more effort needs to go into understanding the impacts of climate change on carbon sequestration potential.

Conservation of biodiversity and natural resources

Primary industries are dependent on ecosystem processes to sustain their productivity and it is these ecosystem processes that are highly vulnerable to climate change. Likewise, the health of natural ecosystems and biodiversity are highly vulnerable to climate change. There are relatively fewer options available for planned adaptation to climate change in natural ecosystems than in heavily managed agricultural systems. Some of these options include expanding reserve systems, improving connectivity, reduction of a variety of threatening processes and assisted migration.

However, there is a limit to the feasibility of these options and given that much primary industry production occurs in relatively intact ecosystems (rangelands, marine ecosystems) through to more fragmented landscapes (broadacre agriculture, plantation forestry) increasing emphasis will be placed on novel ways of conserving biodiversity in the matrix of primary production.

This will lead to some trade-offs and win-win opportunities. For example Ecosystem Based Fisheries Management provides an opportunity to improve the resilience of both marine ecosystems and fisheries production in the face of climate change (Chapter 13). In agricultural landscapes there maybe some trade-offs required, e.g. habitat re-engineering of developed agricultural land to improve biodiversity outcomes. This may not necessarily involve trying to restore habitat to its former condition but rather taking the opportunity to create novel landscapes that might be better suited to some key species under a changed climate. This brings into play related issues such as stewardship arrangements, which are gaining momentum under existing policies for ecosystem management, but they will need to be reframed and reemphasised in the context of climate adaptation.

Building adaptive capacity

Adaptive capacity is the ability of a human system or ecosystem to adjust or respond to climate impacts. Having the necessary adaptive capacity helps primary industries to moderate the impacts of climate variability and change and to take advantage of new opportunities arising from climate change. Adaptive capacity is dependent on having a good balance of five forms of capital: human, social, natural, physical and financial (Nelson *et al.* 2007, Chapter 15) that together provide the ability to cope with the challenges posed by climate change and variability. We are at an early stage in characterising, mapping and enhancing adaptive capacity across our primary industries and this is a key knowledge gap that needs to be addressed if we are to successfully target adaptation options.

A key component in building adaptive capacity is to better understand the attitudinal, social, financial, behavioural, institutional or environmental barriers to adopting adaptation measures (Chapter 15). Howden *et al.* (2007) have suggested a number of approaches to overcome these barriers to build adaptive capacity and to change the decision environment. These approaches were

outlined in the summary (Chapter 16) but more effort needs to be put into the structured implementation of these measures if significant progress in adaptation is to be achieved.

Summary and knowledge gaps

While significant progress is being made on understanding the vulnerability of primary industries to climate change and the exploration of various adaptation options, significant gaps in knowledge and implementation remain. These include:

- improving the skill and reducing the uncertainties of seasonal climate forecasts and climate change projections but in a way that does not reinforce the rather limiting linear approach of action on adaptation being predicated on better climate science;

- developing adaptation technologies that will assist primary industries to maintain their viability in the medium term, e.g. climate-ready crops, improved management systems;

- taking a systems approach to assessing the opportunities and trade-offs in transformational changes to primary industries;

- developing a much more explicit approach to exploring the opportunities in the interactions between adaptation and mitigation;

- identifying mechanisms and processes to develop win-win options between adaptation and biodiversity and ecosystem management goals, and

- putting considerably more effort into the social, economic and policy aspects of building adaptive capacity and overcoming barriers to adoption.

While acknowledging that there are some serious knowledge gaps that have to be filled before comprehensive adaptation strategies can be implemented, it is equally important that the quest for improved information does not delay early actions from taking place. Indeed, there are already indications that opportunities for considered, pre-emptive actions are being lost and that in some areas we are instead having to respond reactively to the initial impacts of climate change (e.g. management of declining water resources,

Chapter 12). One priority, in planning early actions, will be to identify 'no-regrets' options that (1) are robust enough to be effective in coping with a range of plausible climate change impacts and/or (2) have environmental, production or other benefits irrespective of climate change. Such options could be adaptively refined over time and incorporated into more comprehensive adaptation strategies as improved information becomes available.

References

Dessai S, Hulme M, Lempert R, Pielke R Jr (2009). Climate prediction: a limit to adaptation? In *Adapting to Climate Change – Thresholds, Values and Governance*. (Eds W Neil Adger, I Lorenzoni, K O'Brien) Cambridge University Press, Cambridge.

Easterling W, Aggarwal P, Batima P, Brander K, Erda L, Howden M, Kirilenko A, Morton J, Soussana J–F, Schmidhuber J, Tubiello F (2007). Food, fibre and forest products. In *Climate Change 2007: Impacts, Adaptation and Vulnerability. Contribution of Working Group II to the Fourth Assessment Report of the Intergovernmental Panel on Climate Change*. (Eds ML Parry, OF Canziani, JP Palutikof, PJ van der Linden, CE Hanson), pp. 273–313. Cambridge University Press, Cambridge.

Howden SM, Soussana JF, Tubiello FN, Chhetri N, Dunlop M, Meinke HM (2007). Adapting agriculture to climate change. *Proceedings of the National Academy of Sciences* **104**, 19691–19696.

Keogh M, Thompson A (2008). 'Preliminary modelling of the farm-level impacts of the Australian Greenhouse Emissions Trading Scheme.' Research Report. Australian Farm Institute, Surry Hills, NSW.

Kurz WA, Dymond CC, Stinson G, Rampley GJ, Neilson ET, Carroll AL, Ebata T, Safranyik L (2008). Mountain pine beetle and forest carbon feedback to climate change. *Nature* **452**, 987–990.

Nelson R, Kokic P, Elliston L, King J (2005). Structural adjustment: a vulnerability index for Australian broadacre agriculture. *Australian Commodities* **12**, 171–179.

Nelson R, Brown PR, Darbas T, Kokic P, Cody K (2007). The potential to map the adaptive capacity of Australian land managers for NRM policy using ABS data. CSIRO, Australian Bureau of Agricultural and Resource Economics, prepared for the National Land & Water Resources Audit. http://www.nlwra.gov.au/products/pn21327

18

FREQUENTLY ASKED QUESTIONS

CJ Stokes and SM Howden

Climate change: how is the climate changing?

1) How are humans influencing the earth's climate?

There are many natural and human factors that affect the earth's climate, causing it to vary over time. The natural factors include variation in received solar radiation and volcanic activity. However, it is human actions that appear to be most strongly influencing observed climate changes, particularly over the past five decades. This has been via the emissions of greenhouse gases from energy production, industrial activities, agriculture and land use change among other sources.

2) What is the effect of greenhouse gases?

On average, the amount of solar radiation reaching the earth is balanced by the amount of energy radiated back into space by the planet. Greenhouse gases in the atmosphere intercept some of the energy that is radiated from the earth's surface and re-radiate it to the lower atmosphere and earth's surface. At moderate levels, greenhouse gases therefore have a beneficial effect in maintaining warmer surface conditions (raising temperatures by 33°C above the freezing conditions that would otherwise exist) particularly at night. However, when greenhouse gases in the atmosphere increase, the temperature of the earth increases until it radiates sufficient energy to restore the balance between incoming solar energy and escaping radiation. Over the past century, the earth's temperature has risen 0.76°C with much of this being ascribed to human-induced greenhouse gas emissions. Aside from an increase in global temperature, changing the energy balance also alters patterns of circulation in oceans and the atmosphere, thereby changing other aspects of climate such as rainfall.

3) Is it possible to predict exactly how climate will change for a particular region or enterprise?

In short, no. There are too many unknowns and uncertainties for true prediction. For example, we cannot predict scientifically the global response to calls to reduce greenhouse gas emissions – this is a political decision, not amenable to scientific analysis. Equally, models of the ocean-atmosphere-land system have many uncertainties relating to various key processes. Together, these result in a wide range of possible future climate changes, sea level rises and subsequent impacts. Modelled projections of the future climate are further complicated by the fact that changes in climate will vary from location to location. Because of this spatial variation (and the associated increase in complexity and uncertainty regarding the influence of finer scale climate processes) the confidence in climate projections declines as we move from global to regional to enterprise scales. In addition, the limited range of climate variables included in current climate model output may not provide the specific weather measures required for management decisions (e.g. cold indices for stone fruit). Nonetheless, there is adequate information to construct projected scenarios of change which can be used to explore possible impacts and adaptive responses to these impacts.

4) What do projections of future scenarios indicate about likely climate changes?

The effect of increased atmospheric concentrations of greenhouse gases on global temperatures has been understood for over 100 years. There is a high degree of confidence that existing emissions have warmed the globe and that future warming is virtually certain. Warming is expected to be stronger in the interior of Australia and slightly less in coastal regions. Projected changes in rainfall are much less certain, with the range of possible change including the possibility of both wetting and drying trends for many locations in Australia. Nevertheless, there is a strong likelihood of reductions in rainfall in the south of the continent, especially the south-west and possible trends towards wetter conditions in the north (although there is roughly equal likelihood of rainfall reductions). Advances in climate modelling will improve projections (and the range of projected metrics), but considerable uncertainty will remain, especially at management scales. Climate projections are best used as a guide for risk management and as indicators of the range of plausible changes in climate that could occur at specific locations. Making decisions in the face of this uncertainty will be an essential and unavoidable aspect of adapting to climate change.

5) What actions can be taken to limit climate change?

Broad cooperation in reducing greenhouse gas (GHG) emissions will be required to limit exposure to the risks of climate change. This is currently the subject of major policy initiatives. Agriculture accounts directly for about 17% of Australia's GHG emissions with further emissions from land clearing, energy use in farming, processing and transport. It also can be a sink for carbon dioxide in plant biomass and soil carbon. In order of importance, the GHGs emitted by agriculture are methane (from ruminant livestock and savanna burning), nitrous oxide (from fertiliser use, savanna burning and manure management) and carbon dioxide (from land clearing and losses of soil and vegetation carbon stores). Agriculture will participate in broader cooperative efforts to reduce GHG emissions although the exact mechanism is still being established. Past emissions of GHG gases have already committed the planet to further warming of about 0.2°C per decade for several decades. Additional changes and impacts beyond this will depend on the effectiveness of global efforts to reduce GHG emissions. While reducing GHG emissions is essential, some climate change will be unavoidable, so adaptation strategies will be required to deal with the associated impacts.

Impacts: how will climate change affect agriculture?

6) How will rising carbon dioxide levels affect primary industries?

There are some broad generalisations that can be made about the ways in which climate change will affect the plant and animal growth that underpins agriculture. The most certain aspect of climate and atmospheric change is the observed increase of carbon dioxide (CO_2) in the atmosphere and oceans. Carbon dioxide is one of the inputs into photosynthesis, and it is often in limiting amounts. Consequently increasing levels of CO_2 can stimulate plant growth by increasing the efficiency with which plants use light, water and nutrients. However, enhanced plant growth and greater nutrient use efficiency are often accompanied by dilution of nutrients in plant material, reducing diet quality for livestock and human consumption. In the oceans, rising levels of CO_2 are increasing their acidity and reducing the availability of calcium carbonate, which is required by many creatures with calcium carbonate-based structures (many of which are at the base of food chains).

7) How will increasing temperatures affect primary industries?

The aspect of climate change in which there is the next highest level of confidence is rising temperatures. In cool climates, warmer temperatures may benefit some perennial plants by providing longer growing seasons, but will generally have an adverse effect on crops, especially horticulture crops such as stone fruit, which require periods of cold temperatures. For annual crops, higher temperatures tend to accelerate plant development, reducing the time to grow. In hot climates, warmer conditions will most likely have an adverse impact on plant growth through greater heat stress, greater water demand and reduced water use efficiency. For intensive livestock in cooler climates, warmer conditions will reduce the need for heating sheds. However, in hotter climates or hotter seasons, warmer conditions are likely to increase animal heat stress, reducing productivity and increasing health concerns. Warmer conditions are likely to assist the spread of tropical pests, weeds and diseases into more southern parts of Australia. Rising ocean temperatures are likely to threaten coral reefs with more frequent bleaching episodes, cause fish species to migrate towards the poles, reduce the viability of southern kelp forests and threaten many important fisheries that have specific ecological, physiological or temperature requirements, such as abalone, rock lobster and Atlantic salmon.

8) How will changes in rainfall and other aspects of climate change affect primary industries?

There is less confidence in regional projections for other climate variables such as rainfall. Changes in plant growth would be expected to mirror changes in rainfall, although the responses may be magnified such that the percentage change in agricultural yield exceeds the percentage change in rainfall (especially for decreases in rainfall). The projected drying trend in southern Australia, if realised, will create particular challenges for water resource management and cropping. Projected increases in year-to-year variability in rainfall could increase the frequency and severity of drought and flood events, and will be compounded by projected increases in rainfall intensity, which could increase the potential for soil erosion. Shifts in the seasonal distribution of rainfall could impact activities such as viticulture, which requires winter rainfall. Winter and spring rainfall is projected to decline in the south of the country, spring rainfall is projected to decline in the east and autumn rainfall is projected to decline along the west coast. Changes in sea level will increase the risk of coastal flooding for crops such as sugarcane in low-lying areas, could cause intrusion of saltwater into ground water in coastal regions and will alter estuaries, which are important breeding grounds for many marine species. If the frequency or severity of cyclones increases, this could pose further risks to agriculture in coastal areas.

Adaptation: what can we do about it?

9) What is the difference between climate change 'mitigation' and 'adaptation'? Do we need both?

Addressing the challenge of climate change will require taking action both by reducing the amount of greenhouse gas emissions, to lessen our exposure to the risks of climate change (mitigation), and by taking actions to reduce the impacts of climate change that does occur (adaptation). The two are linked in that strong, up-front efforts to mitigate climate change will lessen the subsequent effort required for adaptation (while delaying mitigation will increase the impacts of climate change and the adaptation challenge). While mitigation efforts require global cooperation to be effective, adaptation actions taken by individuals at enterprise and community levels can have local benefits.

10) What determines how vulnerable an enterprise, community or industry will be to climate change?

The potential impact of climate change on a current agricultural activity depends on both the magnitude of climate change in the particular location (the exposure, e.g. the percentage change in rainfall) and how strongly each unit change in climate impacts that activity (the sensitivity, e.g. the change in enterprise income for each 1% change in rainfall). However, since land managers are likely to take actions in response to these changes (adaptation), the potential impacts are likely to be modified. The ultimate vulnerability of communities and enterprises to climate change is therefore strongly dependent not just on potential biophysical impacts but also on the way people respond and their capacity to moderate those impacts.

11) What can be done to prepare agriculture to face the challenge of adapting to climate change?

Successful adaptation to climate change will need both strategic preparation and tactical response strategies. Adaptation measures will have to reflect and enhance current 'best-practices' designed to cope with adverse conditions such as drought. Adoption of these new practices will require, among other things: (1) confidence that the climate really is changing; (2) the motivation to change to avoid risks or seize opportunities; (3) demonstration of the benefits of new adaptation options; (4) support during transitions to new management or new land uses; (5) building capacity for communities to develop and implement adaptation strategies; (6) altered transport and market infrastructure, and (7) an effective system to monitor climate change impacts and human adaptive responses to evaluate and adjust policy and management decisions continuously.

12) Is it possible to prepare a specific adaptation strategy based on what the future climate will be like for a particular region, industry or enterprise?

It is not possible to predict precisely how the climate will change at a particular location (see FAQ 4). Even with advances in climate science, substantial uncertainty over future impacts will likely remain and be unavoidable. However, climate change projections and scenarios of change can provide a guide to the range of plausible impacts and opportunities that are likely to occur. Adaptation approaches will therefore need to take a flexible, risk-based approach that incorporates future uncertainty and provides strategies that will be able to cope with a range of possible changes in local climate. Initial efforts in preparing adaptation strategies should focus on equipping primary producers with alternative adaptation options suitable for the range of uncertain future climate changes and the capacity to evaluate and implement these as needed, rather than focusing too strongly yet on exactly where and when these impacts and adaptations will occur.

13) How will management strategies for coping with climate variability help prepare primary producers to adapt to climate change?

Building capacity to cope with climate variability will serve as a strong foundation in preparing to deal with climate change. There are several reasons for this. (1) In the short term, land managers are more likely to be concerned about the immediate challenge of year-to-year variation, because climate change effects will largely be within the bounds of existing climate variability. (2) Adjusting management for seasonal differences from year to year builds in automatic tracking of longer-term climate trends. (3) These approaches encourage a shift from less flexible management styles to ones where managers adaptively monitor and respond to changes over time, and (4) managing for extreme events is already important in agriculture and will become more so as what we now think of as extreme events become more common under climate change. Combining approaches for dealing with short- to medium-term climate fluctuations (climate variability) with strategies for dealing with longer-term trends (climate change) will capitalise on these synergies and provide a more integrated approach for promoting effective responses to climate challenges.

14) Will it be possible to continue doing the same agricultural activities in their present geographic locations under future climates?

Hopefully, incremental adjustments to current agricultural practices will be sufficient to adapt to climate change in many locations. In such cases, it may be sufficient to support the types of gradual shifts in agricultural practice that are part of the normal management response to continually changing operating conditions. In other cases, new opportunities may emerge that are more profitable and appropriate for the altered climate conditions, even if previous production systems remain viable. However, there are likely to be some cases where climate change will make some production systems and options unviable and adaptation will require transformation to alternative land uses. Such transitions are likely to cause severe disruption to affected communities, so it will be important to identify the locations where such risks are greatest in order to provide the necessary support. As the magnitude of climate change increases, managers' past personal experience is likely to become less relevant as ecosystems, land uses, and management practices change. This will require greater support by, and collaboration with, institutions that can facilitate transitions to new systems better adapted to the changing climate.

15) In what ways can government policies promote climate change adaptation in the agricultural sector and support producers in making the necessary management changes?

At the broad policy level, governments and supporting industry institutions can enhance adaptive capacity by investing in (1) climate-related natural, physical, financial, human and social assets; (2) financial and other incentives that provide social status and/or reward early adopters; (3) climate research; (4) adaptive governance approaches that encourage climate-learning and adaptive management through developing effective environmental, economic and social feedback mechanisms; (5) 'mainstreaming' climate change considerations into appropriate government policies, particularly those related to drought, natural resource management and biodiversity, and (6) addressing other key stressors that can interact with those of climate change. A key policy challenge is to reduce maladaptation and to promote synergies with mitigation and with policies addressing other issues.

16) What are some specific examples of actions that can be taken to prepare Australian agriculture for climate change?

Some adaptation options will have broad relevance and benefit across primary industries. For example, the delivery of information from climate models could be enhanced by providing projections of management- and policy-relevant weather metrics (e.g. cold indices for stone fruit), downscaling regional projections to management scales, and combining information on both climate variability and trend in seasonal and medium-term (decadal) forecasts. Biotechnology and traditional breeding have the potential to develop new varieties better suited to future climates. Nutrient supplies could be adjusted (by fertiliser use, legumes and varietal selection) to maintain the quality of grain, fruit, fibre and forage sources. As water resources become more constrained, measures to increase irrigation efficiency will become more important. Such efforts could be assisted by identifying and establishing alternative less water-intensive production options, and by implementing water markets and water-sharing arrangements.

Improved quarantine, monitoring and control measures will help to control the spread of pests, weeds and diseases. Better agricultural systems models and other research could help to provide assessments of climate change impacts and adaptation options that are more effective and integrate a broader range of management-relevant considerations. Monitoring and evaluation systems that track changes in climate, impacts on agriculture and effectiveness of adaption measures, will help in deciding when alternative options should be implemented and in refining these measures over time. Timely inclusion of climate information in policy and management decisions would be helped by communicating new information as it becomes available and closer collaboration among policy-makers, managers, researchers and extension agencies. Many of these options have close synergies with existing 'best practice' and natural resource management initiatives. Such options should be prioritised as part of 'no regrets' strategies because they will provide immediate and ongoing benefits aside from preparing industries for climate change.

INDEX

Acacia species   142, 262, 263
adaptation,   2–3, 8, 277–79
   challenges   270–73
   community scale   247–48
   incremental   270–71
adaptation priorities   265–67
adaptation strategies,   246, 249–50, 264–65, 278–79
   for aquaculture   215, 218, 219
   for broadacre grazing   160–64
   for cotton   56–62
   for fisheries   215–20
   for forestry   143–48
   for grains   25–35
   for horticulture   126–129
   for livestock industries   177–81
   for rangelands   160–64
   for rice   73–79
   for sugarcane   91–95
   for water resources   196–200
   for winegrapes   107–11
adaptation-mitigation interactions   271–72
adaptive capacity   245–52, 273
adaptive capacity assessments   250–51
agistment   177
agricultural greenhouse gases *see* greenhouse gases
agricultural industrial parks   179
Agricultural Production Systems Simulator (APSIM)   90, 93, 249
animal health and husbandry   157–58, 174, 177
apples   121, 124–25
aquaculture,   205–207, 210, 214–15, 218–20, 263
   adaptation strategies   215, 218, 219
   Australian industry   206–207
   costs and benefits of adaptation   218
   feed   218
   impact of climate change   263–64
   options for dealing with climate variability   214–15
   risk of maladaptation   217–18
   selective breeding   215
atmospheric CO_2 enrichment,
   and horticulture   123–24
   and pastures   175
   and rice   71–72
   and winegrapes   108–109
atmospheric CO_2 increase,   4–5, 276
   and broadacre grazing   155–56, 159

and cotton   51–52
and cropping   22, 23, 25, 26, 28
and forestry   139–41
and horticulture   127
and pastures   155–56, 159, 175
plant responses   4–7
and rangelands   155–56, 159
and winegrapes   103–104, 110
atmospheric water vapour   7
autumn break   27

baselines,
   climate   189–91
   water resources   191
biodiversity conservation   272–73
bioenergy   146, 237
biosequestration   146, 148, 157, 162, 235–40, 260, 261, 272
broadacre grazing,   153–64
   adaptation strategies   160–64
   and animal husbandry   157–58
   and animal nutrition   156
   and atmospheric CO_2 increase   155–56, 159
   Australian industry   153–54
   costs and benefits of adaptation   162
   and fire regimes   156–57
   and forage   154–56
   and greenhouse gas emissions   158
   impact of projected climate change   154–59
   and increasing temperature   155, 159
   management   154, 156, 160–64
   options for dealing with climate variability   159–60
   and pasture growth   155–56, 159
   and pasture species   157
   pests and diseases   157–58, 161
   policy changes   158
   and rainfall change   155, 159
   and regional impacts   158–59
   risks of maladaptation   162
   and seasonal forecasts   159–60
   and soil degradation   157
   and vegetation change   156–57
   and weeds   157, 161
budburst   102, 103, 105, 107, 108, 123, 125, 126

C_3 plants   5, 26, 87, 155, 156, 175, 178
C_4 plants   5, 26, 87, 155, 156, 175, 178

carbon dioxide *see* CO$_2$
Carbon Pollution Reduction Scheme   158, 271–72
carbon sequestration   146, 148, 157, 162, 235–40, 260, 261, 272
carbon sinks   231, 235–40, 272
catchment risk scores   193–94, 195
catchment scale, water resources   193–94
cereals *see* grains
chilling, of fruit   103, 107, 121, 125, 128
citrus fruit   121
climate baselines   189–91
climate change,   2, 13–14, 16, 190, 258, 275–77
    adaptation priorities   265–67
    coping strategies   264–65
    regional impacts   259–64
climate change policy   175–76
climate change prediction   275–76
climate change projections   2, 13–19, 173
climate forecasting systems   30, 76, 79, 90, 126, 129, 145, 159–60, 176, 179, 196, 249
climate models   15, 16, 29, 110, 127, 140–42, 162, 173, 176, 192, 206, 263, 270
climate, future Australian   15–18
climatic events, extreme   16–18, 88, 103, 122–23, 174, 211, 263, 265
CLIMEX   127, 142
CO$_2$ fertilisation   71–72, 87
CO$_2$, ocean levels   7
    see also atmospheric CO$_2$
community scale adaptation   247–48
consumer behaviour   127, 158
cost and benefits of adaptation,
    to aquaculture   218
    to broadacre grazing   162
    to cotton   60–61
    to fisheries   218–19
    to forestry   145–46
    to grains   33–35
    to horticulture   128
    to intensive livestock industries   180
    to rice   77
    to water resources   198
    to winegrapes   110
cotton,   49–62
    adaptation options   56–62
    Australian industry   50
    breeding programs   58, 60
    costs and benefits of adaptation   60–61
    crop maturity   59
    crop rotation   60
    cropping regions   59–60
    cultivar choice   57–58
    decision support tools   56, 59

    diseases   56
    evaporative demand   52
    and extension material   56
    fertiliser application   55, 58–59
    Helicoverpa resistance   52, 55
    impact of projected climate change   50–54, 57
    increase in CO$_2$ 51–52
    nitrogen nutrition   59
    options for dealing with climate variability   54–56
    pests   53–54, 55–56
    planting time variation   58
    production efficiencies   58–59
    quality   60
    season length   59
    and soil fertility   58–59
    and temperature increase   52–54
    water availability   52
    water management   54–55
    water use efficiency   59
    and weeds   52, 56
    yields   57–58
cotton seed   50
cyclones   7, 14, 18, 88, 123, 210, 211, 237

dairy consumption   158
dairy industry, Australian   171–72, 179–80
diseases   23, 31, 56, 103, 104, 107, 123, 124, 126, 127, 144–45, 157–58, 161
drought   14, 17–18, 50, 123, 125, 128, 139, 141, 143, 146, 154, 156, 162, 190, 193, 194
dryland agriculture   194
dryland salinisation   23, 29, 261

East Australian current   210, 212
ecosystem-based fisheries management   216–17
El Niño-Southern Oscillation system (ENSO)   14, 30, 90, 91, 159–60, 176, 190, 196, 212
emission-reduction agreements   258
emissions trading scheme   146, 176, 234, 271
erosion management   28–29
Eucalyptus species   138, 139, 140, 141, 142, 143, 145, 146, 290, 261
evaporation   7, 18, 29–30, 69, 154, 190, 192
extreme climatic events   16–18, 88, 103, 122–23, 174, 211, 263, 265

feed supplements   160–61
feeding, supplementary   179
feedlots   172, 173, 177
financial institutions   32
fire management, and forestry   144
fire regimes, changing   156–57

fire weather risk   18
fire, and greenhouse gas emissions   240
fish stocks   214, 216, 217–18, 219
fish, species distribution   216
fisheries,   7, 205–20
    adaptation strategies   215–20
    Australian industry   205, 207–09
    Australian regions   208, 210–12
    costs and benefits of adaptation   218–19
    ecosystem-based management   216–17
    fuel consumption   218
    habitat change   210–11
    impacts of projected climate change   209–14,
        263–64
    indigenous fishers   209, 213–14, 217
    and introduced species   212
    management   214, 215–17
    options for dealing with climate
        variability   214–15
    pelagic   212–13
    recreational fishers   213–14, 217
    risk of maladaptation   217–18
fishmeal production   218
forage   27, 154–56, 160, 162
forest management   143–48
forestry,   137–48
    adaptation strategies   143–48
    and assessment of climate conditions   145
    and atmospheric CO_2 increase   139–41
    Australian industry   138–39, 146–47
    bioclimatic analysis   141–42
    costs and benefits of climate change   145–46
    and drought   139, 141, 143
    impact of projected climate change   139–42
    establishment strategies   145
    and fertilisation   144
    and fire management   144
    and frost   145
    and genotype selection   143
    modelling analyses of climate change   140–42
    options for dealing with climate
        variability   142–43
    and pests, diseases and weeds   142, 144–45
    plantations   138–39, 141, 144, 145, 146, 147,
        236, 237, 260, 263
    and rainfall change   139
    and rainfall variability   142–43
    risks of maladaptation   145
    and site selection   143, 144
    and species selection   141
    and tree spacing   143–44
    and tree thinning   143
    and water use   144

forests,
    carbon sequestration   146, 231, 235–37
    and greenhouse gas emissions   231
fossil fuel   179, 218, 231
free-air CO_2 enrichment (FACE) experiments   51,
    71, 104, 124, 140
frost   6, 17, 23, 24, 27, 51, 52, 71, 105, 106, 123, 124,
    141, 145, 258
fruit ripening   121
fruit sunburn   106, 121, 124
fruit, and temperature change   121–22
future climate, Australia's   15–18

governance   246, 247, 249, 278
government policies   158, 161, 175–76, 247, 278
grain nitrogen   28, 33
grain protein   22, 26, 28
grainfill   25, 26, 27
grains,   21–35
    adaptation options   25–35
    Australian regions   21–22
    and CO_2 increase   22, 23, 25, 26, 28
    climate monitoring   30–31
    costs and benefits of adaptation   33–35
    crop management   23, 25, 27–28, 30
    and decision-making   23–25
    fertiliser application   28, 33
    and forecasting systems   23–25
    and frost risk   24
    and heat shock   26
    impact of projected climate change   22–23
    land use change, regional   32
    nutrient management changes   28, 33
    options for dealing with climate variability   23–
        25
    pests, diseases and weeds   31
    and plant breeding   26
    planting time variation   27
    and rainfall   22–23, 27
    research and development   31–32
    risk of maladaptation   32–33
    seasonal forecasting   30
    sowing times   23–24, 33
    species change   26–27
    and temperature increase   22, 25, 26, 27
    varietal change   25–26, 33
    yield   22, 25, 26
grasses, rangeland   155–56
grazing, broadacre see broadacre grazing, intensive
    livestock industries
greenhouse gas emissions,   2, 7–8, 231, 258, 276
    and broadacre grazing   158
    and fire   240

and forests   231
by livestock   175–76
options for reducing   231–35
mitigation strategies   230, 231, 241
rangelands   239–40
greenhouse gases   14, 229–41, 275
groundwater supply   188, 193–94

hail   18, 122–23
heat shock   26
heat stress,
in animals   155, 157, 159, 161, 174, 177, 178, 180, 263, 276
in plants   26, 53, 105–106, 125, 127, 276
Helicoverpa spp.   55, 56, 59
Helicoverpa resistance, in cotton   52, 55
horticulture,   119–30
adaptation strategies   126–129
and atmospheric CO_2 increase   127
Australian industry   119–21
and CO_2 enrichment   123–24
costs and benefits of adaptation   128
crop management   124, 126–27
crop quality   128–29
cultivar selection   125, 126, 127, 128
and drought   123, 125, 128
and extreme climatic events   122–23
and farm management   129
and frost   123, 124
impact of projected climate change   121–24
and market behaviour   126, 127–28
modelling   127
options for dealing with climate variability   124–26
and pests, diseases and weeds   123, 124, 126, 127
and rainfall variation   122
risk of maladaptation   128
and seasonal forecasts   126, 129
site selection   124, 126, 128
and temperature changes   121–22
and water balance changes   122
water management   125–26, 127, 129
yield   121, 126
hydrological cycle, increased intensity   7

Improving Water Information Program   191, 193
incremental adaptation   270–71
indigenous fishers   209, 213–14, 217
industrial parks   179
infrastructure   32, 108, 189, 198, 250
integrated catchment management   129, 197
intensive livestock industries,   171–81

adaptation strategies   177–81
animal health and welfare   174
Australian   171–73
costs and benefits of adaptation   180
feeding strategies   177
impact of projected climate change   173–76
management   176, 177, 179–81
options for dealing with climate variability   176–77
and pasture production   174–75
policy analysis   180
research   180–81
risks of maladaptation   179–80
seasonal climate forecasts   176, 179
and water availability   174
IPCC scenarios   14–15
integrated pest management   (IPM)   55, 56
irrigated agriculture   194–95, 198
irrigation,
of cotton   54–55
of grains   22, 27–28, 30, 34
planning   196
of rice   68–69, 74, 75
of winegrapes   106
irrigation allocations   106, 107, 194–95
irrigation systems   28, 54, 77, 78, 106, 125, 177, 195

knowledge gaps   251, 273–74
knowledge generation   250–51
krill   213
Kyoto Protocol   230, 258

landscape resilience   179
Leeuwin current   208, 212, 218, 263
legumes   5, 59, 156, 160 161, 162, 175
livestock industries *see* intensive livestock industries
livestock,
agistment   177
breeding   178, 179, 181
feeding   177, 178, 179
feeds, and methane reduction   232–33
greenhouse gas emissions   175–76
heat stress   155, 157, 159, 161, 174, 177, 178, 180
husbandry   177
management   161
selection   177
shelter   177, 178, 180, 181, 262, 263
water requirements   157
water use   177
welfare   174

long-spined sea urchin   211–12

maladaptation risk,
    in aquaculture   217–18
    in broadacre grazing   162
    in fisheries   217–18
    in forestry   145
    in grain cropping   32–33
    in horticulture   128
    in intensive livestock industries   179–80
    in rice   76–77
    in sugarcane   92–93
    in water resources   198
    in winegrapes   110
marine environment   7, 209–14
marine species, distribution   211–12
meat consumption   158
meat prices   158
methane emission reduction   232–34
micronaire   52, 53, 58, 60
microsporogenesis   69–70
moisture conservation   29–30, 73
Murray–Darling Basin Sustainable Yields
    Project   192, 193

National Land and Water Resources Audit
    (NLWRA)   191, 192, 193
National Water Account   188, 191
native forests   138, 231, 236
natural resource conservation   272–73
nitrification   234–35
nitrogen use efficiency   5, 59
nitrogen fertilisers   23, 28, 55, 56, 58, 93, 176,
    234–35
nitrogen fixation   5
nitrogen leaching   30
nitrous oxide emission reduction   231, 234–35,
    240
northern fisheries region   210–11, 263

ocean, changes   7, 211, 263

pasture species   157, 162, 175, 177, 178, 179,
    232
pastures   154, 155–56, 159, 160, 174–75
pelagic fisheries   212–13, 264
pests   23, 31, 53–54, 55–56, 103, 104, 107, 123,
    124, 126, 127, 144–45, 157–58, 161
photoinhibition   122
Phytophthora cinnamomi   123, 142
pig industry, Australian   172
Pinus species   138, 140, 141, 142, 143, 145, 147,
    260, 261, 262

plant responses,
    to atmospheric CO_2 4–5, 6–7, 124, 139, 140
    to temperature increase   5–7
plantations   138–39, 141, 144, 145, 146, 147, 236,
    237, 260, 263
planting time variation   27
policies, government   278
policy, climate-related changes   158
policy-making   247–50, 251
poultry industry, Australian   172–73
power generation   178–79
prawn fisheries   208, 210–11, 217
priority challenges   270–73

rainfall,   14, 17
    and broadacre grazing   155, 159
    changes   7, 14, 16, 17, 123, 190, 277
    and cropping   22–23
    extreme   122–23
    and forestry   139, 142–43
    and horticulture   122
    and rangelands   158, 159
    and sugarcane   87–88
    and winegrapes   105, 106–107, 109
rangeland productivity   159
rangelands,   154–64
    adaptation strategies   160–64
    and atmospheric CO_2 155–56
    carbon sequestration   239–40
    changing rainfall   158
    and climate change impacts   158–59
    and climatic variability   159–60
    and fire   239–40
    grasses   155–56
    and greenhouse gas emissions   158, 239–40
    management   160–61
    options for dealing with climate
        variability   159–60
    pests, diseases and weed management   161
    and rainfall change   158, 159
    regional variation due to climate change   158–
        59
    and temperature increase   159
    vegetation   156–57
recreational fishers   213–14, 217
regional impacts of climate change   158–59,
    259–64
research   250–51
revegetation   179
rice,   67–79
    adaptation strategies   73–79
    aerobic   75–76, 78–79
    Australian industry   67–68

breeding   75
and CO$_2$ fertilisation   71, 72
combine and sodsowing   75, 77–78
combined temperature and CO$_2$ effects   71–72
costs and benefits of adaptation   77
crop rotations   72–73
cropping area modification   72
fallows   73
flowering   69–71
and frost risk   71
identification of suitable soils   74, 78
impact of projected climate change   68–72
irrigation   68–69, 74, 75
and lower temperatures   70–71, 73
microsporogenesis   69, 70
nitrogen fertilisers   77
nutrient management   73
options for dealing with climate variability   72–73
raised beds   74–75
risk of maladaptation   76–77
and seasonal climate forecasts   79
sowing times   72
stubble retention   73
and temperature increase   69–70
water use   68–69, 72, 73, 74–76
wet-and-dry   75, 76
whole farm planning   74
yield   70, 71–72
risk management   270
rock lobster   212
rural communities   247–48

salinisation   29, 105
salmon   207, 210, 214
seasonal climate forecasting   30, 76, 79, 90, 126, 129, 145, 159–60, 176, 179, 196, 249
seedling growth   140
shelter, livestock   177, 178, 180, 181, 262, 263
site selection   124, 126, 144
small pelagic fishery   213
social capital   248, 250
soil carbon   179, 237–39, 240
soil degradation   157
soil erosion   157
soil moisture, and climate change   18, 156
soil nitrogen   234–35
soils, carbon sequestration   237–39
south-east fisheries region   211–12
southern bluefin tuna   208, 210, 213, 214, 216
southern squid jig fishery   213
sowing times, of grain   23–24, 33
squid   213

storms   18, 210
streamflow changes   191
sugarcane,   85–95
adaptation strategies   91–95
Australian industry   85–86
breeding   95
and CO$_2$ concentrations   87
and C:N ratio   87, 88
costs and benefits of adaptation   93–94
decision-support tools   92
and extreme weather   88
forecasting systems   90
and harvesting   88, 93, 94, 95
impact of projected climate change   86–90
and improved adaptive capacity   95
management strategies   92, 95
marketing   95
and milling   88
modelling of impacts   90, 93–94
options for dealing with climate variability   90–91
plant growth   87–88
policy and institutional change   92
and rainfall change   87–88
and regional impacts of climate change   88–89
risks of maladaptation   92–93
and temperature increase   87
and transport   88, 94, 95
and trash production   93
water supply   92, 93
yields   90
sunburn damage   106, 121, 122
surface water supply   192–93

temperature,
average annual   14, 15
extreme   16
global mean   2
night-time   6
seasonal average change   15
temperature increase,
effect on agriculture   276–77
and broadacre grazing   155, 159
and cotton   52–54
and cropping   22, 25, 26, 27
and horticulture   121–22
plant responses   5–7
projections   16–17
and rangelands   159
and rice   69–70
and sugarcane   87
and winegrapes   102–103, 107–108, 110
thermotolerance   178

tillage   27, 54
trade   32
transformational change   270–71
transition support   249–50
tree mortality   139, 141, 143, 145, 146
tree planting   236, 237

uncertain change, strategies   264–65
uncertainty, managing   270

vapour pressure deficit (VPD)   6, 29–30
vegetation change   156–57
vernalisation   6, 26, 103, 121
vineyard situation   107
vulnerability   246–47, 277

water accounts   191
water allocations   54, 55, 68, 69, 72, 196, 198
water balance changes, and horticulture   122
Water for the Future   191, 200
water management,
    and horticulture   125–26, 127, 129
    and cotton   54–55
    and rice   68–69, 72, 73
    and sugarcane   92, 93
water pricing   196
water reform   191, 199–200
water resources baselines   191
water resources,   187–200
    adaptation strategies   196–200
    at catchment scale   193–94
    costs and benefits of adaptation   198
    and dryland agriculture   194
    government policy/programs   191
    groundwater supply   188, 193–94
    impact of projected climate change   189–95
    infrastructure   189, 198
    integrated catchment management   197
    and irrigated agriculture   194–95
    management   189–91, 193, 195
    options for dealing with climate
        variability   195–96
    planning   189–90, 196
    research   190–91
    and rainfall variability   190–91
    risk management   196–98
    risks of maladaptation   198
    seasonal prediction systems   196
    streamflow changes   191
    surface water supply   192–93

water trading   72, 111, 129, 196, 197, 200
water use efficiency,   5, 6, 188–89
    and cotton   59
    and livestock   177
    and winegrapes   106–107
waterlogging   30
weeds   23, 31, 52, 56, 123, 124, 144–45, 157, 161
western fisheries region   212
winegrapes,   101–11
    adaptation strategies   107–11
    Australian industry   101–102
    breeding programs   107
    and CO_2 increase   103–104, 108–109, 110
    cost and benefits of adaptation   110
    and extreme climatic events   103
    fertiliser application   108–109
    and fire/smoke   105, 110
    and frost   105, 106, 107–108
    fruit development   102–103, 104, 106, 107
    harvesting   103
    and heat stress   105–106
    impact of projected climate change   102–105
    irrigation   104, 108
    management strategies   108, 110
    options for dealing with climate
        variability   105–107
    and pests and diseases   103, 104, 105, 107,
        109–110
    pruning   107–108
    and rainfall   104–105, 109
    rainfall variability   106–107
    risks of maladaptation   110
    and salinisation   105
    and soil salinity   109
    and sunburn   106
    and temperature increase   102–103, 107–108,
        110
    temperature variability   105–106
    and variety selection   107–108
    and vineyard situation   107, 110–11
    water scheduling   108
    water supply   109, 110, 111
    water use   106–107
    use of recycled water   109, 111
    windbreaks   106
    within-canopy berry shading   108
    yield   108
winemaking, changes to   108
winery infrastructure   108
wool   158